# 27th Winter Workshop on Nuclear Dynamics 2011

# (WWND 2011)



Journal of Physics: Conference Series Volume 316

## Winter Park, Colorado, USA
## 6-13 February 2011

ISBN: 978-1-61839-287-9
ISSN: 1742-6588

**Printed from e-media with permission by:**

Curran Associates, Inc.
57 Morehouse Lane
Red Hook, NY 12571

**Some format issues inherent in the e-media version may also appear in this print version.**



Printed by Curran Associates, Inc. (2011)

For permission requests, please contact the Institute of Physics at the address below.

Institute of Physics
Dirac House, Temple Back
Bristol BS1 6BE UK

Phone: 44 1 17 929 7481
Fax: 44 1 17 920 0979

techtracking@iop.org

**Additional copies of this publication are available from:**

Curran Associates, Inc.
57 Morehouse Lane
Red Hook, NY 12571 USA
Phone: 845-758-0400
Fax: 845-758-2634
Email: curran@proceedings.com
Web: www.proceedings.com

# TABLE OF CONTENTS

## PREFACE

These are the proceedings of the 27th edition of the Winter Workshop on Nuclear Dynamics, which was held in Winter Park, Colorado. As in previous years, the unique character of this conference series has allowed us to bring nuclear scientists from very different fields together to discuss their scientific achievements.

At the high-energy frontier very exciting results from the first heavy-ion collisions at the LHC were shown. At the lower energies the RHIC beam energy scan is underway to span the gap between the SPS and RHIC and search for critical phenomena in the nuclear matter phase diagram. The nuclear structure and astrophysics communities have started to construct dedicated facilities at Michigan State University (F-RIB) and the GSI in Germany (FAIR).

Theory is progressing across all these different energy regimes, and the new data, in particular from the LHC and RHIC, are motivating more detailed modeling and a deeper understanding of the underlying physics.

These proceedings of the 27th Winter Workshop on Nuclear Dynamics again provide a snapshot of the status of the field. The articles, many of which are written by some of the most promising young scientists in the field, are conveying the feeling of excitement that permeates all subfields of nuclear science.

**Rene Bellwied (University of Houston)**
**Claude Pruneau (Wayne State University)**

## PEER REVIEW STATEMENT

All papers published in this volume of *Journal of Physics: Conference Series* have been peer reviewed through processes administered by the proceedings Editors. Reviews were conducted by expert referees to the professional and scientific standards expected of a proceedings journal published by IOP Publishing.

# Beam Energy Dependence of Directed and Elliptic Flow Measurement from the STAR Experiment

**Yadav Pandit (for the STAR Collaboration)**
Department of Physics, Kent State University,USA
E-mail: ypandit@kent.edu

**Abstract.** Measurements of anisotropic flow in heavy-ion collisions provide insight into the early stage of the system's evolution. These proceedings present directed and elliptic flow for Au+Au collisions at 39, 11.5 and 7.7 GeV, and for Cu+Cu at 22.4 GeV, measured in the STAR Experiment at RHIC. Differential measurements of directed and elliptic flow of charged particles as a function of centrality, transverse momentum and pseudorapidity are discussed.

## 1. Introduction

The study of collective flow in relativistic nuclear collisions has potential to offer insights into the equation of state of the produced matter [1, 2]. Anisotropic flow is characterized by the Fourier coefficients [3]

$$v_n = \langle \cos n(\phi - \Psi_R) \rangle, \tag{1}$$

where $\phi$ denotes the azimuthal angle of an outgoing particle, $\Psi_R$ is the orientation of the reaction plane and $n$ denotes the harmonic. The reaction plane is defined by the beam axis and the impact parameter [3].

Directed flow, $v_1$, is the first harmonic coefficient of the above Fourier expansion of the final momentum-space azimuthal anisotropy, and it reflects the collective sidewards motion of the particles in the final state. In regions that are closer to beam rapidity ($y$) than to midrapidity, directed flow is imparted very early, at a pre-equilibrium stage of the collision [4, 5], and thus it probes the onset of bulk collective behavior. Both hydrodynamic and nuclear transport models [6, 7] indicate that directed flow is a sensitive signature for phenomena related to a possible phase transition, especially in the general region of beam energy under investigation here [8]. In particular, the shape of $v_1(y)$ in the midrapidity region is of special interest because it has been argued that directed flow may exhibit flatness at midrapidity due to a strong, tilted expansion of the source. Such tilted expansion gives rise to anti-flow or a 3rd flow [9] component. The anti-flow is perpendicular to the source surface, and is in the opposite direction to the repulsive "bounce off" motion of nucleons. If the tilted expansion is strong enough, it can cancel and reverse the motion in the bounce-off direction and results in a negative $v_1(y)$ slope at midrapidity, potentially producing a wiggle-like structure in $v_1(y)$. A wiggle for baryons is a possible signature of a phase transition between hadronic matter and Quark Gluon Plasma (QGP), although QGP is not the only possible explanation [10, 11, 9]. If strong but incomplete baryon stopping is assumed together with strong space-momentum correlations caused by transverse radial expansion, then a wiggle structure might be explained even in a hadronic system [12].

Elliptic flow, $v_2$, is the second harmonic coefficient of the Fourier expansion above. The initial-state spatial eccentricity of the participant zone drives the process whereby the interactions produce an anisotropic distribution of momenta relative to the reaction plane. The elliptic momentum anisotropy saturates quite early in the collision evolution, although a little later than when directed flow is imparted [4, 5]. Elliptic flow can provide information about the pressure gradients in a hydrodynamic description, and about the effective degrees of freedom, the extent of thermalization, and the equation of state of the matter created at early times. Studying the dependence of elliptic flow on system size, number of constituent quarks, and transverse momentum/mass, are crucial to the understanding of the properties of the produced matter [1].

## 2. Methods and Analysis

In this proceedings, we report $v_1$ and $v_2$ measurements by the STAR experiment from $\sqrt{s_{NN}}$ = 39, 11.5 and 7.7 GeV Au + Au and $\sqrt{s_{NN}}$ = 22.4 GeV Cu + Cu collisions. Data were taken from Run 10 (2010) and Run 5 (2005). The STAR Time Projection Chamber (TPC) [16] was used as the main detector for charged particle tracking and second order event plane determination for elliptic flow analysis. The centrality was determined by the number of tracks from the pseudorapidity region $|\eta| < 0.5$. Forward Time Projection Chambers (FTPCs) were also used for charged particle tracking at forward rapidities. Two Beam Beam Counters covering $3.3 < |\eta| < 5.0$ were used to reconstruct the first-order event plane for directed flow analysis. The pseudorapidity gap between BBC and TPC allows us to reduce some of the non-flow effects. Elliptic flow measurements with the standard event plane method are based on the first-order event plane from the BBC as well as the second-order event plane based on tracks in the main Time Projection Chamber [16] of STAR. We also present elliptic flow based on two- and four-particle correlations . We analyzed minimum bias events with the primary collision vertex position along the beam direction ($V_z$) within 30,50 and 70 cm of the center of the detector are selected for this analysis for Au+ Au collisions at 39, 11.5 and 7.7 GeV respectably and 30 cm for Cu+Cu collisions at 22.4 GeV . In order to reject events which involve interactions with the beam pipe and also to minimize beam-gas interactions, the event vertex radius ($\sqrt{V_x^2 + V_y^2}$, where $V_x$ and $V_y$ are the vertex positions along the $x$ and $y$ directions, respectively) is required to be less than 2 cm. For these analyses, tracks which have transverse momenta $p_t > 0.2$ GeV /$c$, pass within 3 cm of the primary vertex, have at least 15 space points in the main TPC acceptance ($|\eta| < 1.0$) or 5 space points in the case of tracks in the FTPC acceptance $2.5 < |\eta| < 4.0$). Also, the ratio of the number of actual space points to the maximum possible number of space points for that track's trajectory was required to be greater than 0.52, which prevents split tracks from being counted as two tracks.The differential measurement of directed and elliptic flow is presented as a function of pseudorapidity ($\eta$), transverse momentum and centrality.

Figure 1, left panel, shows the first-order event plane resolution as a function of centrality for Au + Au collisions at $\sqrt{s_{NN}} = 39$, 11.5 and 7.7 GeV and for Cu + Cu collisions at $\sqrt{s_{NN}} = 22.4$ GeV based on the event plane reconstructed using BBC information. The right panel shows the second-order event plane resolution reconstructed using charged tracks from the TPC. The errors shown are statistical. Generally, the first-order event plane is used for directed flow measurement and the second-order event plane is used for elliptic flow measurement in the standard method of flow analysis. STAR is well suited for these measurements, with very good event plane resolution down to at least 7.7 GeV.

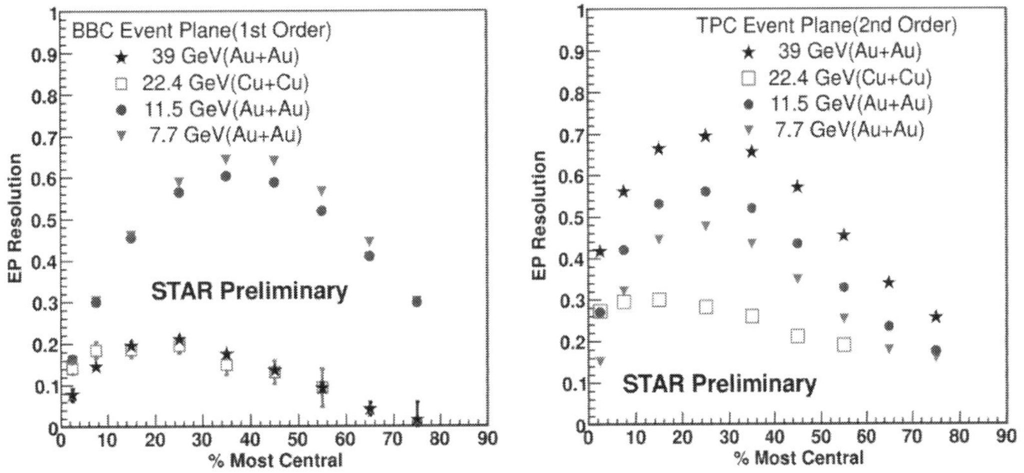


**Figure 1.** The left panel shows the first-order event plane resolution based on information from the STAR BBC as a function of centrality for Au + Au collisions at $\sqrt{s_{NN}}$ = 39, 11.5 and 7.7 GeV and for Cu + Cu collisions at $\sqrt{s_{NN}}$ = 22.4 GeV. The right panel shows the second-order event plane resolution based on the TPC as a function of centrality for Au + Au collisions at $\sqrt{s_{NN}}$ = 39, 11.5 and 7.7 GeV and for Cu + Cu collisions at $\sqrt{s_{NN}}$ = 22.4 GeV. The errors shown are statistical.

## 3. Results and Discussion

### *3.1. Directed flow*

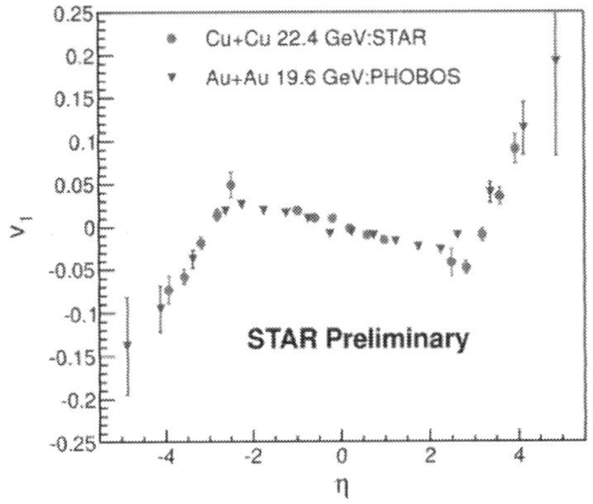


**Figure 2.** Charged hadron $v_1$ vs. $\eta$ for 0–60% centrality Cu+Cu collisions at $\sqrt{s_{NN}}$ = 22.4 GeV obtained from the standard method based on the BBC event plane. The errors shown are statistical. Results are compared to $v_1$ (solid triangles) from 0–40% centrality Au+Au collisions at $\sqrt{s_{NN}}$ = 19.6 GeV [21] from the PHOBOS collaboration.

Figure 2 shows charged hadron $v_1\{BBC\}$ in Cu+Cu collisions for 0–60% centrality at

$\sqrt{s_{NN}}$ = 22.4 GeV, compared to 0–40% centrality Au+Au collisions at $\sqrt{s_{NN}}$ = 19.6 GeV. It is seen that the latter (PHOBOS) results are quite similar, notwithstanding the difference in system size, and the fact that the centrality range and beam energy are not quite the same. At 200 GeV and 62.4 GeV, for 30-60% central collisions, we have previously reported that directed flow does not vary within errors between Au+Au and Cu+Cu [22]. The new finding reported here suggests that this behavior extends to lower energies. Figure 3 shows charged hadron $v_1$ as

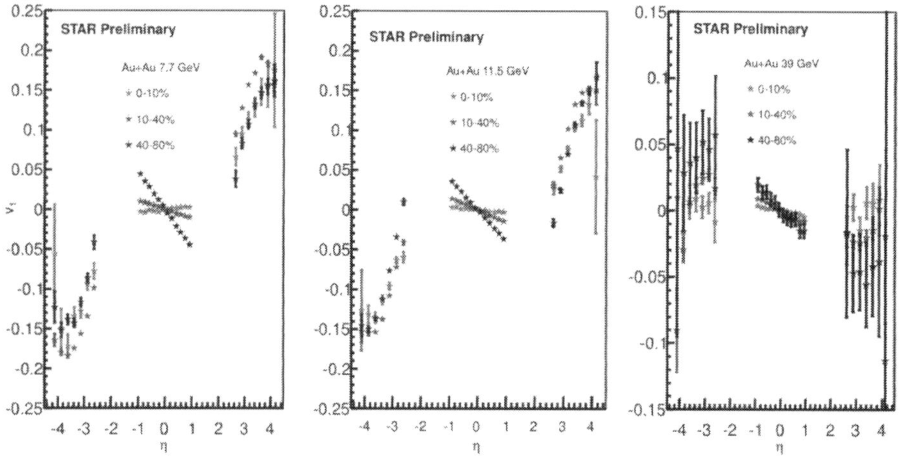


**Figure 3.** Charged hadron $v_1$ as a function of pseudorapidity $\eta$ for central 0–10 %, mid central 10–40% and peripheral 40–80 % Au+Au collisions at $\sqrt{s_{NN}}$ = 39, 11.5 and 7.7 GeV. The errors shown are statistical.

a function of pseudorapidity $\eta$ for central 0–10 %, mid central 10–40% and peripheral 40–80 % Au+Au collisions at $\sqrt{s_{NN}}$ = 39, 11.5 and 7.7 GeV. The errors shown are statistical. We observe centrality dependence of the directed flow signal specially prominent at mid rapidity( $|\eta| < 1.0$) than at forward rapidities. Figure 4 shows charged hadron $v_1$ as a function of transverse momentum $p_T$ for central 0–10 %, mid central 10–40% and peripheral 40–80 % Au+Au collisions at $\sqrt{s_{NN}}$ = 39, 11.5 and 7.7 GeV. The errors shown are statistical. For this plot, particles from the main TPC are used with $|\eta| < 1.0$ . We observe the centrality dependence of directed flow as a function of $p_T$ . We also observe that the directed flow signal crosses zero at $p_T > 1.0 GeV/c$ from negative to positive , which was previously reported at higher RHIC energies [22]

Figure 5 (left panel) shows charged hadron $v_1$ as a function of $\eta$ scaled by the corresponding $y_{\rm beam}$ for Au+Au Collisions at 7.7, 11.5, 39, 62.4 and 200 GeV for 30–60 % central collisions. The new results reported here are the charged hadron $v_1\{\rm BBC\}$ in Au+Au collisions for 30–60% centrality at $\sqrt{s_{NN}}$ = 39,11.5 and 7.7 GeV. On the right, we plot charged hadron $v_1$ as a function of $\eta$ scaled by the corresponding $y_{\rm beam}$ for Cu+Cu collisions at 22.4, 62.4 and 200 GeV. The results for 62.4 and 200 GeV are for 30–60% centrality previously reported by STAR [22]. The new results reported here are the charged hadron $v_1\{\rm BBC\}$ in Cu+Cu collisions for 0–60% centrality at $\sqrt{s_{NN}}$ = 22.4 GeV. We observe that $v_1(\eta/y_{\rm beam})$ shows a beam energy scaling behavior, though not perfect, that has already been observed at higher RHIC energies.

Figure 6 (left) shows charged hadron $v_1$ as a function of $\eta$ - $y_{\rm beam}$ for Au+ Au Collisions at 7.7, 11.5, 39, 62.4 and 200 GeV for 30–60 % central collisions. On the right, we show charged hadron $v_1$ as a function of $\eta - y_{\rm beam}$ for Cu+Cu collisions 22.4, 62.4 and 200 GeV. The results for 62.4 and 200 GeV are for 30–60% centrality, previously reported by STAR [22]. In this frame,

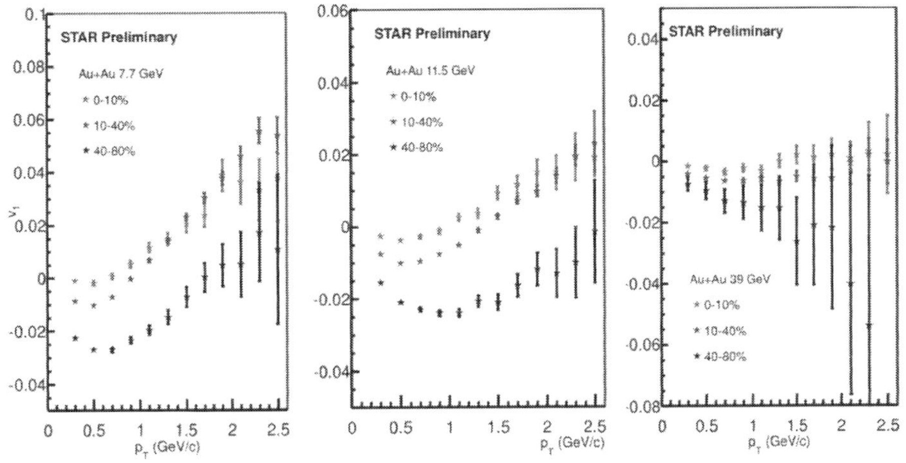


**Figure 4.** Charged hadron $v_1$ as a function of transverse momentum $p_T$ for central 0–10 %, mid central 10–40% and peripheral 40–80 % Au+Au collisions at $\sqrt{s_{NN}} = 39$, 11.5 and 7.7 GeV. The errors shown are statistical. For this plot, particles from the main TPC are used with $|\eta| < 1.0$

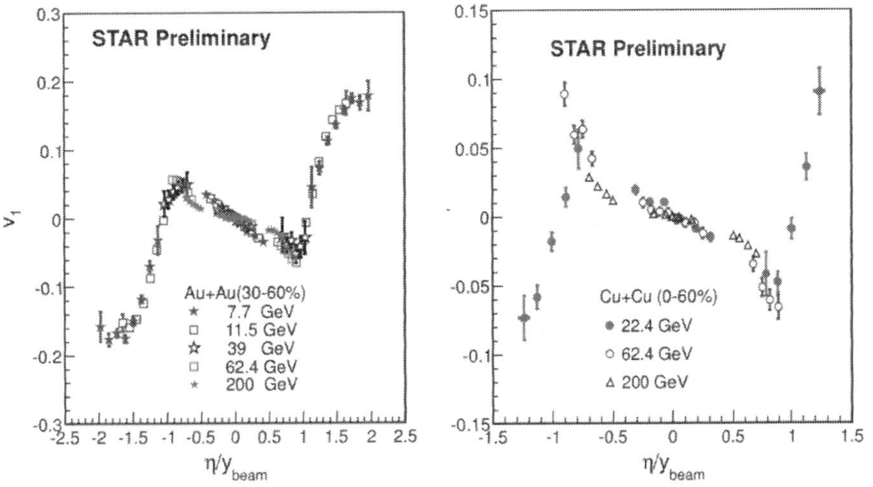


**Figure 5.** The left panel shows charged hadron $v_1$ as a function of $\eta$ scaled by the respective $y_{\rm beam}$ for Au+ Au Collisions at 7.7, 11.5, 39, 62.4 and 200 GeV for 30–60 % central collisions. The right panel shows charged hadron $v_1$ as a function of $\eta$ scaled by the corresponding $y_{\rm beam}$ for Cu+Cu collisions at 22.4, 62.4 and 200 GeV. The results for 62.4 and 200 GeV are for 30–60% centrality, previously reported by STAR [22].

zero on the horizontal axis corresponds to beam rapidity for each of the these beam energies. The data support the limiting fragmentation hypothesis [22] in the region $-2.0 < y_{\rm beam} < -1$ for both system Au+ Au and Cu+ Cu presented here.

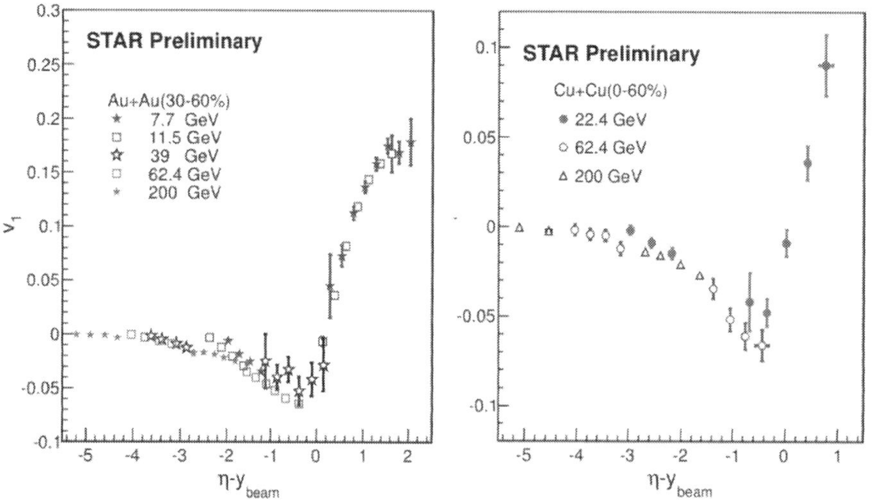


**Figure 6.** The left panel shows charged hadron $v_1$ as a function of $\eta - y_{\text{beam}}$ for Au+Au collisions at 7.7, 11.5, 39, 62.4 and 200 GeV for 30-60 % centrality. The right panel gives charged hadron $v_1$ as a function of $\eta - y_{\text{beam}}$ for Cu+Cu collisions at 22.4, 62.4 and 200 GeV. The results for 62.4 and 200 GeV are for 30–60% centrality, previously reported by STAR [22].

### *3.2. Elliptic flow*

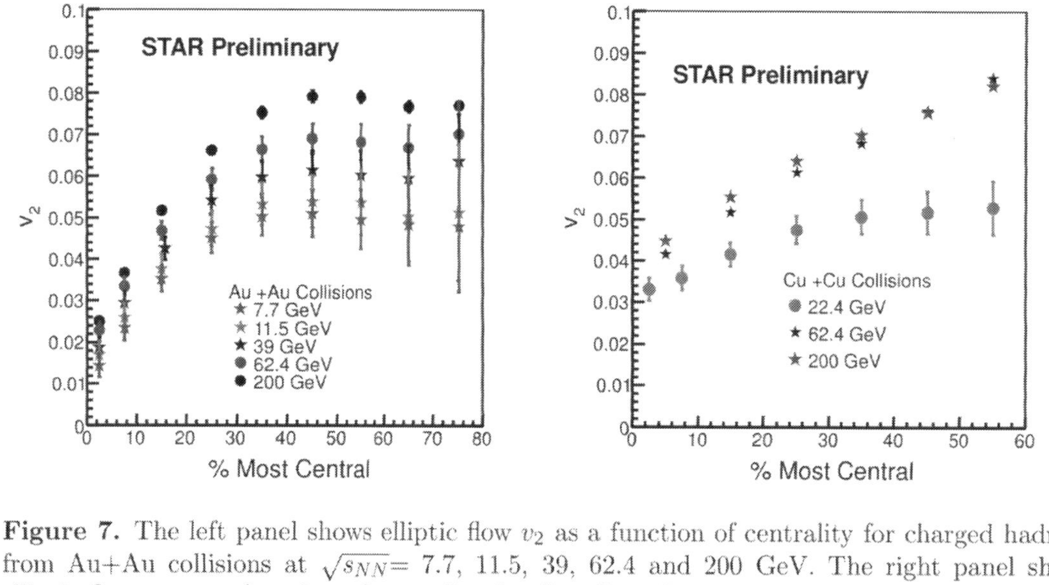


**Figure 7.** The left panel shows elliptic flow $v_2$ as a function of centrality for charged hadrons from Au+Au collisions at $\sqrt{s_{NN}}$= 7.7, 11.5, 39, 62.4 and 200 GeV. The right panel shows elliptic flow $v_2$ as a function of centrality for Cu+Cu collisions at $\sqrt{s_{NN}}$=22.4, 62.4 and 200 GeV. These results are based on the two-particle direct cumulant method with $|\eta| < 1.0$ and $0.2 < p_T < 2.0 GeV$.

Figure 7 (left) shows elliptic flow $v_2$ as a function of centrality for charged hadrons from Au+Au collisions at $\sqrt{s_{NN}}$= 7.7, 11.5, 39, 62.4 and 200 GeV. The right panel presents elliptic flow $v_2$ as a function of centrality for Cu+Cu collisions at $\sqrt{s_{NN}}$=22.4, 62.4 and 200 GeV. These results are based on the two-particle direct cumulant method with $|\eta| < 1.0$ and $0.2 < p_T < 2.0$ GeV. We observe that $v_2$ as a function of centrality increases with beam energy.

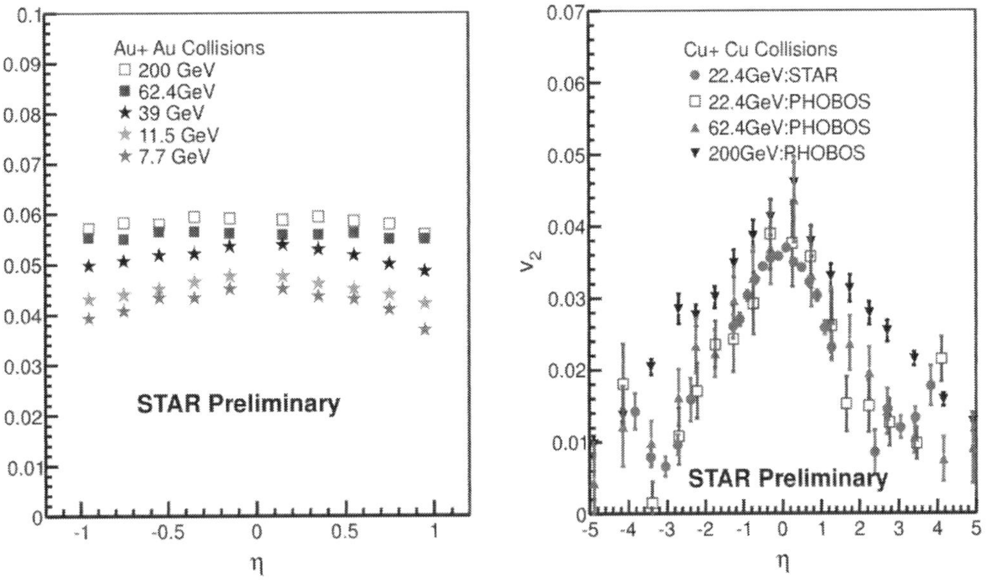


**Figure 8.** The left panel shows $v_2$ as a function of $\eta$ for Au+Au collisions at 7.7, 11.5, 39, 62.4 and 200 GeV for 0–40 % central collisions. On the right, we show the elliptic flow versus $\eta$ for charged hadrons from 22.4 GeV Cu+Cu collisions at 0–60% centrality measured with the $\eta$-subevent method with an $\eta$ gap of 0.3 units. Results are compared with data from the PHOBOS collaboration at 22.4, 62.4 and 200 GeV with 0–40 % centrality.

Figure 8 left panel shows $v_2$ as a function of $\eta$ for Au+Au collisions at 7.7, 11.5, 39, 62.4 and 200 GeV for 0–40 % centrality. On the right, we show elliptic flow versus $\eta$ for charged hadrons from Cu+Cu collisions at 0–60% centrality at $\sqrt{s_{NN}}$ = 22.4 GeV compared with the results from the PHOBOS collaboration at 22.4, 62.4 and 200 GeV with 0–40 % centrality. The new 22.4 GeV Cu+Cu results reported here are based on the $\eta$-subevent method with an $\eta$ gap of 0.3 units. We observe that $v_2$ as a function of $\eta$ increases with beam energy. This energy dependence of $p_T$ integrated elliptic flow as a function of centrality and $\eta$ is driven by the increase in mean $p_T$ with energy.

Figure 9 left panel shows $v_2$ as a function of $p_T$ for Au+Au collisions at 7.7, 11.5 and 39 GeV for 0-80 % centrality. The results reported here are based on the event plane method using full event plane from TPC and are consistent with the result obtained with two particle cumulant methods. On the right, we show elliptic flow versus $p_T$ for charged hadrons from Cu+Cu collisions at 0–60% centrality at $\sqrt{s_{NN}}$ = 22.4 GeV, compared with the previously published results at 200 GeV. Again, the new 22.4 GeV Cu+Cu results reported here are based on the $\eta$-subevent method with an $\eta$ gap of 0.3 units. The error bars include only statistical uncertainties. We observe $v_2$ as a function of $p_T$ obtained using two particle correlation as well as event plane method for Au+Au collisions shows the energy dependence at lower beam energies

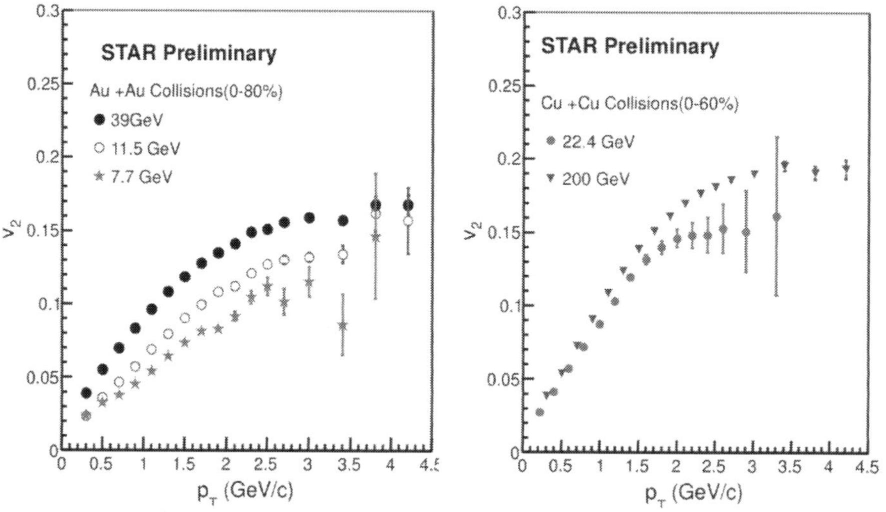


**Figure 9.** The left panel shows $v_2$ as a function of $p_T$ for Au+Au collisions at 7.7, 11.5 and 39 GeV for 0–80 % centrality. The right panel shows elliptic flow versus $p_T$ for charged hadrons from Cu+Cu collisions at 0–60% centrality at $\sqrt{s_{NN}} = 22.4$ GeV measured with the $\eta$-subevent method with an $\eta$ gap of 0.3 units. Results are compared with previously published data at 200 GeV.

upto 39 GeV. However the result obtained using four particle cumulant method [29] shows very small energy dependence.

Figure 10 presents elliptic flow $v_2\{4\}$ for charged hadrons as a function of transverse momentum for various beam energies from 7.7 GeV to 2.76 TeV. The results from 7.7 to 200 GeV are from Au+Au collisions at RHIC and those for 2.76 TeV are from Pb+Pb collisions at the LHC [30]. The red line shows a fit to the results from Au+Au collisions at $\sqrt{s_{NN}}$= 200 GeV. The bottom panel shows $v_2\{4\}$ divided by the fitted line, plotted vs. $p_T$ for all energies. The results are shown for 10–20 % central collisions. It is observed that $v_2\{4\}(p_T)$ for all energies shows similar values beyond $p_T \sim 500$ MeV/$c$. The agreement is within $\sim$ 10 % of the 200 GeV data if we consider only energies of 39 GeV and above. The statistical errors at the energies of 11.5 and 7.7 GeV are larger. This saturation is remarkable considering the very wide energy range involved.

## 4. Summary

In these proceedings, we present STAR results for directed and elliptic flow as a function of transverse momentum, pseudorapidity and centrality for Au+Au collisions at 39, 11.5 and 7.7 GeV and for Cu+Cu collisions at 22.4 GeV. Our findings extend to 22.4 GeV the observation that directed flow is independent of system size (comparing Cu+Cu with Au+Au) at 62.4 and 200 GeV. Our findings also demonstrate that $v_1(\eta/y_{\rm beam})$ shows a beam energy scaling behavior, though not perfect, that has already been established at higher RHIC energies. When viewed in the projectile frame of reference, our results are consistent with the limiting fragmentation hypothesis. Elliptic flow as a function of centrality and pseudorapidity increases with beam energy in both systems studied, i.e., Au+Au and Cu+Cu. The centrality and pseudorapidity dependence are similar to that observed at higher RHIC beam energies. The measured elliptic

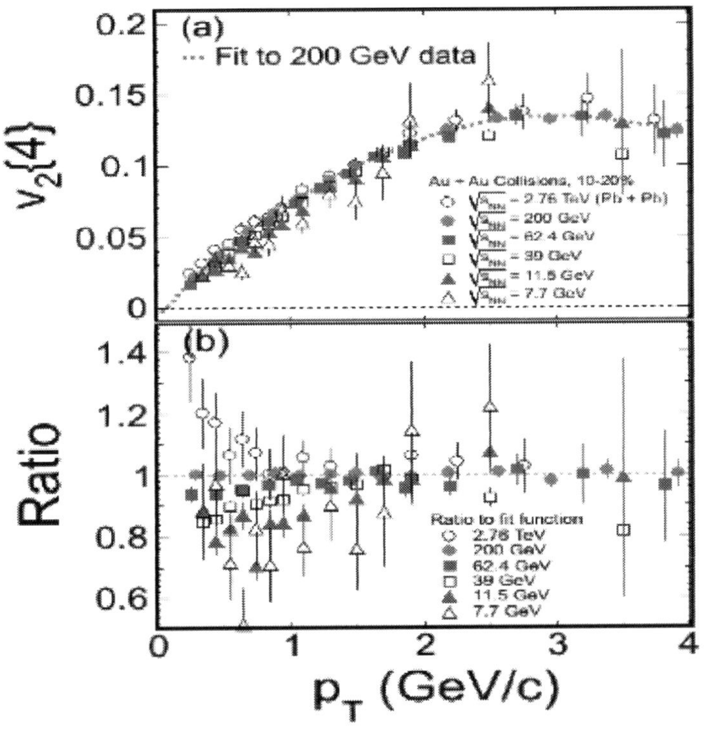


**Figure 10.** Elliptic flow $v_2\{4\}$ for charged hadrons as a function of transverse momentum for various beam energies from 7.7 GeV to 2.76 TeV. The results for 7.7 to 200 GeV are from Au+Au collisions and those for 2.76 TeV are from Pb+Pb collisions [30]. The red line shows a fit to the results from Au+Au collisions at $\sqrt{s_{NN}}$= 200 GeV. The bottom panel shows $v_2\{4\}$ divided by the fitted line, plotted vs. $p_T$ for all energies. The results are shown for 10–20 % central collisions.

flow as a function of transverse momentum increases from 7.7 to 39 GeV but shows saturation above that energy up to 2.76 TeV.

We thank the RHIC Operations Group and RCF at BNL, and the NERSC Center at LBNL and the resources provided by the Open Science Grid consortium for their support. This work was supported in part by the Offices of NP and HEP within the U.S. DOE Office of Science, the U.S. NSF, the Sloan Foundation, the DFG cluster of excellence "Origin and Structure of the Universe", CNRS/IN2P3, RA, RPL, and EMN of France, STFC and EPSRC of the United Kingdom, FAPESP of Brazil, the Russian Ministry of Sci. and Tech., the NNSFC, CAS, MoST, and MoE of China, IRP and GA of the Czech Republic, FOM of the Netherlands, DAE, DST, and CSIR of the Government of India, the Polish State Committee for Scientific Research, and the Korea Sci. & Eng. Foundation and Korea Research Foundation.

## 5. References

[1] BRAHMS Collaboration, I. Arsene *et al.* Nucl. Phys. A **757**, 1 (2005); PHOBOS Collaboration, B. B. Back *et al.* Nucl. Phys. A **757**, 28 (2005); STAR Collaboration, J. Adams *et al.* Nucl. Phys. A **757**, 102 (2005); PHENIX Collaboration, K. Adcox *et al.* Nucl. Phys. A **757**, 184 (2005).
[2] J.-Y. Ollitrault, Phys. Rev. D **46**, 229 (1992).
[3] A. M. Poskanzer and S. A. Voloshin, Phys. Rev. C **58**, 1671 (1998).
[4] H. Sorge, Phys. Rev. Lett. **78**, 2309 (1997).
[5] N. Herrmann, J. P. Wessels, and T. Wienold, Annu. Rev. Nucl. Part. Sci. **49**, 581 (1999).
[6] U. W. Heinz, arXiv:0901.4355.
[7] S. A. Bass *et al.* Prog. Part. Nucl. Phys. **41**, 225 (1998).
[8] B. I. Abelev *et al.* (STAR Collaboration), STAR Note SN0493, 2009.
[9] L. P. Csernai and D. Rohrich, Phys. Lett. B **458**, 454 (1999).
[10] J. Brachmann *et al.*, Phys. Rev. C **61**, 024909 (2000).
[11] H. Stöcker, Nucl. Phys. A **750**, 121 (2005).
[12] R. J. Snellings, H. Sorge, S. A. Voloshin, F. Q. Wang and N. Xu, Phys. Rev. Lett **84**, 2803 (2000).
[13] K. H. Ackermann *et al.*, Nucl. Instr. Meth. A **499**, 624 (2003).
[14] C. A. Whitten, Jr. STAR Collaboration, AIP Conf. Proc. **980**, 390 (2008).
[15] F. S. Bieser *et al.*, Nucl. Instr. Meth. A **499**, 766 (2003).
[16] M. Anderson *et al.*, Nucl. Instr. Meth. A **499**, 659 (2003).
[17] K. H. Ackermann *et al.*, Nucl. Instr. Meth. A **499**, 713 (2003).
[18] B. Christie, S. White, P. Gorodetzky, D. Lazic, STAR Note SN0175, 1994.
[19] M. Anderson, P. M. Dinh, J. Y. Ollitrault, A. M. Poskanzer, S. A. Voloshin, Phys. Rev. C **66**, 014901 (2002).
[20] J. Adams *et al.* (STAR Collaboration), Phys. Rev. C **72**, 014904 (2005).
[21] B. B. Back *et al.* (PHOBOS Collaboration), Phys. Rev. Lett. **97**, 012301 (2006).
[22] B. I. Abelev *et al.* (STAR Collaboration), Phys. Rev. Lett. **101**, 252301 (2008).
[23] Z. W. Lin and C.-M. Ko, Phys. Rev. C **65**, 034904 (2002); L.-W. Chen, C.-M. Ko, J. Phys. G **31**, S49 (2005).
[24] S. A. Bass *et al.*, Prog. Part. Nucl. Phys. **41**, 225 (1998); M. Bleicher *et al.*, J. Phys. G **25**, 1859 (1999).
[25] B. I. Abelev *et al.* (STAR Collaboration), Phys. Rev. C **81**, 044902 (2010).
[26] B. I. Abelev *et al.* (the STAR Collaboration), Phys. Rev. C **75**, 054906 (2007).
[27] R. Nouicer for the PHOBOS Collaboration, J. Phys. G, Nucl. Part. Phys. **34**, S887 (2007).
[28] B. Alver *et al.* (PHOBOS Collaboration), Phys. Rev. Lett. **98**, 242302 (2007).
[29] Nicolas Borghini, Phuong Mai Dinh, and Jean-Yves Ollitrault ,Phys. Rev. C **63**, 054906 (2001).
[30] K. Aamodt et al. (ALICE Collaboration), arXiv:1011.3914 [nucl-ex].

# Rapidity dependence of azimuthal correlations for p+p and d+Au at $\sqrt{s} = 200\text{GeV}$

**Xuan Li**[1,2] **for the STAR Collaboration**

1) School of Physics, Shandong University, Jinan, 250100, China
2) Brookhaven National Lab, Bldg. 510A, Physics Department, upton, NY, 11973, USA

E-mail: xuanli@rcf.rhic.bnl.gov

**Abstract.** Forward di-jet production in d+Au collisions at RHIC provides sensitivity to Bjorken-x between 0.001 and 0.02 for the nuclear gluon density . The STAR experiment at RHIC has continuous azimuthal coverage with electromagnetic calorimeters for pseudo-rapidity from -1 to 4. We have measured forward+mid-rapidity $\pi^0 - \pi^0$ and $\pi^0 - h$ azimuthal correlations and forward+forward rapidity $\pi^0 - \pi^0$ azimuthal correlations. Both analyses use the Forward Meson Spectrometer (FMS) triggered data in RHIC 2008 run p+p and d+Au collisions at $\sqrt{s_{NN}}$=200 GeV. The end-cap electromagnetic calorimeter (EEMC) at STAR covers pseudo-rapidity from 1.07 to 2. This provides us sensitivity to the nuclear gluon density at intermediate x values. Preliminary results on the $\pi^0 - \pi^0$ and $\pi^0$ + jet-like azimuthal correlations with both the FMS and the EEMC will be discussed.

## 1. Introduction

Deep inelastic scattering experiments show that the nucleon gluon density increases rapidly as Bjorken-x ($x_{BJ}$) decreases[1]. The increase at low $x_{BJ}$ is due to partons splitting into more partons. As the gluon density increases, two partons recombining into one becomes increasingly likely. When parton recombination compensates parton splitting, the number of partons will stop increasing and parton saturation is realized. At a given $x_{BJ}$, the nuclear (mass number A) gluon density is approximately $A^{1/3}$ times the nucleon gluon density[2]. Consequently, saturation in the nuclear gluon density is expected to be realized at higher $x_{BJ}$ region than for the nucleon gluon density at fixed $Q^2$.

Forward inclusive $\pi^0$ production probes asymmetric partonic scattering and primarily is from a large $x_F$ quark scattering from a low $x_{BJ}$ gluon. The inclusive production measured in p+p collisions at $\sqrt{s} = 200\text{GeV}$ is consistent with NLO perturbative Quantum Chromodynamics (pQCD) descriptions. Inclusive production measured in d+Au collisions at $\sqrt{s} = 200\text{GeV}$ are better described by Color Glass Condensate (CGC) models[3, 4]. The lower x region can be selectively probed through forward di-jet correlations[5]. The shape and magnitude of back-to-back azimuthal correlation is related to the parton density distribution[6, 7].

With RHIC 2008 run p+p and d+Au collisions, we have measured both forward+mid-rapidity and forward+forward rapidity azimuthal correlations to probe low $x_{BJ}$ nuclear parton distributions. From the forward+mid-rapidity $\pi^0$-$\pi^0$ and $\pi^0 - h$ measurements, the back-to-back azimuthal correlation peaks look similar in p+p and d+Au data. Lack of significant broadening from p+p to d+Au indicates the forward+mid rapidity correlations are not near the

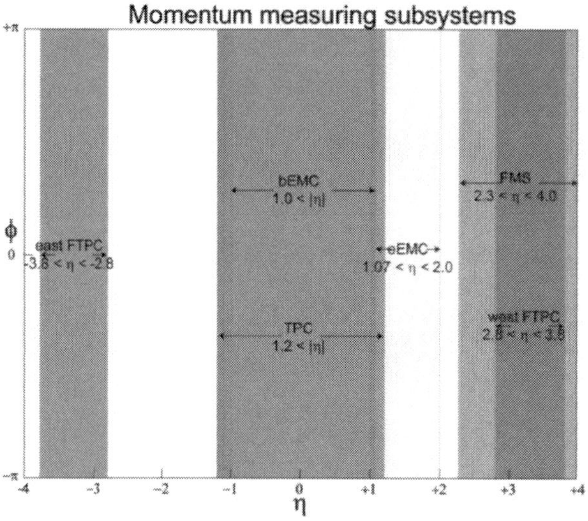


**Figure 1.** *STAR electromagnetic calorimetry $\eta$ $\phi$ coverage.*

saturation region[8]. Broadening from p+p to d+Au has been observed in the forward+forward rapidity back-to-back azimuthal correlations, especially in the central dAu data which show good agreement with the CGC predictions[6]. The observed suppression of the away-side peak in central d+Au forward di-pion events is consistent with gluon saturation expectations.

With $\pi^0$ measured in the forward region, the transition to saturation can be probed through varying the pseudo-rapidity of the associated particle (or jet-like clusters). Azimuthal angle correlations between particles detected at forward rapidity associated with particles at near-forward rapidity probe the intermediate $x_{BJ}$ region to look for the sharpness of the transition from low to high partonic densities[9].

## 2. Experiment setup

The Solenoid Tracker at RHIC (STAR) has nearly continuous full azimuth acceptance of electromagnetic calorimeters in $-1.0 < \eta < 4.0$ (See Figure 1) . The Barrel ElectroMagnetic Calorimeter (BEMC: $-1.0 < \eta < 1.0$), the End-cap ElectroMagnetic Calorimeter (EEMC: $1.07 < \eta < 2.0$) and the Forward Meson Spectrometer (FMS: $2.3 < \eta < 4.0$) comprise the STAR electromagnetic calorimeters. The electromagnetic calorimeters together with the Time Projection Chamber (TPC: $-1.0 < \eta < 1.0$) provide a broad $\Delta\eta \times \Delta\varphi$ coverage for correlation measurements.

In the 2008 run, RHIC provided high luminosity p+p and d+Au collisions. The p+p runs provide reference for di-jet correlations in d+Au runs. The FMS, facing the deuteron beam for the d+Au runs, can probe the gluon density in the nuclear down to $x_{BJ} \approx 10^{-4}$. A high-tower trigger was used for the FMS data taken. The forward+near-forward rapidity azimuthal correlations use neutral pions measured in the FMS and neutral pions or jet-like candidates detected in the EEMC.

## 3. Azimuthal correlation analysis

Non-linear effects on particles produced when a dilute system (e.g. the deuteron) interacts with dense nuclear targets have been predicted by many models. One model which is based on pQCD calculations predicts the partons interact coherently with multiple nucleons changing from a $2 \rightarrow 2$ partonic elastic scattering process to $2 \rightarrow$ *many* process[10]. With higher gluon density in d+Au interactions than in p+p, CGC models predict broadening of back-to-back azimuthal correlations and their suppression (i.e. mono-jets) due to $2 \rightarrow many$ partonic scattering[7]. The CGC signatures in the low $x_{BJ}$ range get enhanced in the large pseudo-rapidity region. With a forward pseudo-rapidity $\pi^0$ triggered in the FMS, we look at the coincident $\pi^0$ in the EEMC. To minimize particle selection bias effect, the FMS $\pi^0$ and EEMC jet-like candidate correlations are also studied.

An electromagnetic calorimeter records the deposited energy of particles passing through it. A threshold bounded cluster finder was developed to identify clusters' energy deposition from connected towers of the EEMC(BEMC). The threshold is tuned to discriminate energy deposition of particles produced by collisions from electronic noise and background. Assuming all clusters are photons, neutral pions are identified from cluster pair mass distributions (Figure 2). The ratio of the leading tower energy to the cluster energy is selected to be $> 0.9$ to suppress background such as hadronic showers. To reduce the detector edge bias, there are fiducial volume selections: $1.1 < \eta < 1.9$ for EEMC and $-0.9 < \eta < 0.9$ for BEMC. Energy sharing between cluster pairs is defined by the energy difference over the cluster pair energy sum. The energy sharing ratio is chosen to be $< 0.7$ to reduce non-$\pi^0$ background. The mass distributions of cluster pairs are in agreement with previous BEMC $\pi^0$ results both in p+p and d+Au runs. Figure 2 shows the invariant masses of EEMC di-clusters with FMS triggered $\pi^0$ in p+p and d+Au data. $\pi^0$'s are evident in the EEMC di-clusters. To emphasize $\pi^0$ signals for azimuthal correlations $mass < 0.2GeV/c^2$ cuts are applied. The FMS $\pi^0$ + BEMC $\pi^0$ azimuthal correlations with the threshold bounded cluster finder are quantitatively consistent with previous results[8]. To compare with forward+mid rapidity results, we select FMS $\pi^0$ $p_t^{FMS} > 2.5GeV/c$ and EEMC $\pi^0$ $1.5GeV/c < p_t^{EEMC} < p_t^{FMS}$. The azimuthal correlation gives the probability to find a $\pi^0$ in the EEMC when a $\pi^0$ is found in the FMS. Presently, background and efficiency corrections are not applied. The azimuthal correlations between FMS $\pi^0$ and EEMC $\pi^0$ are fit with a constant plus a gaussian function. The width of the back-to-back peak is shown in Figure 3. Low statistics in these correlations are caused by cluster pair kinematics. Unlike the BEMC situation, comparison between data and simulation shows that most of the $\pi^0$ candidates are in single clusters in the EEMC. The studies to get more $\pi^0$'s using EEMC single clusters with the help of shower maximum detector (SMD) in EEMC[11] to separate photons is ongoing.

To reduce fluctuations caused by parton fragmentation, jet-like clusters are better substitutes than neutral pions to reflect the behavior of scattered partons. The STAR calorimeters are primarily sensitive to electromagnetic shower energy and only partially sensitive to hadronic shower energy. Consequently, a jet cone algorithm developed from the threshold bounded cluster finder identifies jet-like clusters rather than jets. Initially, we defined a jet-like cluster by using cluster as seed, cone radius $R = 0.5$ ($R : \sqrt{\Delta\eta^2 + \Delta\varphi^2}$), and only one iteration. Later on, with more studies including comparisons between data and simulation, we improved the algorithm by using jet cone radius $R = 0.6$ and multiple iterations to get stable jet-like clusters. Triggered neutral pions are detected in the FMS with transverse momentum $p_t^{FMS} > 2.5GeV/c$ and $2.5 < \eta < 4.0$. Associated jet-like clusters are detected in the EEMC with transverse momentum $1.5GeV/c < p_t^{EEMC} < p_t^{FMS}$ and $1.1 < \eta < 1.9$. Jet-like clusters with $mass > 0.2GeV/c^2$ are chosen for the correlation analysis to suppress clusters that are only a single $\pi^0$, and hence are not explicitly jet-like. The cross section for forward particle productions drops rapidly as $p_t$ grows, so lower $p_t$ cuts are chosen to compare data and simulation. From the jet-like invariant mass distribution (Figure 4), data and simulation are consistent both in p+p and d+Au collisions. The consistency between azimuthal correlations in data and simulation was also checked. The coincident probability (efficiency uncorrected) of azimuthal angle differences between FMS $\pi^0$ and EEMC jet-like candidates are shown in Figure 5 with FMS $\pi^0$ $p_t^{FMS} > 2.5GeV/c$ and EEMC jet-like candidate $1.5GeV/c < p_t^{EEMC} < p_t^{FMS}$. The scales of the peak in the back-to-back correlations ($\Delta\varphi = \pi$) are higher in d+Au interaction than that in p+p. Comparisons

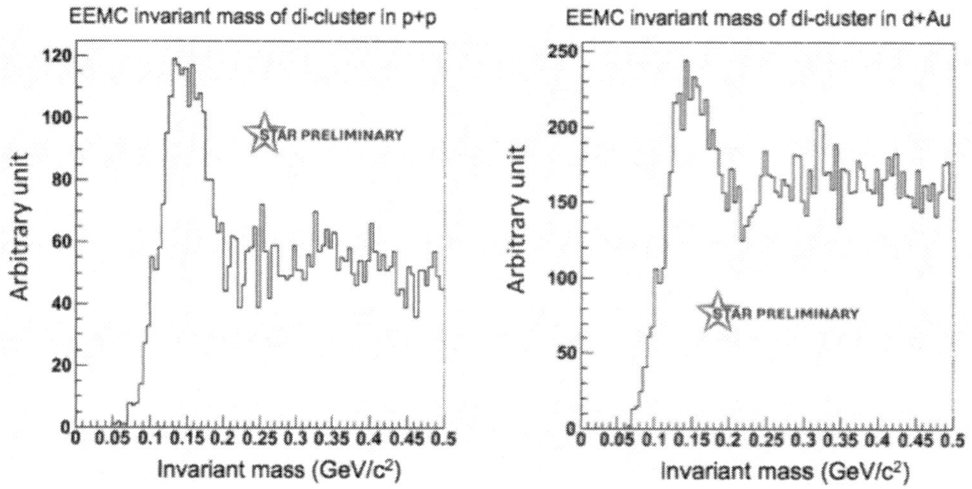


**Figure 2.** *Invariant masses of EEMC di-clusters ($1.5GeV/c < p_t^{EEMC} < p_t^{FMS}$) associated with forward $\pi^0$ ($p_t^{FMS} > 2.5GeV/c$) with further event selections mentioned in the text for p+p (left) and d+Au (right) data.*

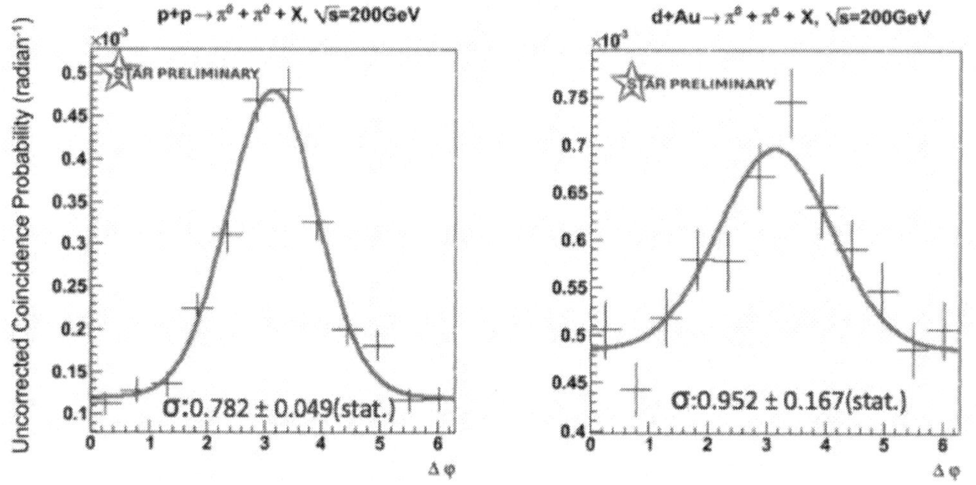


**Figure 3.** *Uncorrected coincidence probability versus azimuthal angle difference between a forward $\pi^0$ ($p_t^{FMS} > 2.5GeV/c$) and a near-forward rapidity $\pi^0$ ($1.5GeV/c < p_t^{EEMC} < p_t^{FMS}$). The red lines stand for the fitting functions which are a constant plus a Gaussian function. The left plot shows the correlations in p+p data and d+Au results are shown on the right plot.*

of azimuthal correlations between p+p and d+Au indicate higher pedestal in d+Au. The width of the peak centered at $\Delta\varphi = \pi$ in d+Au is significantly broader than that in p+p ($\sigma_{dAu} - \sigma_{pp} = 0.131 \pm 0.016$). This indicates that the system in d+Au collisions is approaching saturation at the small Bjorken-x region $0.003 < x_{BJ} < 0.02$. Systematic studies using a random seed in detector coverage and repeating same jet-like clusters algorithm show similar correlation results. Underlying events are also studied with a random seed and a single iteration of the jet finder. These studies show that the jet-like clusters pick up more events in the peak rather than the pedestal compared to the underlying events.

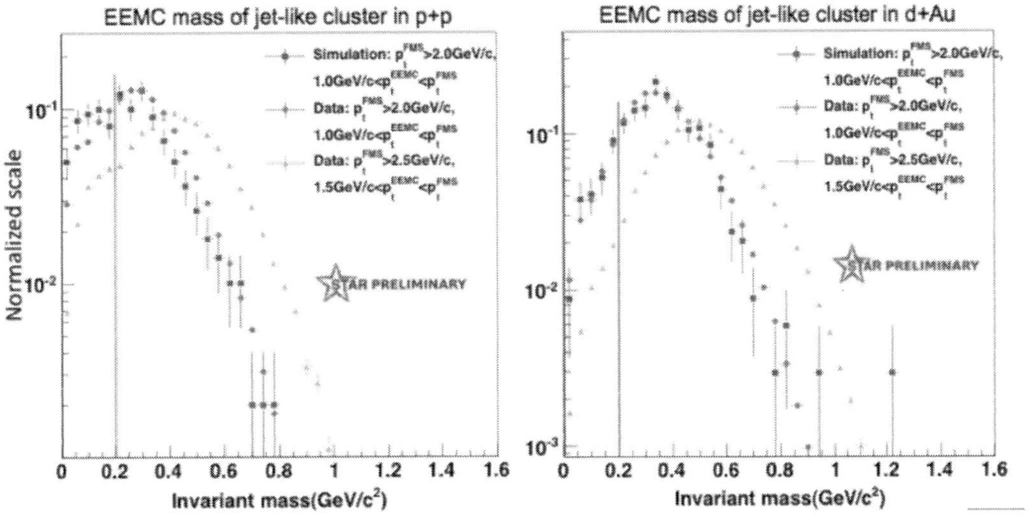


**Figure 4.** *Comparison of data and simulation for EEMC jet-like cluster invariant masses (normalized). With FMS* $\pi^0$ $p_t^{FMS} > 2.0GeV/c$ *and the EEMC jet-like cluster* $1.0GeV/c < p_t^{EEMC} < p_t^{FMS}$*, blue distributions stand for simulation and red distributions stand for data. The green points show the results in data with FMS* $\pi^0$ $p_t^{FMS} > 2.5GeV/c$ *and the EEMC jet-like cluster* $1.5GeV/c < p_t^{EEMC} < p_t^{FMS}$*. The invariant masses of EEMC jet-like clusters in p+p are shown in the left plot and the d+Au results are shown in the right plot. Red lines indicate* $0.2GeV/c^2$ *cut for invariant masses of jet-like clusters.*

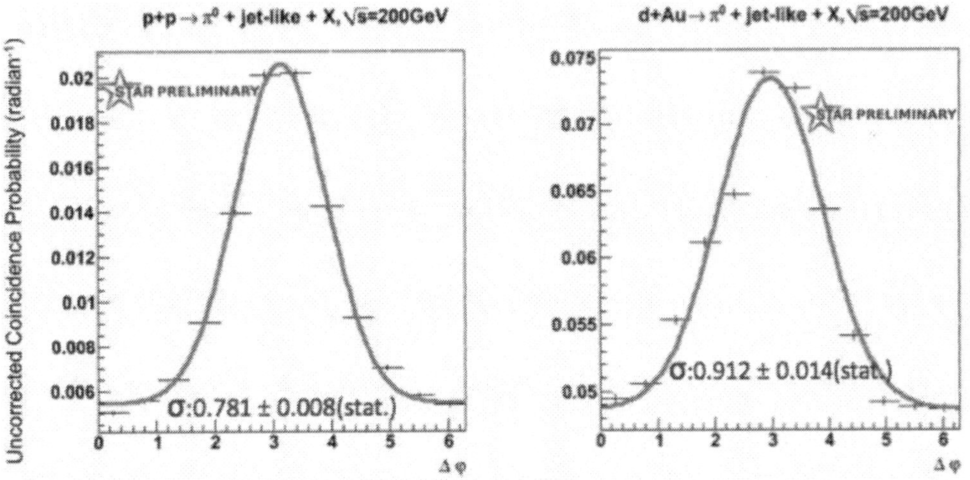


**Figure 5.** *Uncorrected coincidence probability versus azimuthal angle difference between a forward* $\pi^0$ $(p_t^{FMS} > 2.5GeV/c)$ *and a near-forward rapidity jet-like candidate* $(1.5GeV/c < p_t^{EEMC} < p_t^{FMS})$*. The red lines stand for the fitting functions which are a constant plus a Gaussian function. The left plot shows the correlations in p+p data and d+Au results are shown on the right plot.*

## 4. Conclusions and Outlook

RHIC 2008 run had d+Au collisions at $\sqrt{s} = 200$GeV and the STAR FMS provided opportunities to study forward particle triggered di-hadron (hadron+jet-like) azimuthal correlations to probe the low $x_{BJ}$ region. The FMS-EEMC correlations fill in the region between forward+mid rapidity correlations and forward+forward rapidity correlations to provide a fuller picture of the saturation boundary. Azimuthal correlations between a neutral pion in the FMS and jet-like clusters in the EEMC are found to have a broader back-to-back peak in d+Au collisions than in p+p collisions. Combined with earlier studies, this suggests a smooth transition from low to high parton densities.

Back-to-back correlation broadening effect from p+p to d+Au is expected to be stronger at lower $p_t$ in saturation models. Azimuthal correlations in FMS+EEMC with lower $p_t$ cuts are being studied. The forward+forward correlations in d+Au collisions shows dependency on centrality selections[12]. The STAR Beam-Beam Counter(BBC) multiplicity provides a measure of centrality of the event. Therefore, the azimuthal correlations in d+Au interaction can be divided into peripheral and central part with different multiplicity selections. Comparison between d+Au central collisions and d+Au peripheral collisions will be studied.

## 5. References

[1] Rizvi E 2003 Talk presented at the international euro physics conference on high energy physics
[2] Kowalski H and Teaney D 2003 *Phys. Rev. D* **68** 114005 (*Preprint* hep-ph/0304189)
[3] Adams J *et al.* 2006 *Phys. Rev. Lett.* **97** 152302
[4] Kharzeev D, Kovchegov Y V and Tuchin K 2004 *Phys. Lett. B* **599** 23–31
[5] Guzey V, Strikman M and Vogelsang W 2004 *Phys. Lett. B* **603** 173–183
[6] Braidot E 2011 *Nucl. Phys. A* **854** 168–174
[7] Kharzeev D, Levin E and McLerran L 2005 *Nucl. Phys. A* **748** 627–640
[8] Braidot E 2009 *Nucl. Phys. A* **830** 603–606
[9] Iancu E and Venugopalan R 2003 The color glass condensate and high energy scattering in qcd (*Preprint* hep-ph/0303204v3)
[10] Qiu J and Vitev I 2004 *Phys. Rev. Lett* **93** 262301
[11] Allgower C E *et al.* 2003 *Nucl. Instrum. Meth. A* **499** 740
[12] Braidot E 2010 Suppression of forward pion correlations in d+au interactions at star (*Preprint* hep-ph/1005.2378)

# Probing the QCD Critical Point with Higher Moments of Net-proton Multiplicity Distributions

**Xiaofeng Luo for the STAR Collaboration**

Department of Modern Physics, University of Science and Technology of China, Hefei 230026, China

Nuclear Science Division, Lawrence Berkeley National Laboratory, Berkeley, CA 94720, USA

E-mail: xfluo@lbl.gov

**Abstract.** Higher moments of event-by-event net-proton multiplicity distributions are applied to search for the QCD critical point in the heavy ion collisions. It has been demonstrated that higher moments as well as moment products are sensitive to the correlation length and directly connected to the thermodynamic susceptibilities computed in the Lattice QCD and Hadron Resonance Gas (HRG) model. In this paper, we will present measurements for kurtosis ($\kappa$), skewness ($S$) and variance ($\sigma^2$) of net-proton multiplicity distributions at the mid-rapidity ($|y| < 0.5$) and $0.4 < p_T < 0.8$ GeV/$c$ for Au+Au collisions at $\sqrt{s_{\mathrm{NN}}}$=19.6, 39, 62.4, 130 and 200 GeV, Cu+Cu collisions at $\sqrt{s_{\mathrm{NN}}}$=22.4, 62.4 and 200 GeV, d+Au collisions at $\sqrt{s_{\mathrm{NN}}}$=200 GeV and p+p collisions at $\sqrt{s_{\mathrm{NN}}}$=62.4 and 200 GeV. The moment products $\kappa\sigma^2$ and $S\sigma$ of net-proton distributions, which are related to volume independent baryon number susceptibility ratio, are compared to the Lattice QCD and HRG model calculations. The $\kappa\sigma^2$ and $S\sigma$ of net-proton distributions are consistent with Lattice QCD and HRG model calculations at high energy, which support the thermalization of the colliding system. Deviations of $\kappa\sigma^2$ and $S\sigma$ for the Au+Au collisions at low energies from HRG model calculations are also observed.

## 1. Introduction

One of the main goals of heavy ion collision is to explore the phase structure of hot, dense nuclear matter [1]. Finite temperature Lattice QCD calculations demonstrate that with vanishing baryon chemical potential ($\mu_B = 0$), the phase transition from the hadronic phase to the Quark Gluon Plasma (QGP) phase is a smooth crossover [2]. The corresponding transition temperature is about $170 - 190$ MeV [3, 4]. At large $\mu_B$ region, the phase transition between hadronic phase and QGP phase is of first order [5] and with a second order end point at the boundary towards the crossover region, which is so called QCD Critical Point (CP) [6]. Although many efforts have been made by theorist and experimentalist to search for the CP, its location or even the existence is still not confirmed yet [7]. The first principle Lattice QCD calculation at finite $\mu_B$ are difficult due to the fermion sign problem. Several techniques, such as Re-weighting, Image baryon chemical potential and Taylor expansion [8] *etc.*, have been developed to overcome those problems and make the Lattice QCD calculable at finite $\mu_B$ region. However, large uncertainties are still there.

Experimentally, heavy ion collision provides us a good opportunity to search for the CP. To access the QCD phase diagram, we can tune the chemical freeze out temperature ($T$) and baryon chemical potential ($\mu_B$) by varying the colliding energy. As the characteristic signatures of the

CP in a static and infinite medium is the divergence of the correlation length ($\xi$) and increase of non-Gaussian fluctuations. Thus, non-monotonic signals of CP are expected to be observed if the evolution trajectory ( $T, \mu_B$ ) in the QCD phase diagram of the system pass nearby the critical region and the signals are not washed out by the expansion of the colliding system.

Due to the finite size effect, rapid expansion and critical slowing down *etc.*, the typical correlation length ($\xi$) developed in the heavy ion collision near the QCD critical point is a small value about $2-3$ fm [9]. Recently, model calculations reveal that higher moments of conserved quantities distributions are proportional to the higher power of the correlation length [10, 11], such as fourth order cumulant $<(\delta N)^4> -3 <(\delta N)^2>^2 \sim \xi^7$ , where $\delta N = N - M$, $N$ is the particle multiplicity in one event and $M$ is the averaged particle multiplicity of the event sample. On the other hand, the higher moments as well as moment products of conserved quantities distributions are also directly connected to the corresponding susceptibilities in Lattice QCD [12, 13] and HRG model [14] calculations, for e.g. the third order susceptibility of baryon number ($\chi_B^{(3)}$) is related to the third cumulant ($<(\delta N_B)^3>$) of baryon number distributions as $\chi_B^{(3)} = <(\delta N_B)^3>/VT^3$; $V, T$ are volume and temperature of system respectively. It has been found that the volume independence baryon number susceptibility ratio can be directly connected to the moment products of baryon number distributions as $\kappa\sigma^2 = \chi_B^{(4)}/\chi_B^{(2)}$ and $S\sigma = \chi_B^{(3)}/\chi_B^{(2)}$, which allows us to compare the theoretical calculations with experimental measurements. Theoretical calculations also demonstrate that the experimental measurable net-proton number (proton minus anti-proton number) event-by-event fluctuations can reflect the baryon number and charge fluctuations [15]. Thus, higher moments of the net-proton multiplicity distributions are applied to search for the QCD critical point in the heavy ion collisions [16, 17, 18]. When approaching the QCD critical point, the moment products $\kappa\sigma^2$ and $S\sigma$ will show large deviation from its Poisson statistical value. The skewness is expected to change its sign when system evolution trajectory in the phase diagram cross phase boundary [19].

In year 2010, RHIC Bean Energy Scan (BES) program [20] was carried out to map the first order QCD phase boundary and search for the QCD critical point by tuning the colliding energy from 39 GeV down to 7.7 GeV with the corresponding $\mu_B$ coverage about $100-410$ MeV. With the large uniform acceptance and good capability of particle identification STAR detector, it provides us very good opportunities to find the QCD critical point with sensitive observable, if the existence of QCD critical point is true.

## 2. Experimental Method

The data presented in the paper are obtained using the Solenoidal Tracker at RHIC (STAR). Those are Au+Au collisions at $\sqrt{s_{\rm NN}}$=19.6 (year 2001), 39, 62.4 (year 2004), 130 and 200 GeV (year 2004), Cu+Cu collisions at $\sqrt{s_{\rm NN}}$=22.4, 62.4, 200 GeV, d+Au at $\sqrt{s_{\rm NN}}$=200 GeV (year 2003) and p+p collisions at $\sqrt{s_{\rm NN}}$=62.4 (year 2006), 200 GeV (year 2009). The main subsystem used in this analysis is a large, uniform acceptance cylindrical Time Projection Chamber (TPC) covering a pseudo-rapidity range of $|\eta| < 1$ and $2\pi$ azimuthal coverage. As a primary tracking device, it can measure the trajectories and momenta of particles with transverse momenta above 0.15 GeV/$c$. To ensure the purity and similar efficiency, the protons and anti-protons are identified with the ionization energy loss ($dE/dx$) measured by TPC of STAR detector within $0.4 < p_T < 0.8$ GeV/$c$ and mid-rapidity ($|y| < 0.5$). Several track quality cuts are also used to select the tracks with good quality in each event. We require the distance of closest approach (dca) to the primary vertex of proton (antiproton) track less than 1 cm to suppress the contamination from secondary protons (antiproton). To study the centrality dependence of higher moments, centralities are determined by the uncorrected charged particle multiplicities ($dN_{ch}/d\eta$) within pseudo-rapidity $|\eta| < 0.5$ measured by TPC. By comparing measured $dN_{ch}/d\eta$ with the Monto Carlo Glauber model results, we can obtain the average number of participant $N_{part}$, impact parameter $b$ and number of binary nucleon-nucleon collisions $N_{bin}$ for each centrality class.

## 3. Analysis Method

In this section, we will introduce you how to perform the higher moments analysis with experimental data. First, we will show you the definition of the cumulants and various moments used in our analysis, such as standard deviation ($\sigma$), skewness ($S$) and kurtosis ($\kappa$). Then, we will discuss about the *Centrality Bin Width Effect* (CBWE).

### *3.1. Moments and Cumulants of Event-by-Event Fluctuations*

In statistics, probability distribution functions can be characterized by the various moments, such as mean ($M$), variance ($\sigma^2$), skewness ($S$) and kurtosis ($\kappa$). Before introducing the above moments used in our analysis, we would like to define cumulants, which are alternative methods to the moments of a distribution. The cumulants determine the moments in the sense that any two probability distributions whose cumulants are identical will have identical moments as well.

Experimentally, we measure net-proton number event-by-event wise, $N_{p-\bar{p}} = N_p - N_{\bar{p}}$, which is proton number minus antiproton number, in the mid-rapidity ($|y| < 0.5$) and within the transverse momentum $0.4 < p_T < 0.8$ GeV/$c$. In the following, we use $N$ to represent the net-proton number $N_{p-\bar{p}}$ in one event. The average value over whole event ensemble is denoted by $< N >$, where the single angle brackets are used to indicate ensemble average of an event-by-event distributions.

The deviation of $N$ from its mean value are defined by

$$\delta N = N - < N > . \tag{1}$$

Then, we can define the various order cumulants of event-by-event distributions as:

$$C_{1,N} =< N >, C_{2,N} =< (\delta N)^2 >, C_{3,N} =< (\delta N)^3 >, C_{4,N} =< (\delta N)^4 > -3 < (\delta N)^2 >^2 . \tag{2}$$

An important property of the cumulants is their additivity for independent variables. If $X$ and $Y$ are two independent random variables, then we have $C_{i,X+Y} = C_{i,X} + C_{i,Y}$ for $i$th order cumulant. This property will be used in our study.

Once we have the definition of cumulants, various moments can be denoted as:

$$M = C_{1,N}, \sigma^2 = C_{2,N}, S = \frac{C_{3,N}}{(C_{2,N})^{3/2}}, \kappa = \frac{C_{4,N}}{(C_{2,N})^2} \tag{3}$$

And also, the moments product $\kappa\sigma^2$ and $S\sigma$ can be expressed in term of cumulant ratio.

$$\kappa\sigma^2 = \frac{C_{4,N}}{C_{2,N}}, S\sigma = \frac{C_{3,N}}{C_{2,N}}. \tag{4}$$

With above definition of various moments, we can calculate various moments and moment products with the measured event-by-event net-proton fluctuations for each centrality.

### *3.2. Centrality Bin Width Effect (CBWE) Correction*

The centralities in this analysis are determined by the uncorrected charged particle multiplicity ($N_{ch}$) measured at middle pseudo-rapidity ($|\eta| < 0.5$) by the TPC of the STAR detector. Before calculating various moments of net-proton distributions for one centrality, such as 0−5%, 5−10% , we should consider the so called *Centrality Bin Width Effect* (CBWE) arising from the impact parameter fluctuations due to the finite centrality bin width. This effect must be corrected, as it may cause different centrality dependence. To formulate and demonstrate the centrality bin width effect, we write the event-by-event net-proton distributions in one centrality:

$$P(N) = \sum_i \omega_i f^{(i)}(N), (\sum_i \omega_i = 1), \tag{5}$$

where the $\omega_i$ and $f^{(i)}(N)$ are the weighted and net-proton distributions for $i$th impact parameter in one centrality, respectively. From eqs. (2)-(5) and (11), we can calculate the various order cumulants for the distribution $P(N)$ as below:

$$C_{1,N} = \sum_i \omega_i C^i_{1,N} = \sum_i \omega_i < N >_i \tag{6}$$

$$C_{2,N} = (\sum_i \omega_i C^i_{2,N}) + C^{'}_{2,C^i_{1,N}} \tag{7}$$

$$C_{3,N} = (\sum_i \omega_i C^i_{3,N}) + C^{'}_{3,C^i_{1,N}} + 3 \times C^{'}_{1,C^i_{1,N},1,C^i_{2,N}} \tag{8}$$

$$\begin{aligned} C_{4,N} &= (\sum_i \omega_i C^i_{4,N}) + C^{'}_{4,C^i_{1,N}} + 4 \times C^{'}_{1,C^i_{1,N},1,C^i_{3,N}} \\ &+ 6 \times C^{'}_{1,(C^i_{1,N})^2,1,C^i_{2,N}} - 12 \times (C^{'}_{1,C^i_{1,N}})(C^{'}_{1,C^i_{1,N},1,C^i_{2,N}}) \\ &- 3 \times (C^{'}_{2,C^i_{1,N}})^2 + 3 \times C^{'}_{2,C^i_{2,N}}, \end{aligned} \tag{9}$$

where the $C^i_{k,N}$ $(k = 1,2,3,4)$ are the $k^{th}$ order cumulant for net-proton distribution $f^{(i)}(N)$; the $C^{'}_{k,X}$ $(X = C^i_{m,N}, k, m = 1,2,3,4)$ are $k^{th}$ order cutmulant for random variable $X = C^i_{m,N}$ under the discrete probability distribution $Prob(X)$=$\omega_i$; the $C^{'}_{1,X,1,Y}$=$< XY > - < X >< Y >$ $(X = C^i_{k,N}, k = 1,2,3,4)$ are the first order joint cumulant for random variable $X, Y$ under the discrete probability distribution $Prob(X,Y)$=$\omega_i$. We find that the higher order cumulants $C_{k,N}$ $(k = 1,2,3,4)$ can be expressed by the addition of two parts, one is the weighted average of the same order cumulant of each sub-distribution $f^{(i)}(N)$, and the other part is the cumulant of lower order cumulant under the discrete weighted distributions, which stems from the fluctuation of impact parameters within the centrality and results in the CBWE.

Experimentally, the smallest centrality bin is determined by a single uncorrected reference multiplicity value measured by TPC. Generally, we usually report our results for a wider centrality bin, such as $0-5\%$,$5-10\%$,...etc., to supress the statistical fluctuations. To eliminate the centrality bin width effect, we calculate the various moment for each single $N_{ch}$ within one wider centrality bin and weighted averaged by the number of events in each $N_{ch}$.

$$\sigma = \frac{\sum_r n_r \sigma_r}{\sum_r n_r} = \sum_r \omega_r \sigma_r, \tag{10}$$

$$S = \frac{\sum_r n_r S_r}{\sum_r n_r} = \sum_r \omega_r S_r, \tag{11}$$

$$\kappa = \frac{\sum_r n_r \kappa_r}{\sum_r n_r} = \sum_r \omega_r \kappa_r, \tag{12}$$

where the $n_r$ is the number of events in $r^{th}$ refmult and the corresponding weight $\omega_r = n_r / \sum_r n_r$.

## 4. Results

In this section, we will present beam energy and system size dependence for the various moments $(M, \sigma, S, \kappa)$ as well as moment products $(\kappa\sigma^2, S\sigma)$ of net-proton distributions for Au+Au collisions at $\sqrt{s_{\rm NN}}$ = 19.6, 39, 62.4, 130, 200 GeV, Cu+Cu collisions at $\sqrt{s_{\rm NN}}$ = 22.4, 62.4, 200 GeV, d+Au collisions at $\sqrt{s_{\rm NN}}$ = 200 GeV and p+p collisions at $\sqrt{s_{\rm NN}}$ = 62.4, 200 GeV. First, we will show the typical event-by-event net-proton multiplicity distributions from different colliding systems. Then we studied the centrality as well as energy dependence of various moments and moment products. The systematic errors are estimated by varying the following requirements for $p(\bar{p})$ tracks: DCA, track quality reflected by the number of ot points used in track reconstruction and the $dE/dx$ selection criteria for $p(\bar{p})$ identification. The statistical and systematic error are shown separately by lines and brackets, respectively.

### *4.1. Event-by-Event Net-proton Multiplicity Distributions*

Event-by-event net-proton multiplicity distributions for Au+Au collisions at $\sqrt{s_{\rm NN}}$ = 39 GeV measured within $0.4 < p_T < 0.8$ GeV/$c$ and $|y| < 0.5$ are shown in Fig. 1. Going from peripheral to central collisions, it is found that the distributions become wider and more symmetric for central collisions.

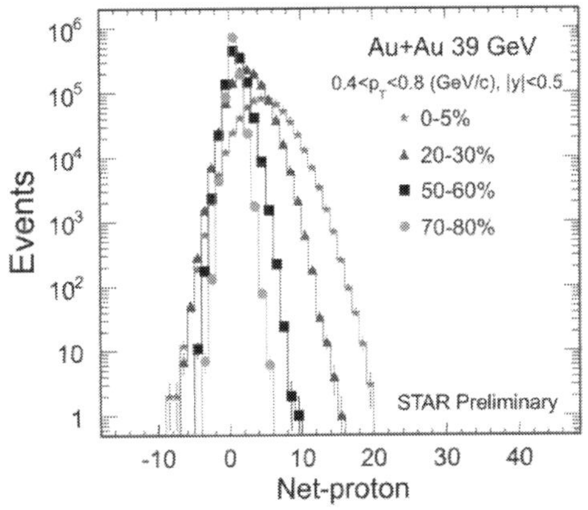


**Figure 1.** Typical event-by-event net-proton multiplicity distributions for Au+Au collisions at $\sqrt{s_{\rm NN}}$ = 39 GeV.

### *4.2. Centrality Dependence of Moments and Moment Products of Net-proton Distributions*

The centrality $(N_{part})$ dependence for various moments $(M, \sigma, S, \kappa)$ of net-proton multiplicity distributions from Au+Au collisions at $\sqrt{s_{\rm NN}}$=39, 62.4 and 200 GeV, Cu+Cu collisions at $\sqrt{s_{\rm NN}}$=22.4, 62.4 and 200 GeV, d+Au collisions at $\sqrt{s_{\rm NN}}$=200 GeV and p+p collisions at $\sqrt{s_{\rm NN}}$=62.4 and 200 GeV are shown in Fig. 2 and Fig. 3. We find that $M$ and $\sigma$ increase with $N_{part}$ monotonically, while $S$ and $\kappa$ decrease with $N_{part}$. The centrality and energy dependence of $S$ and $\kappa$ indicate that the net-proton distributions become more symmetric for more central collision and higher energies.

The dashed lines in the Fig. 2 and Fig. 3 represent the expectations from Central Limit Theorem (CLT) when assuming the superposition of many identical and independent particle emission sources in the system [17, 18]. In Fig. 2 and 3, the centrality dependence of various moments can be well described by the dashed lines expected from CLT. Especially in Fig. 3, the various moments of p+p and d+Au collisions follow the CLT lines of Cu+Cu collision at the

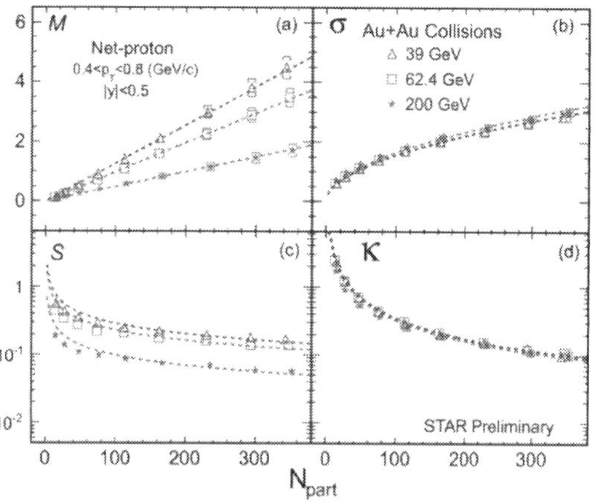


**Figure 2.** Centrality dependence of various moments of net-proton multiplicity distributions for Au+Au collisions at $\sqrt{s_{\rm NN}}$ =39, 62.4, 200 GeV. The dashed lines shown in the figure are expectation lines from Central Limit Theorem (CLT).

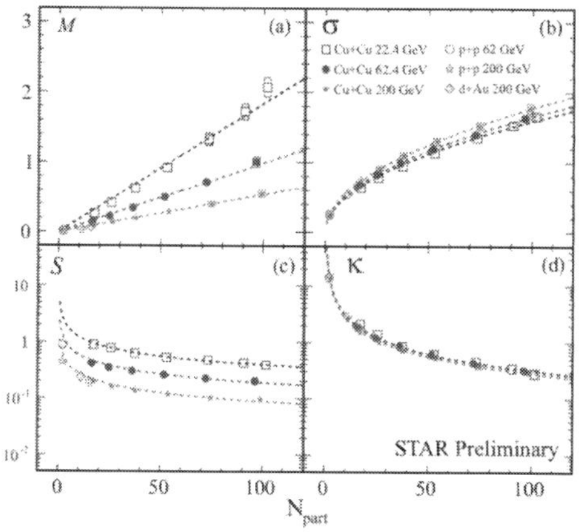


**Figure 3.** Centrality dependence of various moments of net-proton multiplicity distributions for Cu+Cu collisions at $\sqrt{s_{\rm NN}}$ = 22.4, 62.4 and 200 GeV, d+Au collisions at $\sqrt{s_{\rm NN}}$ = 200 GeV and p+p collisions at $\sqrt{s_{\rm NN}}$ = 62.4 and 200 GeV. The dashed lines shown in the figure are expectation lines from Central Limit Theorem (CLT).

corresponding energy very well. This also supports the identical independent emission sources assumption. Fig. 4 shows the centrality dependence of moment products $S\sigma$ and $\kappa\sigma^2$ of net-

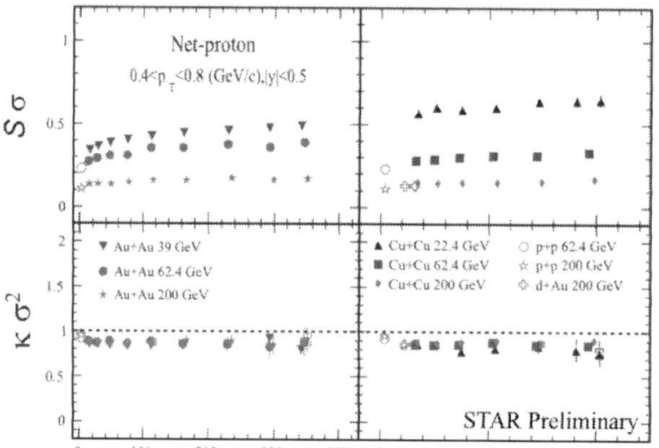


**Figure 4.** Moment products ($\kappa\sigma^2$ and $S\sigma$) of net-proton distributions for Au+Au, Cu+Cu, d+Au and p+p collisions.

proton distributions, which are directly related to the baryon number susceptibility ratio in Lattice QCD and HRG models as $\kappa\sigma^2 = \chi_B^{(4)}/\chi_B^{(2)}$ and $S\sigma = \chi_B^{(3)}/\chi_B^{(2)}$, for p+p, Cu+Cu and Au+Au collisions at various colliding energies. $S\sigma$ shows a weak increase with centrality, while the $\kappa\sigma^2$ shows no centrality dependence.

*4.3. Energy Dependence of the Moment Products ($S\sigma$ and $\kappa\sigma^2$)*

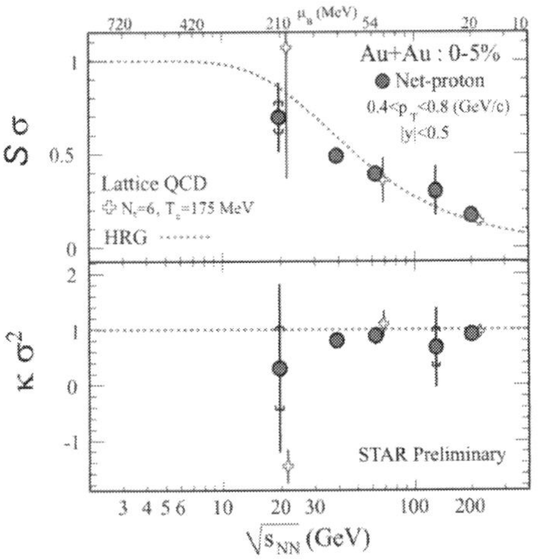


**Figure 5.** Energy dependence of moment products ($\kappa\sigma^2$ and $S\sigma$) of net-proton distributions for central Au+Au collisions (0 – 5%, 19.6 GeV: 0 – 10%, 130 GeV: 0 – 6%). The red dashed lines denote the HRG model calculations, in which the $S\sigma = \tanh(\mu_B/T)$ and $\kappa\sigma^2$=1. The empty markers denote the results calculated from Lattice QCD [4].

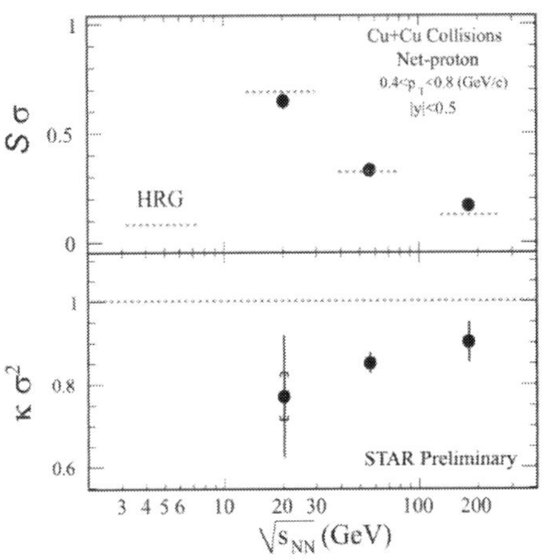


**Figure 6.** Energy dependence of moment products ($\kappa\sigma^2$ and $S\sigma$) of net-proton distributions for Cu+Cu central collisions (0 – 10%, 22.4 GeV: 0 – 5%). The dashed lines shown in the figures are from HRG model calculations, in which $S\sigma = \tanh(\mu_B/T)$ and $\kappa\sigma^2$=1. The $\mu_B$ and $T$ values are taken from [21].

In Fig. 5, we show the energy dependence of the $S\sigma$ and $\kappa\sigma^2$ of net-proton distributions for most central Au+Au collisions (0 – 5%, 19.6 GeV: 0 – 10%, 130 GeV: 0 – 6%). Lattice QCD [4] and HRG model [14] calculations are also shown for comparison. Lattice QCD results are obtained with time extent $N_\tau$ = 6 and phase transition temperature at $\mu_B$=0, $T_c$ = 175 MeV. The red dashed lines of the HRG model in the upper panel and lower panels are evaluated by $S\sigma = \tanh(\mu_B/T)$ and $\kappa\sigma^2 = 1$, respectively, where the $\mu_B/T$ ratio at chemical freeze-out is parameterized as a function of colliding energy based on reference [22]. The corresponding baryon chemical potential ($\mu_B$) at chemical freeze-out for each energy is shown in the upper band of the Fig. 5. We find that the moment products ($\kappa\sigma^2$ and $S\sigma$) of Au+Au collisions at $\sqrt{s_{NN}}$=200, 130, 62.4 GeV are consistent with Lattice QCD and HRG model calculations. The $S\sigma$ and $\kappa\sigma^2$ for Au+Au collisions at $\sqrt{s_{NN}}$ = 39 GeV deviates from HRG model calculations. Surprisingly, $\kappa\sigma^2$ from Lattice QCD calculations at $\sqrt{s_{NN}}$ = 19.6 GeV show a negative value [4]. However, due to the limited statistics, the statistical errors of experimental data are large at 19.6 GeV. The STAR is running to take more data at 19.6 GeV in year 2011. Those deviations could be linked to the chiral phase transition [23] and presence of QCD critical point [24]. Recent linear $\sigma$ model calculations demonstrate that the forth order cumulant of the fluctuations for $\sigma$ field will be universally negative, when the QCD critical point is approached from cross-over side [24]. It will cause the measured $\kappa\sigma^2$ as well as kurtosis ($\kappa$) of net-proton distributions to be smaller than their Poisson expectation values.

Fig. 6 shows the energy dependence of $\kappa\sigma^2$ and $S\sigma$ for Cu+Cu central collisions. The red dashed lines in the figure are obtained from the HRG model by using the formula $S\sigma = \tanh(\mu_B/T)$, where the $\mu_B$ and $T$ are from thermal model fits of the particle ratios. We find that our experimental data is consistent with HRG model expectations for $S\sigma$ of net-proton distributions. While the $\kappa\sigma^2$ deviates from HRG model calculations and monotonically decrease as the collision energy decreases.

### *4.4. Charged Particle Density ($dN_{ch}/d\eta$) Scaling of $S\sigma$ and Evidence of Thermalization in Heavy Ion Collisions*

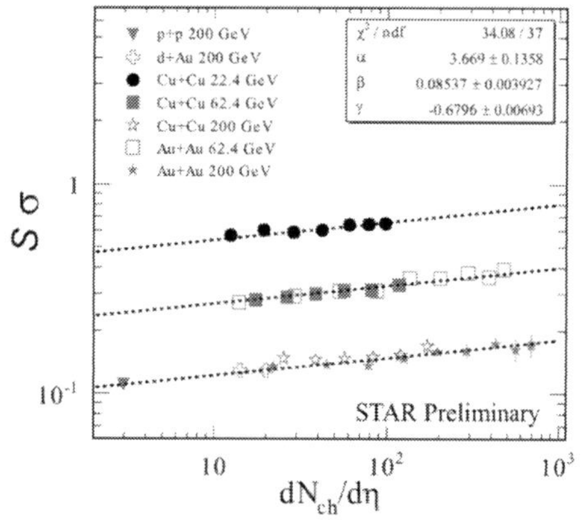


**Figure 7.** $S\sigma$ of net-proton distributions as a function of charged particle density at mid-rapidity ($dN_{ch}/d\eta$) for various colliding systems. The dashed lines in the figures is the fitting lines.

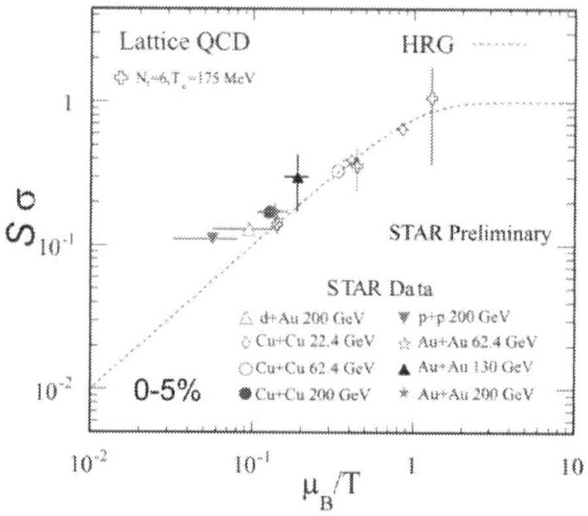


**Figure 8.** $S\sigma$ of net-proton distributions as a function of baryon chemical potential over temperature ratio ($\mu_B/T$) for most central collisions ($0-5\%$, 130 GeV: $0-6\%$, d+Au : $0-20\%$). The red dashed line in the figures is the result of the HRG model. Lattice QCD results with $N_\tau = 6$ and $T_c = 175$ MeV are also shown in the figure.

The $S\sigma$ of net-proton distributions for various colliding systems including Au+Au collisions at $\sqrt{s_{NN}} = 62.4$ and 200 GeV, Cu+Cu collisions at $\sqrt{s_{NN}} = 22.4$, 62.4 and 200 GeV, d+Au and p+p collisions at $\sqrt{s_{NN}} = 200$ GeV, as a function of $dN_{ch}/d\eta$ are shown in the Fig. 7 with double logarithm axis. It is obvious that for a fixed colliding energy, such as $\sqrt{s_{NN}}$=200 GeV, the moment products of $S\sigma$ of net-proton distributions for different system size the p+p, d+Au, Cu+Cu and Au+Au collisions have power law dependence on the charged particle density ($dN_{ch}/d\eta$). Thus, we fit the $S\sigma$ of net-proton distributions for various colliding systems with double power law formula:

$$S\sigma(dN_{ch}/d\eta, \sqrt{s_{\rm NN}}) = \alpha \times (dN_{ch}/d\eta)^{\beta} \times (\sqrt{s_{\rm NN}})^{\gamma} \tag{13}$$

The fitting results are shown in the Table. 1. The $S\sigma$ can be well described by the power law

**Table 1.** Fitting parameters for $S\sigma$ as a function of $dN_{ch}/d\eta$

| Parameters | Value | Approx. |
|---|---|---|
| $\chi^2/ndf$ | 34.08/37 | 0.92 |
| $\alpha$ | 3.669±0.1358 | $\frac{11}{3}$ |
| $\beta$ | 0.0853±0.003927 | $\frac{1}{12}$ |
| $\gamma$ | -0.6796±0.00693 | $-\frac{2}{3}$ |

formula $S\sigma = \frac{11}{3} \times (\frac{1}{s^4}\frac{dN_{ch}}{d\eta})^{\frac{1}{12}}$, where the $s$ is the square of the center of mass energy. For high energy heavy ion collisions the temperature is approximately constant and the ratio $\mu_B/T << 1$, thus we have the approximation $\mu_B/T \sim \tanh(\mu_B/T) = S\sigma = \frac{11}{3} \times (\frac{1}{s^4}\frac{dN_{ch}}{d\eta})^{\frac{1}{12}}$. This denotes the relation between $\mu_B/T$ and the charged particle density and colliding energy for high energy nuclear collisions.

Multiplicity fluctuations and inclusive yields are two basic properties in high energy heavy ion collisions. For a thermal system, both the fluctuations and yields should be described by the thermodynamic parameters ($\mu_B$ and $T$), which completely determine the properties of the thermal system. The fluctuation observable $S\sigma$ of most central net-proton distributions versus thermodynamic parameter $\mu_B/T$ ratios, which are extracted from the thermal model fit of the particle ratio, is shown in the Fig. 8 for various colliding systems. Lattice QCD calculations with $N_\tau = 6$ and $T_c = 175$ MeV and the HRG model relation $S\sigma = \tanh(\mu_B/T)$ are also plotted in Fig. 8 for comparison. We find that high energy heavy ion collisions, such as Au+Au and Cu+Cu collisions, are consistent with Lattice QCD and HRG model calculations, while the elementary p+p collision deviate from the HRG model calculations. In addition to the perfect description of the particle yields by the thermal model, the agreement of higher order fluctuations with thermal model predications provide further evidence that the colliding system has achieved thermalization in most central high energy heavy ion collision.

## 5. Summary

The higher moments of net-proton multiplicity distributions measured in heavy ion collision experiment have been applied to search for the QCD critical point, due to the high sensitivity to the correlation length ($\xi$). The beam energy and system size dependence for higher moments $(M, \sigma, S, \kappa)$ as well as moment products ($\kappa\sigma^2$, $S\sigma$) of net-proton multiplicity distributions have been presented with a broad energy range and different system sizes, which include Au+Au collisions at $\sqrt{s_{NN}}$ = 200, 130, 62.4, 39 and 19.6 GeV, Cu+Cu collisions at $\sqrt{s_{NN}}$ = 200, 62.4 and 22.4 GeV, d+Au collisions at $\sqrt{s_{NN}}$ = 200 GeV, p+p collisions at $\sqrt{s_{NN}}$ = 200 and 62.4 GeV. The moment product $\kappa\sigma^2$ shows no centrality dependence while $S\sigma$ shows a weak centrality dependence. The energy dependence is studied by comparing the results from the Au+Au 200 GeV to those from the BES energies. The moment products $\kappa\sigma^2$ and $S\sigma$ of net-proton distributions from most central Au+Au collisions are consistent with Lattice QCD and HRG model calculations at high energy (200, 130, 62.4 GeV) while deviating from (smaller than) HRG model calculations at $\sqrt{s_{NN}}$ = 39 GeV. Lattice QCD calculations show negative value for $\kappa\sigma^2$ at $\sqrt{s_{NN}}$ = 19.6 GeV. The deviations could potentially be linked to chiral phase transitions and QCD critical point. But the experimental data is with large error bar due to the limited statistics. Fortunately, this ambiguity can be clarified soon by 19.6 GeV data taken in year 2011 with higher statistics. Recent model calculations show that the $\kappa\sigma^2$ value will always be smaller than its Poisson statistical value 1, when QCD critical point is approached from the high energy cross-over side.

On the other hand, the mutual agreements between the $\mu_B/T$ extracted from thermal model fits of particle ratio and from the event-by-event fluctuations observable $S\sigma$ of net-proton distributions provides further evidence of thermalization of the hot dense matter created in the heavy ion collisions. Further, the $\kappa\sigma^2$ and $S\sigma$ of net-proton distributions are consistent with Lattice QCD and HRG model calculations at high energy, which support the thermalization of the colliding system. The deviations of the $\kappa\sigma^2$ and $S\sigma$ of net-proton distributions for Au+Au central collisions from HRG model predications at low energies are not well understood. This may result from the non-applicability of grand canonical ensemble or the appearance of QCD critical point and chiral phase transitions at low energies. However, it should be further investigated.

**Acknowledgement**
We thank S. Gupta, F. Karsch, K. Rajagopal, K. Redlich, M. Stephanov for enlightening discussions. We thank the RHIC Operations Group and RCF at BNL, the NERSC Center at LBNL and the Open Science Grid consortium for providing resources and support. This work was supported in part by the Offices of NP and HEP within the U.S. DOE Office of Science, the U.S. NSF, the Sloan Foundation, the DFG cluster of excellence 'Origin and Structure of the Universe' of Germany, CNRS/IN2P3, FAPESP CNPq of Brazil, Ministry of Ed. and Sci. of the Russian Federation, NNSFC, CAS, MoST, and MoE of China, GA and MSMT of the Czech Republic, FOM and NWO of the Netherlands, DAE, DST, and CSIR of India, Polish Ministry of Sci. and Higher Ed., Korea Research Foundation, Ministry of Sci., Ed. and Sports of the Rep. Of Croatia, and RosAtom of Russia.

## References

[1] J. Adams *et al.* (STAR Collaboration), *Nucl. Phys. A* **757** (2005) 102.
[2] Y. Aoki *et al.*, *Nature* **443** 675 (2006).
[3] Y. Aoki *et al.*, *Phys. Lett. B* **643** (2006) 46; M. Cheng *et al.*, *Phys. Rev. D* **74** (2006) 054507.
[4] S. Gupta, X. Luo, B. Mohanty, H. G. Ritter, N. Xu, arXiv: 1105.3934.
[5] S. Ejiri, *Phys. Rev. D* **78** (2008) 074507 ; E. S. Bowman, J. I. Kapusta, *Phys. Rev. C* **79** (2009) 015202.
[6] M. A. Stephanov, *Prog. Theor. Phys. Suppl.* **153** (2004) 139; Z. Fodor and S. D. Katz, *Journal of High Energy Physics* **0404** (2004) 50; R. V. Gavai, S. Gupta, *Phys. Rev. D* **78** (2008) 114503.
[7] M. A. Stephanov, *Int. J. Mod. Phys. A* **20** (2005) 4387; B. Mohanty, *Nucl. Phys. A* **830** (2009) 899c.
[8] Z. Fodor, S. D. Katz, *Phys. Lett. B* **534** (2002) 87; Philippe de Forcrand and O. Philipsen, *Nucl. Phys. B* **642** (2002) 290; R. V. Gavai, S. Gupta, *Phys. Rev. D* **68** (2003) 034506.
[9] M. A. Stephanov *et al.*, *Phys. Rev. D* **60** (1999) 114028; B. Berdnikov and K. Rajagopal, *Phys. Rev. D* **65** (2000) 105017.
[10] M. A. Stephanov, *Phys. Rev. Lett.* **102** (2009) 032301.
[11] C. Athanasiou *et al.*, *Phys. Rev. D* **82** (2010) 074008.
[12] R. V. Gavai and S. Gupta, *Phys. Lett. B* **696** (2011) 459.
[13] M. Cheng, *et al.*, *Phys. Rev. D* **79** (2009) 074505.
[14] F. Karsch and K. Redlich, *Phys. Lett. B* **695** (2011) 136.
[15] Y. Hatta and M. A. Stephanov, *Phys. Rev. Lett.* **91** (2003) 102003.
[16] X. Luo *et al.*, arXiv:1105.5049.
[17] M. M. Aggarwal *et al.* (STAR Collaboration), *Phys. Rev. Lett.* **105** (2010) 022302.
[18] X. Luo *et al.*, *J. Phys. G: Nucl. Phys.* **37** (2010) 094061.
[19] M. Asakawa *et al.*, *Phys. Rev. Lett.* **103** (2009) 262301.
[20] B. I. Abelev *et al.* (STAR Collaboration), *Phys. Rev. C* **81** (2010) 024911; STAR Internal Note - SN0493, 2009.
[21] M. M. Aggarwal *et al.*, *Phys. Rev. C* **83** (2010) 34910.
[22] J. Cleymans *et al.*, *Phys. Rev. C* **73** (2006) 034905.
[23] B. Friman, arXiv:1103.3511.
[24] M. A. Stephanov, arXiv:1104.1627.

27th Winter Workshop on Nuclear Dynamics (WWND 2011) IOP Publishing
Journal of Physics: Conference Series **316** (2011) 012004 doi:10.1088/1742-6596/316/1/012004

# Forward Transverse Single Spin Asymmetries at PHENIX

**Anselm Vossen for the PHENIX Collaboration**
Indiana University, CEEM, 2401 Milo B. Sampson Lane, Bloomington, IN 47408 USA
E-mail: avossen@indiana.edu

**Abstract.** Recent measurements of single transverse spin asymmetries in proton-proton collisions measured by the PHENIX experiment at RHIC are presented. The focus is on the single particle left-right asymmetry $A_N$ for $\pi^0$ at $\sqrt{s} = 200$ GeV and $\sqrt{s} = 62.4$ GeV and the measurement of di-hadron correlations at $\sqrt{s} = 200$ GeV which are produced by the fragmentation of a transversely polarized quark via the Interference Fragmentation Function (IFF) $H_1^{<}$ and thus provide a probe for the quark transversity distribution function.

## 1. Introduction

The single transverse spin asymmetry $A_N$ for pions in proton-proton collisions was first discovered more than 30 years ago at the ZGS. Even though it is a fairly simple observable, the left right asymmetry of count-rates with respect to the spin direction of the polarized proton in the process $p + p^{\uparrow} \rightarrow \pi + \mathrm{X}$, where $\pi$ can be charged or neutral, our understanding of the mechanism(s) leading to this asymmetry is still not complete. The asymmetry persists all the way up to RHIC energies with a center of mass energy of up to 200 GeV. At these energies it is reasonable to use a factorized picture to explain the asymmetries. If transverse momentum dependence of parton distribution functions (PDFs) and fragmentation functions (FFs) is introduced, there are two mechanisms which can generate left-right asymmetries at leading order in the hard part and leading twist in the soft part of the interaction. The first one, called the Sivers effect, is related to the transverse momentum dependence of PDFs. Proposed first by Sivers [1], an azimuthal asymmetry is generated by the spin dependence of the intrinsic transverse momentum of the quarks within the transversely polarized nucleon described by the Sivers function $f_{1T}^{\perp}(x, k_T^2)$. Here $x$ is the fractional momentum of the nucleon carried by the parton and $k_T$ its intrinsic transverse momentum. The Sivers function is naive time reversal odd, therefore a phase shift in the amplitude is needed for the asymmetry to survive. In models for the Sivers function the phase shift is generated by a final state interaction. In a simple model by Burkhard [3], the Sivers function is connected to quark orbital angular momentum, a missing part of the spin puzzle [3]. The idea is, that orbital angular momentum will lead to a red- or blueshift of the quark that participates in the hard scattering. Due to spin-orbit coupling the OAM will be spin dependent. Since the PDFs are strongly dependent on the momentum fraction $x$ of the nucleon that the quark carries, this shift leads to a spin dependent asymmetry in impact parameter space. The attractive strong force will then translate the impact parameter space asymmetry into a momentum space asymmetry of the produced hadrons, a mechanism which Burkhard termed "Chromodynamic lensing". The second effect leading to single transverse spin asymmetries is known as the Collins effect. Here a spin dependent transverse momentum dependence of the fragmentation function is assumed. The spin dependent transverse momentum dependence of the fragmentation of a transversely polarized quark into unpolarized mesons can be described by the Collins fragmentation function $H_1^\perp(z, \kappa_T^2)$ [2], with $z$ being the fractional momentum of the fragmenting parton the hadron carries and $\kappa_T$ the intrinsic transverse momentum of the produced hadron with respect to the parton. This function is chiral odd and couples to the transversity distribution function $h_1(x)$, which is also a chiral odd quantity. In fact, since in QCD chiral odd amplitudes are heavily suppressed, using a chiral odd fragmentation function like the Collins FF or the Interference Fragmentation Function [4] have so far been the only successful ways to access the transversity distribution. Therefore $h_1(x)$ is the least well known of the three leading twist PDFs which are equally important to describe the spin structure in a collinear picture [5].

Both, Sivers and Collins effect have been very successful at describing azimuthal asymmetries in semi-inclusive deep inelastic scattering (SIDIS). Here the two effects can be separated because the parton kinematics can be reconstructed from the scattered lepton. Given universality and factorization, the SIDIS results can be used to explain the asymmetries in p+p collisions. In the collinear case a benchmark for the validity of a factorized approach has been the agreement of the measured cross-sections for pions with the theoretical predictions using PDFs and FFs. Using leading log resummation this agreement works well down to a center of mass energy of 62.4 GeV for $\pi^0$'s [6]. However, if intrinsic transverse momentum is taken into account, it was shown that TMD factorization does not work without modification by process dependent soft factors [11]. The most prominent example is probably the sign flip of the Sivers effect in Drell-Yan and SIDIS [7]. At the time of this writing it is not clear, if the universality for TMD PDFs can be saved. If factorization is broken, it will be up to the experiments to determine how large the effect is.

At large transverse momenta, the phase shift necessary to produce the SSAs can be generated in the hard part by including an additional parton exchange with the nucleon. Since the spin asymmetry is now dependent on the interaction with two partons in the polarized nucleon, the nucleon structure has to be described by twist three correlation functions [8, 9]. It has been shown that the observed asymmetries are well described by a fit based on a model for the twist three correlation functions [10]. If one assumes that contributions from the Collins effect are small, both, Sivers and twist three functions should describe the asymmetries in the kinematic regions in which they are thought to be valid. For Sivers, these are large momentum transfers and small transverse momenta, so there are two momentum scales involved, whereas the twist three mechanism has only one hard scale. In the kinematic region of overlap, it has been shown explicitly for the Drell Yan process that both lead to the same result [12]. Since twist three effects have an intrinsic $1/p_T$ dependence, it is expected that $A_N$ exhibits this behavior at high $p_T$. However this has not been observed yet. Additional interest in $A_N$ has arisen from the discovery that using a relation between the quark gluon correlation function $T_{q,F}$ and the first moment of the Sivers function, there is a sign mismatch between $T_{q,F}$ extracted from p+p collisions and the moment of the Sivers function extracted from fits to SIDIS data [13]. The reason is not understood yet, but may hint either at an incomplete understanding or a strong kinematic dependence of the Sivers function. It appears thus that the measurement of the transverse SSA $A_N$ provides access to a multitude of effects, which also makes it interpretation difficult. This article presents the PHENIX measurement for the transverse spin asymmetry $A_N$ for neutral pions at $\sqrt{s} = 62.4$ GeV and $\sqrt{s} = 200$ GeV in the forward region. The forward and backward regions in this context are defined with respect to the momentum of the polarized proton. Hadrons scattered into the forward region come from fragmenting valence quarks and thus lead to a strong analyzing power. We look at various kinematic dependences, especially at the $p_T$ dependence to find the onset of the $1/p_T$ behavior. By showing all relevant kinematic dependences of the analyzing power, we hope to learn more about the origin of $A_N$ and give data for the check of theoretical model calculations. In addition to the measurement of $A_N$ we also present the measurement of the di-hadron correlations in the central region. Using recent results by the Belle collaboration, using di-hadron correlations which probe the product of transversity distribution to the interference fragmentation function, might lead to the first extraction of transversity from proton-proton data. Thus a new kinematic regime can be accessed, a crucial step towards determining the tensor charge. In addition, the extraction uses only collinear factorization which is different for the Collins effect. Other advantages compared to using the Collins effect is that for the relevant invariant mass range of the produced pion pairs the evolution is known, jet reconstruction is not necessary and the asymmetries are not Sudakov suppressed [14].

## 2. The PHENIX Detector

The PHENIX detector is described in detail in [15]. It is located at the Relativistic Heavy Ion Collider (RHIC) at BNL. The data presented here was taken in the years 2006 and 2008. In 2006, PHENIX sampled a delivered luminosity of $4.6 \times 10^{-3}\text{pb}^{-1}$ at $\sqrt{s} = 62.4$ GeV with a transverse polarization of 48%. We also sampled 0.7 $\text{pb}^{-1}$ at $\sqrt{s} = 200$ GeV with 54% polarization. In 2008 we had only 200 GeV running with a delivered integrated luminosity of 1.1 $\text{pb}^{-1}$ but a lower transverse polarization of 46%. The results for the forward transverse SSA $A_N$ presented here are from $\pi^0$'s detected in the Muon Piston Calorimeters (MPC) [16], located 220 cm on both sides of the interaction region. The calorimeter uses $PbWO_4$ crystals, due to the small Moliere radius of 2.0 cm. The crystals have a surface area of $2.2 \times 2.2$ $\text{cm}^2$ and a radiation length of 20 $X_0$. They are read out by avalanche photo diodes (APDs). With the two detectors a pseudorapidity range of $3.1 < |\eta| < 3.9$ is covered. In the central region $-0.35 < \eta < 0.35$ PHENIX is intrumented by two arms each covering $\pi/2$ of the azimuth. The detectors used for di-hadron analysis are the pad- and wire-chambers for unidentified charged tracks and the electromagnetic calorimeters in each arm to reconstruct neutral pions.

## 3. $A_N$ measurement for reconstructed pions in MPC at $\sqrt{s} = 62.4$ GeV

Figure 1 shows the results for $A_N$ measured in the 62.4 GeV run in 2006. We divided the dataset into two of roughly equal size by selecting pions with $\eta > 3.5$ and $\eta < 3.5$. In the backward region, where sea-quarks in the polarized proton are probed, the asymmetries are very small. This is expected, since both, Sivers function and transversity are only significant in the valence quark region. For hadrons scattered in the forward direction with respect to the polarized proton momentum the asymmetries rise with $x_F$ for the same reason. At high $x_F$ for $\eta < 3.5$ the asymmetries reach a maximum. Compared with the data points at $\eta > 3.5$, for fixed $x_F$ the $p_T$ has to be higher to reach lower $\eta$. This is also evident in the average $p_T$ for each bin noted in the figure. Thus differences between the two different pseudorapidity bins could be attributed to the $p_T$ dependence of the asymmetries which leads to falling asymmetries at high $p_T$ as expected in the twist three framework.

## 4. $A_N$ from clusters in the MPC at $\sqrt{s} = 200$ GeV

With the given crystal size of the MPC, the two photon decay of the $\pi^0$ leads to merged clusters for pion energies > 20 GeV. Translated into the variable $x_F = \frac{2p_L}{\sqrt{s}}$ this means that at 62.4 GeV clusters will be merged for $x_F > 0.65$ and for 200 GeV already at $x_F > 0.2$. At leading order in the parton picture $x_F$ can be interpreted as the difference in the momentum fractions $x$ the partons participating in the hard scattering carry. Figure 4 shows the distribution of $x$ for the participating partons in the central and forward region of PHENIX. From these distributions it also becomes obvious why only forward measurements can access valence quarks at high $x$. At the position of the MPC, the pion energy at which the clusters are merging corresponds to

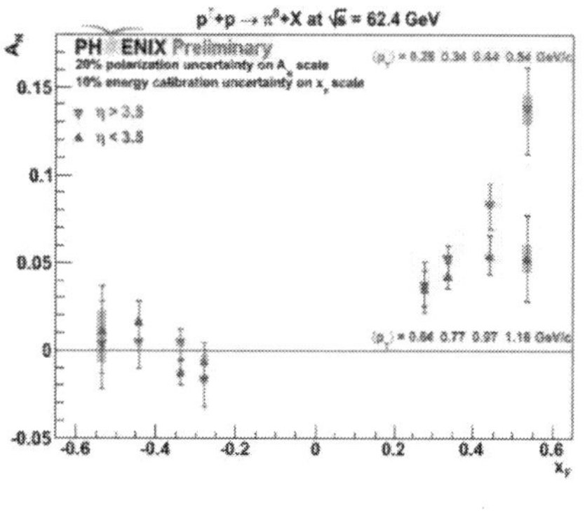


**Figure 1.** Results for $A_N$ at $\sqrt{s}$ = 62.4 GeV in two different bins of $\eta$

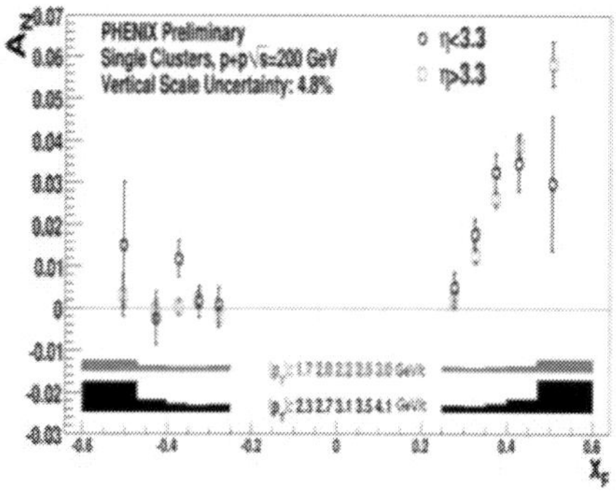


**Figure 2.** Results for $A_N$ at $\sqrt{s}$ = 200 GeV in two different bins of $\eta$

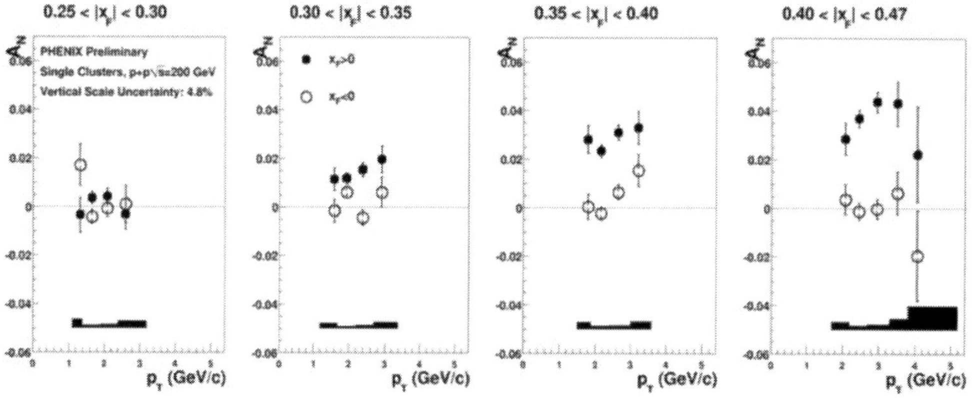


**Figure 3.** $p_T$ dependence of $A_N$ in the MPC at $\sqrt{s} = 200$ GeV for different $x_F$ bins

transverse momenta $p_T$ of greater than 2 GeV/c. Since the high $p_T$ region is of special interest to investigate the predicted $1/p_T$ shape of $A_N$ we decided to analyze asymmetries for clusters with an energy greater than 25 GeV and determine the cluster content from Monte Carlo simulations. We used the Pythia [17] generator with k-factors adjusted, so that the cross sections for pions and direct photons measured at RHIC are truthfully reproduced. The simulation of the MPC was done in Geant3 [18]. The result is shown in figure 6. A majority of the clusters originate from neutral pions, at low $p_T$ more than 90%. At higher $p_T$ the fraction of clusters coming from direct photons at LO rises and the fraction of pion induced clusters falls to 60%. The resulting values for the asymmetry $A_N$ for the 200 GeV data are shown in figure 2. Again, we divided the data in two different $\eta$ ranges, here at $\eta = 3.3$ to provide for two datasets of roughly equal size. The asymmetries follow the same trend as the results from reconstructed pions at 62.4 GeV. In fact the datapoints are compatible with the ones at 62.4 GeV if scaled with $\sqrt{s}$ i.e. by using the dimensionless variable $x_F$. Surprisingly, as shown in fig. 5 this is true for asymmetries down to 19.4 GeV as measured by the E704 collaboration [19], even though at these energies the frameworks for the mechanisms to produce the SSAs introduced above, are no longer valid. Figures 3 and 7 explore the $p_T$ behavior at 200 GeV some more. At high $p_T$ a trend for falling asymmetries can be observed. However, due to the large error bars and the contamination of the cluster asymmetries by other sources of photons the trend is not significant.

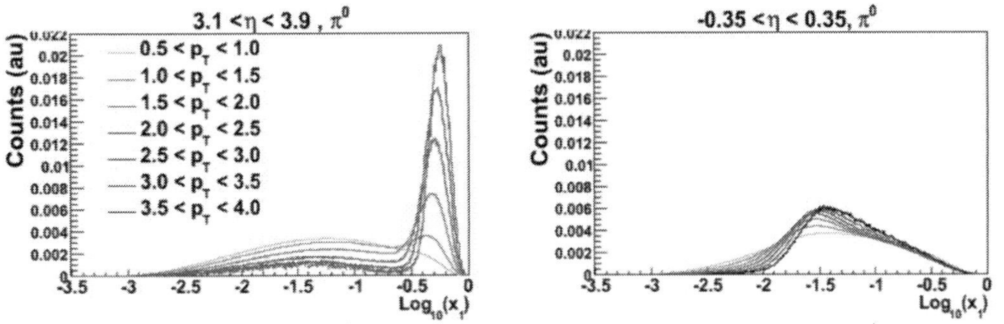


**Figure 4.** Distribution of the Bjorken scaling variable $x$ of the parton fragmenting into the detected hadron for $A_N$ measured forward in the MPC and in the central arm. Curves for different bins in $p_T$ are shown

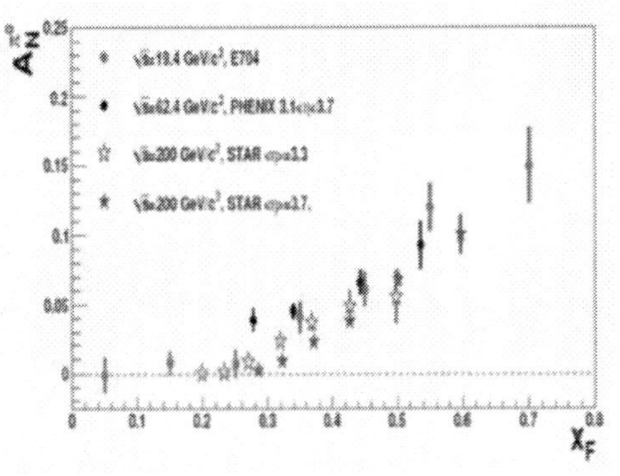


**Figure 5.** World results for $A_N$ at different $\sqrt{s}$. The asymmetries seem only to dependent on $x_F$

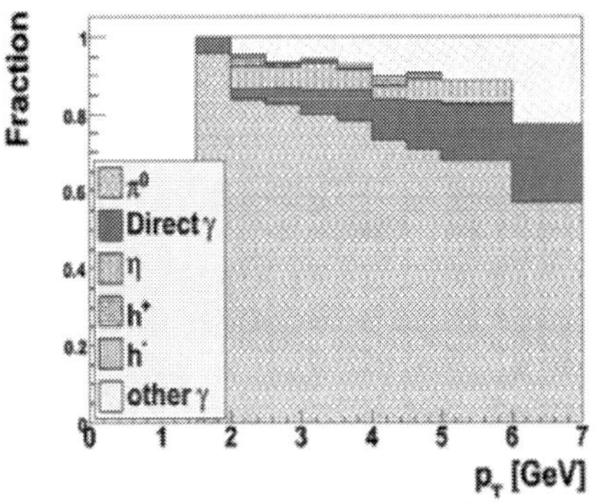


**Figure 6.** Results for the decomposition of the clusters used to compute $A_N$. Most clusters come from $\pi^0$'s but at high $p_T$ the fraction of direct photon grows. The contributions from charged hadrons ($h^+$,$h^-$) is neglible

## 5. Di-hadron correlation measurements in the central region to access transversity

Using di-hadron correlations to access transversity in proton-proton collisions has first been proposed by Baccheta and Radici [20]. In events where two hadrons with different charges are produced within a jet cone, the azimuthal asymmetry of the difference vector $\vec{R}$ between the two hadron momentum vectors around the momentum sum vector with respect to the spin axis is measued. Since the sign of the asymmetry is fixed by the charge ordering, it is not possible to compute this asymmetry for particles with equal charge. At PHENIX this requirement means, that the analysis is restricted to the central region where charged particle tracking and neutral particle detection is available. We use $\pi^0$'s reconstructed in the EMCs and unidentified charged track reconstructed in the pad and drift chambers. A minimum $p_T$ of 1 GeV/c is required. In order to select tracks from the same jet, we use pairs in the same arm. As a reminder, each arm covers $-0.35 < \eta < 0.35$ and $\pi/2$ in azimuth. Figures 8 and 9 show the results in bins of the invariant mass $m_{\mathrm{Inv}}$ of the hadron pair and the combined transverse momentum, respectively. We choose this binning because the recent results of the Belle collaboration for the

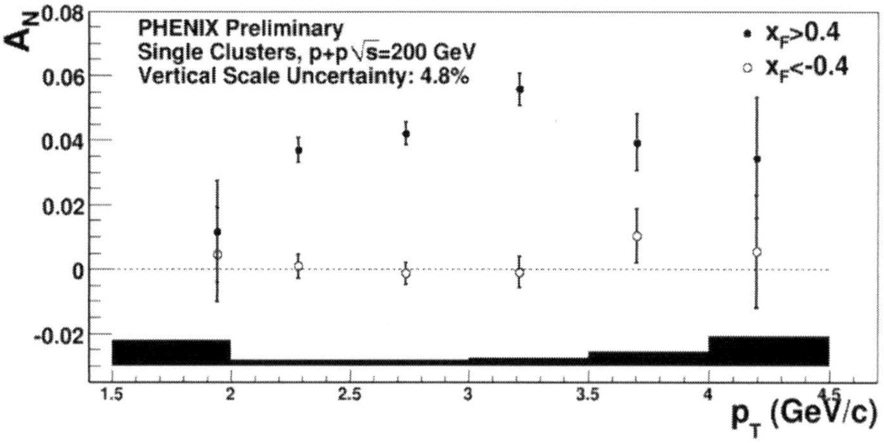


**Figure 7.** $p_T$ dependence of the 200 GeV cluster $A_N$ measured in the MPC

IFF of $(\pi^+/\pi^-)$ pairs shows a strong dependence on the invariant mass of the hadron pair [21]. Since the measured asymmetry is a product of transversity and IFF $A_{RS} \propto h_1(x) \cdot H_1^{\sphericalangle}(z, m_{\text{Inv}})$ we can expect a strong dependence on $x$, $m_{\text{Inv}}$ and the fractional momentum of the fragmenting parton $z$ the produced hadron pair carries. However, $x$ and $z$ are not accessible. The fractional momentum $x$ is correlated to $p_T$, but, as seen in fig. 4 at small rapidities it is not possible to select valence quarks with a $p_T$ cut. The measured asymmetries are small, however they exhibit the linear behavior expected in this range for $m_{\text{Inv}}$. The slope for $\pi^0/h^-$ is opposite to $h^+/\pi^0$. SIDIS results have shown large asymmetries for $\pi^+/\pi^-$ pairs [22], which is not observed in p+p. This can be explained by u-quark dominance in SIDIS and the opposite sign for u- and d-quark transversity. This might also cause the sign difference in the slopes for $\pi^0$ combinations with positive and negative hadrons. But since the fragmentation function has not been measured yet and there are no SIDIS results available for these combinations, this interpretation is speculative.

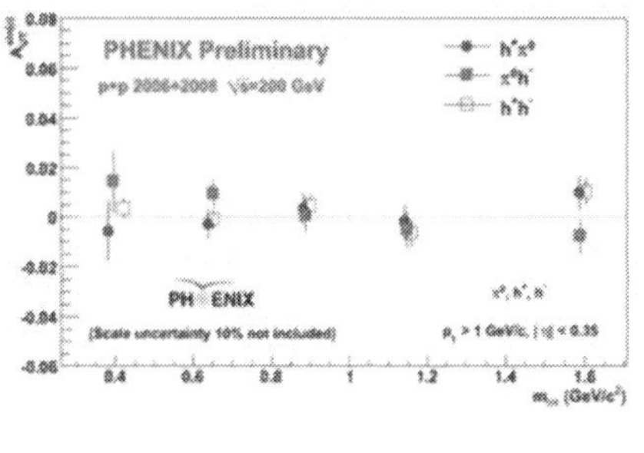


**Figure 8.** Results for the di-hadron asymmetry $A_{\text{RS}}$ in $m_{\text{Inv}}$ binning of the produced hadron pair

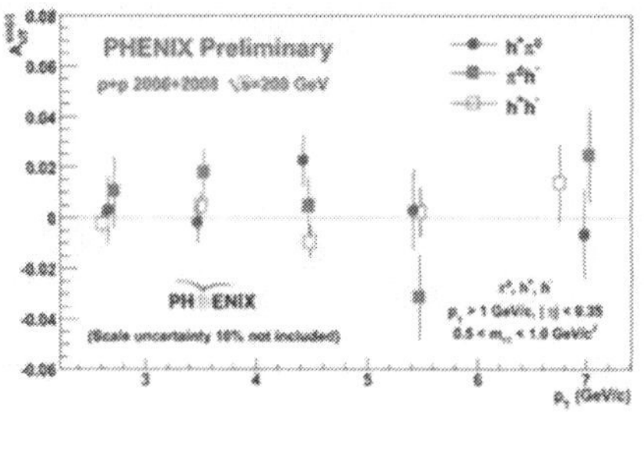


**Figure 9.** Results for the di-hadron asymmetry $A_{\text{RS}}$ in $p_T$ binning of the produced hadron pair

## 6. Summary and Outlook

Transverse spin asymmetries in proton proton collisions provide insights into the transverse spin structure of the proton in a kinematic regime not accessible in SIDIS. Thus measuring these asymmetries is also important in the context of the predicted sign flip of the Sivers function in SIDIS and DY. Due to the presence of two color charges in the initial state they also challenge our understanding of TMD factorization. At PHENIX we measured transverse spin asymmetries in the hard scattering regime. We investigated multiple kinematic dependencies to help identify the underlying mechanism. A focus is the shape of $A_N$ at high $p_T$. To disentangle Collins and Sivers effect contributions and measure transversity in proton-proton collisions, we measured the di-hadron asymmetry $A_{RS}$ in the central region. In the future these measurements will be extended into the PHENIX muon arms in which muons from charged hadron decay can be measured. Since they cover a more forward region of $1.2 < |\eta| < 2.4$ and the full azimuth, it is expected that the magnitude of the asymmetry will be larger.

## References

[1] D. Sivers, *Phys. Rev.* D **43** 83-90, 1990
[2] J.C. Collins, *Nucl. Phys. B* **396** 161, 1993
[3] M. Burkardt and D.S. Hwang, *Phys. Rev.* D **69** 074032, 2004
[4] J.C. Collins, S.F. Heppelmann and G.A. Ladinsky, *Nucl. Phys.* B **420** 565, 1994
[5] R.L. Jaffe and X. Ji, *Phys. Rev. Lett.* **71** 2547-2550, 1993
[6] The PHENIX collaboration, *Phys. Rev.* D **79** 012003, 2009
[7] J.C. Collins, *Phys. Lett.* B **536** 43, 2002
[8] A.V. Efremov and O.V. Teryaev, *Phys. Lett.* B **150**, 383, 1985
[9] J. Qiu and G. Sterman, *Phys. Lett.* **67** 2264, 1991
[10] J. Qiu and G. Sterman, *Phys. Rev,* D **59** 014004, 1999
[11] T.C. Rogers and P.J. Mulders, Phys. Rev. D **81** 094006, 2010
[12] X. Ji, J.-W. Qiu, W. Vogelsang and F. Yuan, *Phys. Rev. Lett.* **97** 08002, 2006
[13] Z.-B. Kang, J.-W. Qiu, W. Vogelsang and F. Yuan, *Phys. Rev.* D **83** 094001, 2011
[14] D. Boer, *Nucl. Phys.* B **806** 23-67, 2009
[15] The PHENIX collaboration, *Nucl. Inst. Meth.* A **499** 469-479, 2003
[16] J. Koster, Ph.D. Thesis, University of Illinois at Urbana-Champaign, 2010
[17] T. Sjöstrand et. al. , *Computer Phys. Commun.* **135** 238, 2001
[18] CERN, *CERN Program Library Long Writeup* W **5013**
[19] D. Adams et. al., *Phys. Lett.* B **264** 462-466, 1991
[20] A. Bacchetta and M. Radici, *Phys. Rev.* D **70** 094032, 2004
[21] A. Vossen, R. Seidl et. al., *Preprint* arXiv:1104.2425v1
[22] A. Airapetian et. al., *JHEP* 0806 017, 2008

Journal of Physics: Conference Series **316** (2011) 012005 doi:10.1088/1742-6596/316/1/012005

# Electromagnetic interactions of heavy ions at small impact parameters and lepton pair productions

**M.Cem Güçlü**[1],**M. Yılmaz Şengül**[1, 2]
[1]Istanbul Technical University, Faculty of Science and Letters, 34469, Maslak, Istanbul,Turkey
[2]Kadir Has University, Faculty of Science and Letters, 34083, Cibali-Fatih, Istanbul, Turkey

E-mail: guclu@itu.edu.tr

**Abstract.** In peripheral collisions, the strong electromagnetic fields are produced since the fully stripped heavy ions moves at ultra-relativistic velocities. Lepton pairs, especially the electron-positron pairs are the result of this collisions. At the Relativistic Heavy Ion Collider (RHIC) and the Large Hadron Collider (LHC), virtual photons produce many particles such as electron-positron pairs, heavy lepton (muons and tauons) pairs, magnetic monopoles, W-pairs, b-quark pairs and possibly the Higgs. At small impact parameters, the colliding nuclei make peripheral collisions, photon fluxes are very large and these are responsible for the multiple photonuclear interactions. Free pair productions, bound free pair productions, and nuclear Coulomb excitations are important examples of such interactions, and these processes play important roles in the beam luminosity at RHIC and LHC. Here we obtained bound free electron-positron pair production with nuclear breakup for heavy ion collisions at RHIC and LHC.

## Introduction

Several years ago, it was suggested that ultra-relativistic heavy ion collisions provide very large fluxes of coherent photons. These photons can be real or virtual and these photons are the source for the electromagnetic production of various particles. These particles can be electron and heavy lepton (muons and tauons) pairs, magnetic monopoles, W-pairs, b-quark pairs and possibly the Higgs. Most of this particles occur at colliding beams of heavy nuclei at energies per nucleon in the range 3-$10^4$ GeV, corresponding to the experiments at AGS, CERN, RHIC facility and LHC facility [1].

There is many questions about on the electromagnetic and electroweak interactions in strongly dynamical hadronic systems. For example, the dilepton production and $J/\psi$ suppression can be a diagnostic of the formation of the quark-gluon plasma. If we know the electromagnetic pair production processes in detail, we can separete it from the hadronic processes. We know that coherent production of electron-positron pair is proportional to the $Z^4$ where Z is the charge of the colliding ions. Therefore the strong field of the colliding ions can help us to calculate the other processes such as W bosons and the Higgs particles [1].

[1] guclu@itu.edu.tr

In peripheral collisions of the ions, it is known that a large number of lepton pairs can be produced owing to the Lorentz contracted high electromagnetic fields that occur in course of the collisions. In ultrarelativistic heavy ion colliders, such as RHIC and LHC, there are two dominant processes that restrict the luminosity of the ion beams: bound free electron-positron pair production (BFPP) and giant dipole resonance (GDR). In both processes, each ion is acted on by the Lorentz contracted electromagnetic field of the other ion. In the BFPP, an electron is captured by one of the colliding ions and leads to the loss of the ion from the beam. This leads to a change in the charge of the ion and causes the ion to fall out of the beam. In the GDR, the Coulomb force dissociates the nucleus where the protons and neutrons oscillate against each other. Then the neutrons fall out of the beam and these neutrons are detected in the forward zero degree calorimeter. Each of these processes causes the limitation of the beam lifetime for the collisions of Au+Au at RHIC or Pb+Pb at LHC [2].

The other intereting processes is the production of antihydrogen from the collisions of anti-proton and heavy nucleus. Anti-hydrogen in flight was first detected at CERN Low Energy Antiproton Ring (LEAR) using the process

$$\overline{p} + Z \rightarrow \overline{H} + e^{-} + Z \tag{1}$$

here, $Z = 54$ (Xe). This process is first suggested by Munger *et al.*[3] and they also calculated the cross section with equivalent photon approximation (EPA). It s a similar pair production mechanism with electron capture process. The calculations were done in high energy limit however the applicability of the EPA to this problem is questionable since the anti-hydrogen production were done in low energy region [4].

The Weizsacker-Williams method were used mostly for the calculation of the cross section of all these particles. The handicap of this method is, the photons must be "on-shell" , however in heavy ion collisions this condition is not always realized. Generally, the virtual photons carry a transverse momentum which is not taken into consideration in the Weizsacker-Williams method [5].

Heavy ion collisons are used to calculate the lepton pair production cross section as a model. This model can also be applied to the W-pair and Higgs production, where the electromagnetic potentials are taken as classical. In this method, the electromagnetic fields that are arose in interaction Lagrangian replaced by their classical term. With this way, the coherent pair production mechanism can be used as the excitations of quantum fields using perturbation theory. For the two photon process, the analytical integrals for the calculation of the cross section can be obtained in momentum space and this analytical integrals can be calculated by Monte Carlo Method [5].

The observation of the Higgs boson would give us a chance to explain the mass splittings in a unified theory of weak and electromagnetic interactions. It is thoght that, at relativistic heavy ion collisions, the coherent two photon mechanism can be applied to produce an intermediate- mass Higgs boson. At coherent pair production mechanism, the cross section behaves as the fourth power of the nuclear charge, relative to the corresponding mechanism at p-p colliders. The coherent pair production at heavy ion collisons provide that the two photon pair production mechanism occurs at distances of far away from the nucleus. Therefore the Higgs boson can be separated from the usual hadronic background. The Weizsacker-Williams method underestimate the coherent production cross sections since the nuclear form factors play important role in the calculation and the form factors are not well identified in this method [6].

First heavy ion collisions were done in Brookhaven National Laboratory (BNL) at the Relativistic Heavy Ion Collider (RHIC) in the mid of 2000 that gave an important contribution to the heavy ion physics. Before RHIC, these type of interactions has been observed only with cosmic rays. Many type of collisions have been studied in the laboratory and their first target was to study on central collisions. The main aim of these collisons at RHIC is to observe the quark-gluon plasma phase that is the signature of a new state of hadronic matter [7].

In this work, a physics of the peripheral collisions are taken into consideration at RHIC and LHC energy regimes. In the collisions, the protons in the nucleus are coherent and the electromagnetic fields that are surrounded by the ion is so strong and acts for a short time. As seen in figure 1, Fermi suggests that the time dependent electromagnetic field can be replaced with the radiation frequency of the field. And this photon spectrum is calculated by the "Weizsacker-Williams method" [7].

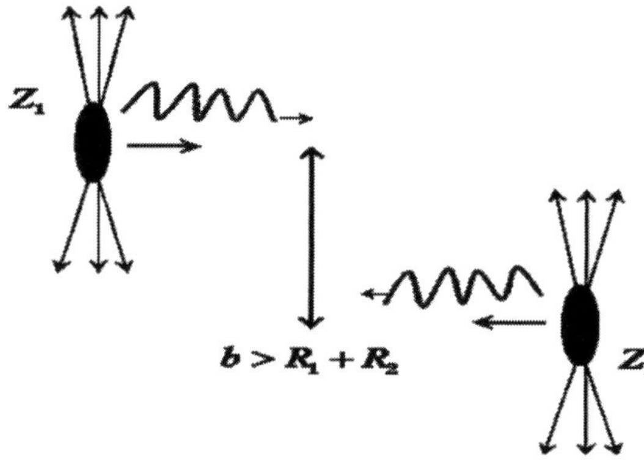


**Figure 1.** Fast nucleus Ze that surrounded by the strong electromagnetic field. The cloud of quasireal photons.

Photon-photon collisions can be observed in hadron-hadron collisions. Relativistic heavy ion collisions is a general method for two photon physics and suggested many years ago. Landau and Lifschitz in 1934 and Racah in 1937 had studied the electron-positron pair production in nucleus-nucleus collisons. After this calculations, many years later the Higgs boson pair production with photon-photon collision is also suggested. Therefore it is clear that strong field of the ultra-relativistic heavy ions at RHIC and LHC are the source of the production of some exotic particles [7].

**Formalism**

In this work, we calculate the cross section of BFPP with the GDR of the ions in lowest-order QED as shown in figure 2. We use the semiclassical approximation in the calculation and the Monte Carlo method to obtain the exact results in Ref. [2]. Bound-free pair production accompanied by giant dipole resonance can be represent as;

$$Z_1 + Z_2 \rightarrow (Z_1^* + e^-)_{1s_{1/2},...} + Z_2^* + e^+. \qquad (2)$$

In the bound free pair production accompanied by giant dipole resonance, the electron is captured by one of the colliding ions that is nuclear excited which leads to the loss of the (one-electron) excited ion from the beam. In our calculations, Monte Carlo techniques were used, and the integrands were tested on about 10 million randomly chosen "positions" to ensure sufficient convergence of our theoretical results. The total numerical error in the computations is estimated to be less than or approximately 5%. And then we calculate the impact parameter dependence cross section of BFPP accompanied by GDR.

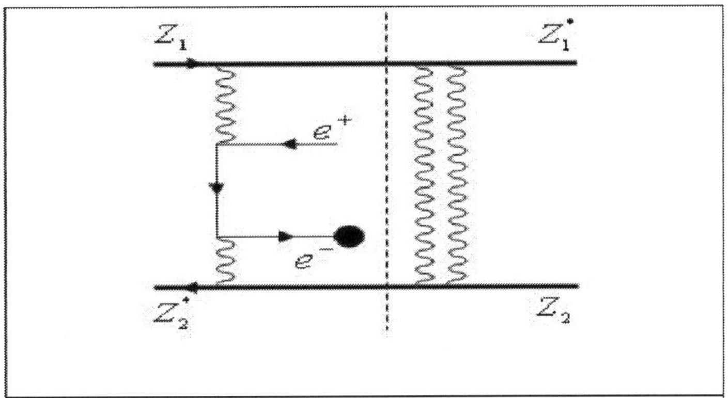


**Figure 2.** Dominant Feynman diagram for two photon reactions BFPP accompanied by GDR in a relativistic heavy ion collision.

The cross sections for the BFPP can be expressed as

$$\sigma = \frac{|N_+|^2}{4\beta^2}\frac{1}{\pi}\left(\frac{Z}{a_H}\right)^3 \sum_{\sigma_q} \int \frac{d^3q\,d^2p_\perp}{(2\pi)^5}\left(\boldsymbol{A}^{(+)}(q:\vec{p}_\perp) + \boldsymbol{A}^{(-)}(q:\vec{q}_\perp - \vec{p}_\perp)\right)^2 \tag{3}$$

where

$$\boldsymbol{A}^{(+)}(q:\vec{p}_\perp) = F(-\vec{p}_\perp:\omega_1)F(\vec{p}_\perp - \vec{q}_\perp:\omega_2)\tau_q(\vec{p}_\perp:+\beta) \tag{4.a}$$

and

$$\boldsymbol{A}^{(-)}(q:\vec{q}_\perp - \vec{p}_\perp) = F(\vec{p}_\perp - \vec{q}_\perp:\omega_2)F(-\vec{p}_\perp:\omega_1)\tau_q(\vec{q}_\perp - \vec{p}_\perp:-\beta) \tag{4.b}$$

are some proper products of the transition amplitudes and scalar parts of the fields as associated with ions $a$ and $b$ [8].

We found an expression for the impact parameter dependence cross section of free-pair production in Ref. [9]. We used the same method to calculate the impact parameter dependence BFPP cross section. Finally, we reach the impact parameter dependence cross section as

$$P_{BFPP}(b) = \sigma_{BFPP}\frac{a}{2\pi(a^2+b^2)^{3/2}}. \tag{5}$$

In this article, we calculate bound free electron-positron pair production accompanied by GDR, and this process can be seen in figure 2. In addition to the lepton pairs, the nuclei also exchange some photons and this may break up the nuclei. After the GDR, if the nucleus emits only one neutron we show this as $(1n)$ [2].

The probability of one Coulomb excitation is

$$P_{C(1n)}(b) = S/b^2 \tag{6}$$

where

$$S = \frac{2\alpha^2 Z^3 N}{A m_N \omega_{GDR}} \approx 5.45 \times 10^{-5} Z^3 N A^{-2/3} fm^2 \tag{7}$$

here $m_N$ is the nucleon mass and the neutron, proton, and mass numbers of the ions are *N*, *Z*, and *A*, respectively. The excitation probability is inversely proportional to the energy $\omega \approx 80 MeV A^{-1/3}$ of the GDR state [2].

The total cross section for BFPP with mutual nuclear excitation can be written as [2];

$$\sigma_{BFPP}^{GDR} = 2\pi \int_{b_{\min}}^{\infty} db\, b\, P_{BFPP}(b) P_{C(1n)}^2(b). \tag{8}$$

**Results and Discussions**

In Table 1, some values of tagged and untagged cross sections are calculated for RHIC and LHC energies for free pair production and bound free pair production. We calculated untagged bound free pair production cross sections and the results are 94.5 b for RHIC and 202 b for LHC collisions. Here we did not include the nuclear excitation processes and these values are shown in the first column in the table. When we include the nuclear excitation processes (mainly the GDR), the results are more than two orders of magnitude smaller than the untagged calculations.

**Table 1.**Integrated cross sections for Au + Au collisions at RHIC energies and for Pb + Pb collisions at LHC energies for free pair production and bound free pair production.

| | $\sigma$ *(barn)* | $\sigma^{1n1n}$ *(barn)* |
|---|---|---|
| Au+Au (RHIC-Free) | $34\text{x}10^4$ | 1.63 |
| Pb+Pb (LHC-Free) | $2.1\text{x}10^5$ | 10.2 |
| Au+Au (RHIC-BFPP) | 94.5 | $4.5\text{x}10^{-3}$ |
| Pb+Pb (LHC-BFPP) | 202 | $9.7\text{x}10^{-3}$ |

This calculations are done by using the Monte-Carlo techniques. Ten million random numbers are used for each variable and the estimated error is about five percent.

**Conclusions**

In this work, we obtained the cross sections of producing electron-positron pairs from the peripheral heavy ion collisions. We have also used the same method for the bound-free pair production processes. This calculations are well agree with the experimental results that were done at SPS and at STAR with some kinematical restrictions. We have also obtained the impact parameter dependence cross sections valid all impact parameters. By using this result, we calculated BFPP with GDR cross sections for $1n1n$ excitations. We have also obtained some results for the future experiments that may be done at LHC.

By using the same procedure and the same method, we can also obtain the cross sections of producing heavy lepton (muons and tauons) pairs, W-pairs, b-quark pairs and possibly the Higgs boson. We will show this calculations in our future works.

**Acknowledgments**
This work is partially supported by Istanbul Technical University and by Kadir Has University.

**References**

[1] Bottcher C and Strayer M R 1990 *J.Phys. G:Nucl. Part. Phys.* **16** 975.
[2] Sengul M Y and Guclu M C 2011 *Phys. Rev. C* **83** 014902.
[3] Munger C T, Brodsky S J and Schmidt I 1994 *Phys. Rev. D* **49** 3228.
[4] Bertulani C A and Baur G 1998 *Phys. Rev. D* **58** 034005.
[5] Wu J S, Bottcher C and Strayer M R 1990 *Phys. Lett. B* **252** 37.
[6] Wu J S, Bottcher C, Strayer M R and Shakin C M 1991 *Phys. Rev. C* **43** 2422.
[7] Baur G, Hencken K, Trautmann D, Sadovsky S and Kharlov Y 2002 *Phys. Rep.* **364** 359.
[8] Sengul M Y, Guclu M C, and Fritzsche S 2009 *Phys. Rev. A* **80** 042711.
[9] Guclu M C 2000 *Nucl. Phys. A* **668** 149.

# Di-lepton production in p+p collisions at $\sqrt{s} = 200$ GeV from STAR

**Bingchu Huang (for the STAR Collaboration)**
Department of Modern Physics, University of Science and Technology of China, Hefei 230026, China
Brookhaven National laboratory, Upton NY 11973, USA
E-mail: huangbc@bnl.gov

**Abstract.** The di-electron mass continuum in the mass range $0 < M_{ee} < 3.3$ GeV/$c$ has been measured at midrapidity ($|y| < 1$) by the STAR experiment in p+p collisions at $\sqrt{s} = 200$ GeV. We will compare the measured di-lepton continuum to a hadronic cocktail simulation. In addition, the omega yields via di-leptonic decays have been measured for $0 < p_T < 1.6$ GeV/$c$.

## 1. Introduction

Di-leptons production is a crucial probe of strongly interacting matter with high energy density matter[1] created in Relativistic Heavy Ion Collider(RHIC). Since di-leptons are not affected by the strong interaction, and can escape from the hot dense matter, di-leptons can probe the whole time evolution and dynamics of the heavy ion collisions. In-medium properties of low-mass vector mesons can be studied via their di-lepton decays. The observation of in-medium modifications of vector mesons may relate to the possibility of Chiral symmetry restoration. In the intermediate mass range($1.1 < M_{l\bar{l}} < 3.0$ GeV/$c^2$), the di-leptons production are directly linked to the thermal radiation of the QGP [2, 3]. The di-electron production measurements in p+p collisions provide a crucial baseline for the future measurements in Au+Au collisions at STAR [4]. Benefit from the newly installed Time-of-Flight detector (TOF) at STAR, we are able to identify electrons with high purity, high acceptance, and high efficiency [5] from low to intermediate transverse momentum ($p_T$) [6]. We combine the TOF and Time Projection Chamber (TPC) [7] for electron identification (eID). The EMC was used for eID at high $p_T$(¿2 GeV/$c$). The di-electron continuum which come from 200 GeV p+p collisions in 2009 is presented in this article. We also present the cocktail simulations and compare the results with the data. The $\omega \to e^+e^-$ yields were obtained and compared to previously published results.

## 2. Data analysis and Results

### *2.1. Detector and data sample*

In this analysis, we have three major detectors that are used for the eID: the TPC, the TOF, and the Barrel Electron-magnetic Calorimeter (EMC) [8]. The TPC is the primary tracking device of the STAR detector providing energy loss, momenta and path-lengths measurements of particles created in the collisions. The energy loss is used for particle identification [9, 10]. At STAR, 72% of the TOF system was installed in year 2009, it could extend the identification of $\pi$ and $K$ up to 1.6 GeV/$c$ and the $p(\bar{p})$ up to 3 GeV/$c$ [11, 12]. Each track included in the

Published under licence by IOP Publishing Ltd

di-electron analysis was required to pass cuts of the velocity($\beta$) from the TOF and the $dE/dx$ from the TPC. Electrons can be clearly identified from low to intermediate $p_T$ by combining the TOF and TPC. In addition, the EMC is used for triggering on high energy photons and electrons based on the energy deposited in the detector.

The p+p data used in this analysis were 107 million events that were required to have a collision vertex within 50 cm from TPC center along the beam line. The electron candidates were obtained with 99% purity by applying velocity and $p_T$ dependent $dE/dx$ cuts on tracks with $p_T > 0.2$ GeV/$c$ and $|\eta| < 1$. The velocity was required as $|1/\beta - 1| < 0.03$. The di-electron invariant mass distribution was generated by pairing all possible $e^+$ and $e^-$ tracks from the same event, which is marked as unlike-sign distribution. In order to understand the combinatorial and correlated background, we reconstructed the backgrounds by both mixed-event technique and like-sign technique. In the like-sign technique, the invariant mass was calculated by pairing two tracks with same charge from the same events. The mixed-event background was reconstructed by two opposite charge sign electrons from different events. The two events were only mixed if they had collision vertices within 5 cm of each other in the beam line direction.

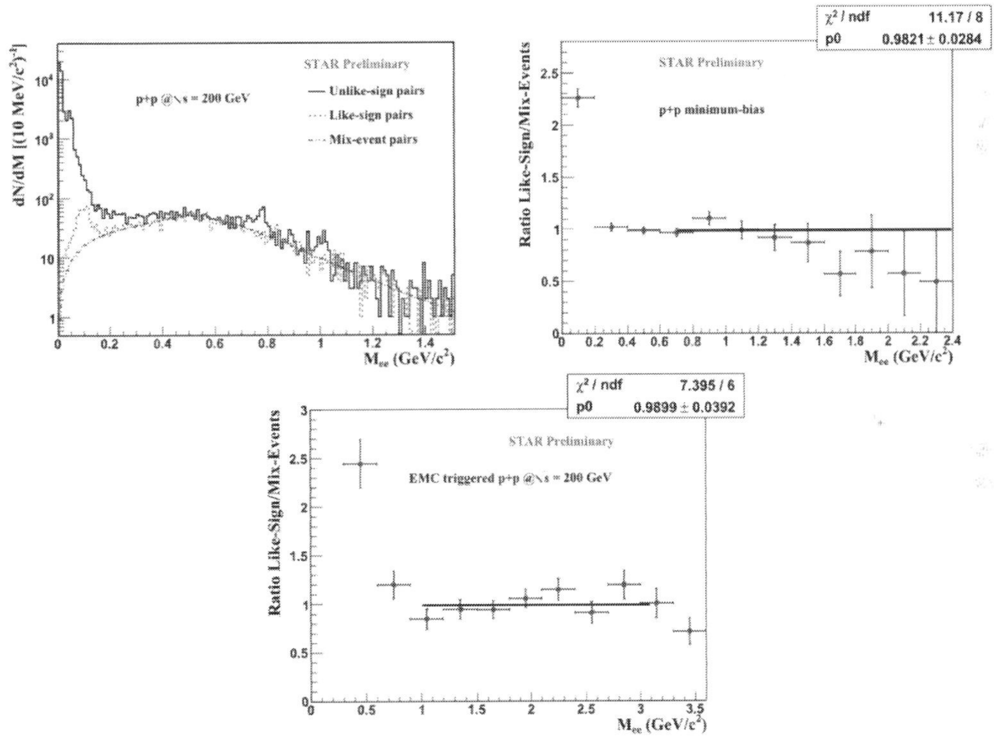


**Figure 1.** (top-left panel) The electron-pair invariant mass distributions for unlike-sign pairs, like-sign, and mixed-event background in minimum-bias p+p collisions. (top-right panel) The ratio of like-sign over mixed-event distributions in minimum-bias p+p collisions. (bottom panel) The ratio of like-sign over mixed-event distributions in EMC triggered p+p collisions.

### *2.2. Background subtraction*

Figure. 1 (top-left panel) shows the invariant mass distribution for unlike-sign pairs, like-sign, and mixed-event background. The mass range of the mixed-event distribution normalized to the like-sign distribution is $0.4 - 1.5$ GeV/$c^2$. Correlated background are at low mass region, which comes from double Dalitz decays followed by a conversion of the decay photon, or two-photon conversions. Such background included in the like-sign distribution but not in the mixed-event background. We subtracted like-sign background at $M_{ee} < 0.7$ GeV/$c^2$, and subtracted normalized mixed-event background at $M_{ee} > 0.7$ GeV/$c^2$. As shown in the ratio plot in Fig. 1 (top-right panel), the shape of like-sign and mixed-event distributions matched reasonably well. The ratio like-sign over mixed-event distributions is around unity for $p_T > 0.2$ GeV/$c$ [13]. To study the difference of like-sign and mixed event distribution at high mass, we made the two backgrounds by using EMC triggered events which triggered on at least one electron track which has $p_T > 2$ GeV/$c$. We found that the shape of like-sign and mixed-event distributions matched in the mass region of $1-3$ GeV/$c^2$, as shown in the bottom panel of Fig. 1. This indicates that the jet contribution is negligible within the STAR acceptance. PYTHIA simulations to estimate the jet contributions are on-going. The di-electron continuum after background subtraction is shown in Fig. 2. The errors are statistical only. The systematic uncertainties on the normalization and background subtraction are still under study.

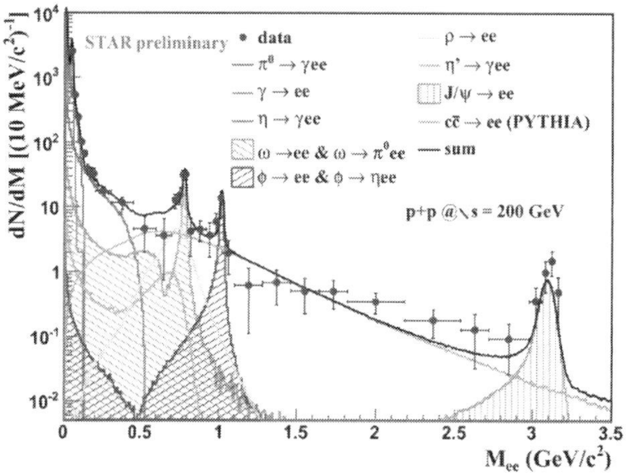


**Figure 2.** The comparison for di-electron continuum between data and simulation in 200 GeV minimum-bias p+p collisions. The di-electron continuum from simulations with different source contributions are also shown. The statistical errors are shown as bars.

### *2.3. Cocktail simulation*

The dominant sources of di-electron signals are hadron decays. For example, light flavor hadrons, $\pi^0$, $\eta$, and $\eta'$ Dalitz decays: $\pi^0 \to \gamma e^+e^-$, $\eta \to \gamma e^+e^-$, and $\eta' \to \gamma e^+e^-$; vector meson decays: $\omega \to \pi^0 e^+e^-$, $\omega \to e^+e^-$, $\rho^0 \to e^+e^-$, $\phi \to \eta e^+e^-$, $\phi \to e^+e^-$, and $J/\psi \to e^+e^-$. Heavy flavor hadron semi-leptonic decays: $c\bar{c} \to e^+e^-$ and $b\bar{b} \to e^+e^-$; and Drell-Yan contributions most dominant at high mass region. We fit the invariant yields of measured mesons with the Tsallis blast-wave functions [14], and use the fit functions as inputs of the simulations. We decayed the input mesons into di-electrons GEANT using STAR year 2009 geometry. The same cuts were applied in the simulation as well as in the data. The different contribution sources had same number of events and rapidity ranges as used in the data analysis. The gamma conversion $\gamma \to e^+e^-$ was not rejected and its contribution to di-electron continuum was simulated with the same method. The Dalitz decays of $\pi^0 \to \pi^0 e^+e^-$, $\omega \to \pi^0 e^+e^-$ and $\eta' \to \gamma e^+e^-$ were obtained by using the Kroll-Wada expression [15]. The $\rho^0 \to e^+e^-$ line shape was convoluted with the Boltzmann phase space factor [16, 17]. Fig. 2 shows the total contribution from the simulation as black solid curve which is found to be consistent with the data over all measured mass region. The uncertainties of the simulations are about 30%, mainly coming from the uncertainties of the invariant yield of input sources. The invariant yield of $\pi^0$ is taken as the average of $\pi^+$ and $\pi^-$ [12, 18]. The yields of $\phi$ [19] and $\rho^0$ [17] are from STAR while $\eta$ [20], $\omega$ [21] and $J/\psi$ [22] are from PHENIX. In this simulation, the $c\bar{c}$ cross section was an input and constrained by the measurement from STAR [6, 23]. The $\chi^2/NDF$ of the comparison between data and simulation is 36/30 in the mass region of 0.1-3.2 GeV/$c^2$. In the intermediate mass region, di-electron continuum is dominated by the $c\bar{c}$ contribution. It is unable to distinguish STAR's measured charm cross section from PHENIX's [6, 24] with current data precisions.

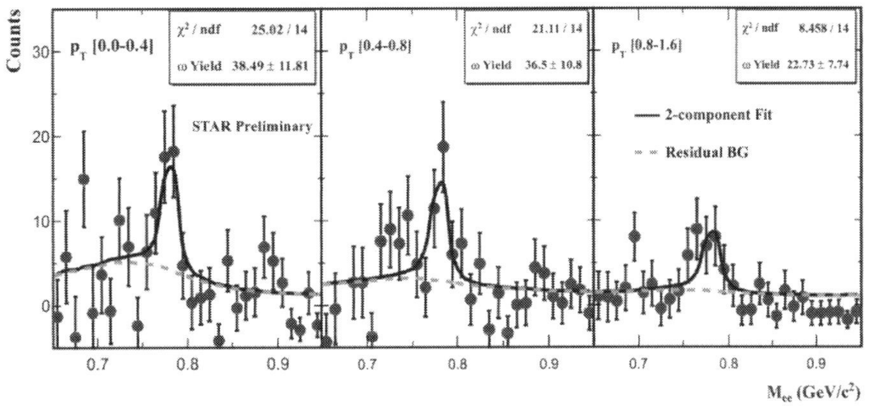


**Figure 3.** The $M_{ee}$ distribution after the mixed-event background subtraction at $0.65 < M_{ee} < 0.95$ GeV/$c^2$ in p+p collisions at $\sqrt{s} = 200$ GeV. The fit is to obtain the $\omega \to e^+e^-$ raw yields in three different $p_T$ bins shown in the plot. The fit range is $0.7 < M_{ee} < 0.85$ GeV/$c^2$. The dashed green line represents the residual background. Errors on data points are statistical.

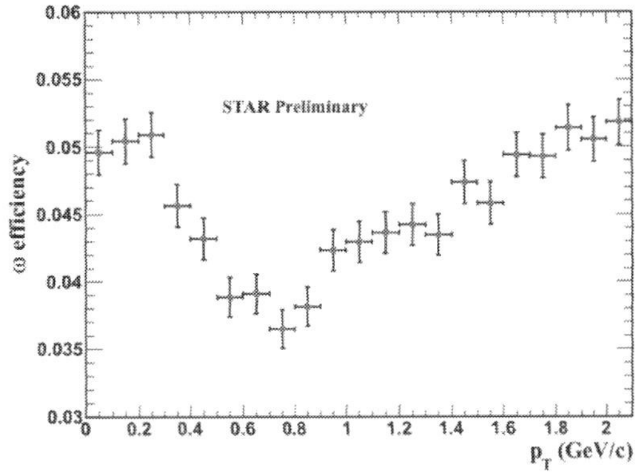


**Figure 4.** The efficiency including STAR acceptance for $\omega \to e^+e^-$ at $|y| < 1$ in p+p collisions at $\sqrt{s} = 200$ GeV.

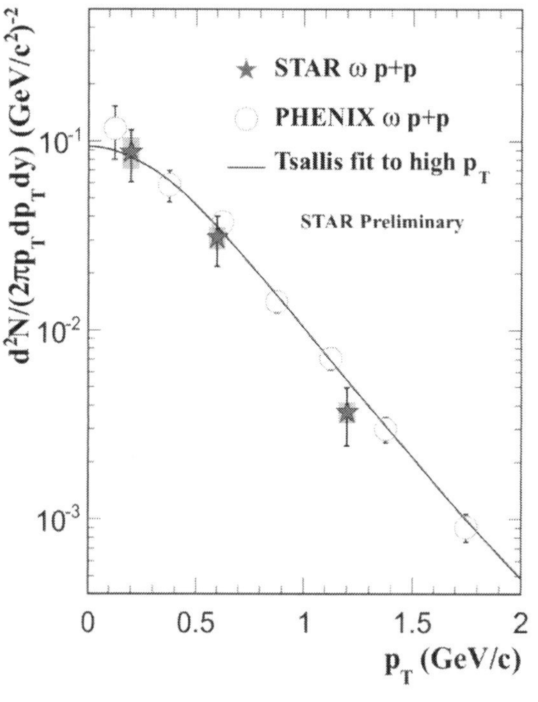


**Figure 5.** The $\omega \rightarrow e^+e^-$ invariant yields, divided by its branch ratio, in non-singly diffractive p+p collisions at $\sqrt{s}$ = 200 GeV. The open circles represent PHENIX published results [25]. The bars are statistical errors and boxes are systematic uncertainties. The line represents the yields of $\omega$ from Tsallis blast-wave function fit to high $p_T$ $\omega$ yields through hadronic decays in 200 GeV p+p collisions.

### *2.4. ω yields*

To extract the $\omega \rightarrow e^+e^-$ signal, we used two components to fit the distribution of $M_{ee}$ at $0.7 < M_{ee} < 0.85$ GeV/$c^2$: the $\omega \rightarrow e^+e^-$ signal and the residual background. The residual background shape and magnitude were from simulation. The systematic uncertainties of the $\omega \rightarrow e^+e^-$ raw yields were derived by the magnitude of the background and it is in the order of 30% due to the uncertainties of the input invariant yields. Figure 3 shows the fit to the $M_{ee}$ distribution at $0.7 < M_{ee} < 0.85$ GeV/$c^2$ in three different $p_T$ bins in p+p collisions at $\sqrt{s}$ = 200 GeV. Combining the TPC tracking, TOF acceptance and response, and the $dE/dx$ cut efficiency, the total efficiency including the STAR acceptance for $\omega \rightarrow e^+e^-$ at $|y| < 1$ is shown in Fig. 4. The $\omega \rightarrow e^+e^-$ invariant yields are shown as stars in Fig. 5 in non-singly diffractive p+p collisions at $\sqrt{s}$ = 200 GeV. The open circles represent PHENIX published results [25]. The bars are statistical errors and boxes are systematic uncertainties, which were dominated by the fit and efficiency correction factors. Our measured $\omega$ yields through di-leptonic decays are consistent with previously published results [25] and the Tsallis function fit to the high $p_T$ $\omega$ yields [14] reconstructed from their hadronic decays.

## 3. Summary

The di-electron continuum is measured in 200 GeV p+p collisions at STAR. The cocktail simulations are consistent with the data in 200 GeV p+p collisions. The di-electron mass distribution provide a reference for the future Au+Au. The newly installed TOF system enables this study. The $\omega$ invariant yields are consistent with the previously published results.

## References

[1] J. Adams *et al.*, Nucl. Phys. A **757**, 102 (2005).
[2] R. Rapp and J. Wambach, Adv. Nucl. Phys. **25**, 1 (2000).
[3] Electromagnetic Probes at RHIC II (Working Group Report), G. David, R. Rapp and Z. Xu, Phys. Rept. **462**, 176 (2008).
[4] K. H. Ackermann *et al.*, Nucl. Instr. Meth. A **499**, 624 (2003).
[5] B. Bonner *et al.*, Nucl. Instr. Meth. A **508**, 181 (2003); M. Shao *et al.*, Nucl. Instr. Meth. A **492**, 344 (2002); J. Wu *et al.*, Nucl. Instr. Meth. A **538**, 243 (2005).
[6] J. Adams *et al.*, Phys. Rev. Lett. **94**, 062301 (2005).
[7] M. Anderson *et al.*, Nucl. Instr. Meth. A **499**, 659 (2003).
[8] M. Beddo *et al.*, Nucl. Instr. Meth. A **499**, 725 (2003).
[9] H. Bichsel, Nucl. Instr. Meth. A **562**, 154 (2006).
[10] Y. Xu *et al.*, Nucl. Instr. Meth. A **614**, 28 (2010).
[11] M. Shao *et al.*, Nucl. Instr. Meth. A **558**, 419 (2006).
[12] J. Adams *et al.*, Phys. Lett. B **616**, 8 (2005).
[13] A. Adare *et al.*, Phys. Rev. C **81**, 034911 (2010).
[14] Z. Tang *et al.*, Phys. Rev. C **79**, 051901 (2009); M. Shao *et al.*, J. Phys. G **37**, 085104 (2010).
[15] N. M. Kroll and W. Wada, Phys. Rev. 98, 1355 (1955).
[16] R. Rapp, Nucl. Phys. A 725, 254 (2003); E. V. Shuryak and G. Brown, Nucl. Phys. A 717, 322 (2003).
[17] J. Adams *et al.*, Phys. Rev. Lett. **92**, 092301 (2004).
[18] J. Adams *et al.*, Phys. Lett. B **637**, 161 (2006).
[19] J. Adams *et al.*, Phys. Lett. B **612**, 181 (2005).
[20] S.S. Adler *et al.*, Phys. Rev. C **75**, 024909 (2007).
[21] S.S. Adler *et al.*, Phys. Rev. C **75**, 051902 (2007).
[22] A. Adare *et al.*, Phys. Rev. Lett. **98**, 232002 (2007).
[23] W. Xie *et al.*, for the STAR Collaboration, these proceedings.
[24] A. Adare *et al.*, Phys. Rev. Lett. **97**, 252002 (2006).
[25] S.S. Adler *et al.*, Phys. Rev. D **83**, 052004 (2011).

27th Winter Workshop on Nuclear Dynamics (WWND 2011) IOP Publishing
Journal of Physics: Conference Series **316** (2011) 012007 doi:10.1088/1742-6596/316/1/012007

# Inclusive $e^+e^-$ pair production in p+p and p+Nb collisions at $E_{kin} = 3.5$ GeV

**Michael Weber**
Physik Department E12, TU München, Germany
E-mail: michael.weber@ph.tum.de

**for the HADES collaboration**

**G. Agakishiev[6], D. Belver[15], A. Belyaev[6], A. Blanco[2], M. Böhmer[9], P. Cabanelas[15], S. Chernenko[6], A. Dybczak[3], E. Epple[8], L. Fabbietti[8], O. Fateev[6], P. Finocchiaro[1], P. Fonte[2,b], J. Friese[9], I. Fröhlich[7], T. Galatyuk[7,c], J. A. Garzón[15], M. Golubeva[11], D. González-Díaz[d], F. Guber[11], M. Gumberidze[13], T. Hennino[13], R. Holzmann[4], P. Huck[9], A. Ierusalimov[6], A. Ivashkin[11], M. Jurkovic[9], B. Kämpfer[5,e], T. Karavicheva[11], I. Koenig[4], W. Koenig[4], B. W. Kolb[4], G. Korcyl[3], G. Kornakov[15], R. Kotte[5], A. Kozuch[3,f], A. Krása[14], F. Krizek[14], R. Krücken[9], H. Kuc[3,13], W. Kühn[10], A. Kugler[14], A. Kurepin[11], A. Kurilkin[6], P. Kurilkin[6], V. Ladygin[6], S. Lang[4], K. Lapidus[8], T. Liu[13], L. Lopes[2], M. Lorenz[7], L. Maier[9], A. Mangiarotti[2], J. Markert[7], V. Metag[10], B. Michalska[3], J. Michel[7], C. Müntz[7], L. Naumann[5], Y. C. Pachmayer[7], M. Palka[7], Y. Parpottas[12], V. Pechenov[4], O. Pechenova[7], J. Pietraszko[7], W. Przygoda[3], B. Ramstein[13], A. Reshetin[11], A. Rustamov[4], A. Sadovsky[11], P. Salabura[3], A. Schmah[a], J. Siebenson[8], Yu.G. Sobolev[14], S. Spataro[g], H. Ströbele[7], J. Stroth[7,4], C. Sturm[4], A. Tarantola[7], K. Teilab[7], P. Tlusty[14], M. Traxler[4], R. Trebacz[3], H. Tsertos[12], T. Vasiliev[6], V. Wagner[14], M. Weber[9], J. Wüstenfeld[5], S. Yurevich[4], Y. Zanevsky[6]**

[1]Istituto Nazionale di Fisica Nucleare - Laboratori Nazionali del Sud, 95125 Catania, Italy
[2]LIP-Laboratório de Instrumentação e Física Experimental de Partículas, 3004-516 Coimbra, Portugal
[3]Smoluchowski Institute of Physics, Jagiellonian University of Cracow, 30-059 Kraków, Poland
[4]GSI Helmholtzzentrum für Schwerionenforschung GmbH, 64291 Darmstadt, Germany
[5]Institut für Strahlenphysik, Forschungszentrum Dresden-Rossendorf, 01314 Dresden, Germany
[6]Joint Institute of Nuclear Research, 141980 Dubna, Russia
[7]Institut für Kernphysik, Goethe-Universität, 60438 Frankfurt, Germany
[8]Excellence Cluster 'Origin and Structure of the Universe' , 85748 Garching, Germany
[9]Physik Department E12, Technische Universität München, 85748 Garching, Germany
[10]II.Physikalisches Institut, Justus Liebig Universität Giessen, 35392 Giessen, Germany
[11]Institute for Nuclear Research, Russian Academy of Science, 117312 Moscow, Russia

[12]Department of Physics, University of Cyprus, 1678 Nicosia, Cyprus
[13]Institut de Physique Nucléaire (UMR 8608), CNRS/IN2P3 - Université Paris Sud, F-91406 Orsay Cedex, France
[14]Nuclear Physics Institute, Academy of Sciences of Czech Republic, 25068 Rez, Czech Republic
[15]Departamento de Física de Partículas, Univ. de Santiago de Compostela, 15706 Santiago de Compostela, Spain

[a] also at Lawrence Berkeley National Laboratory, Berkeley, USA
[b] also at ISEC Coimbra, Coimbra, Portugal
[c] also at ExtreMe Matter Institute EMMI, 64291 Darmstadt, Germany
[d] also at Technische Univesitt Darmstadt, Darmstadt, Germany
[e] also at Technische Universität Dresden, 01062 Dresden, Germany
[f] also at Panstwowa Wyzsza Szkola Zawodowa , 33-300 Nowy Sacz, Poland
[g] also at Dipartimento di Fisica Generale, Università di Torino, 10125 Torino, Italy

**Abstract.** We report on recent data of $e^+e^-$ pair emission in proton nucleus collisions at energies above the light vector meson production thresholds. Invariant mass distributions for the p+Nb system at $E_{kin} = 3.5\ GeV$ are compared to data from elementary p+p reactions at the same beam energy. We observe a constant $\pi^0/\omega$ yield ratio for both systems but an excess in the mass region above the $\pi^0$ mass. Furthermore we present here the normalization procedure that was applied to p+Nb collisions by measuring the production of negative pions in the HADES acceptance.

## 1. Introduction

The spectroscopy of $e^+e^-$ pairs offers a promising opportunity to study hadron properties inside a strongly interacting medium. Electrons and positrons interact only electromagnetically and their kinematics stays almost undistorted while propagating through matter. Systematic studies in different environments, characterized by different temperatures and densities, are therefore possible, if the respective hadron h has a direct decay branch into $e^+e^-$ pairs: $h \longrightarrow e^+e^-$. For the light vector mesons with the same quantum numbers as virtual photons the branching ratio is approximately $10^{-5}$. Background sources of dielectrons like external conversion of real photons, Dalitz decays of hadrons, misidentification of pions as electrons etc. have to be well under control and understood.

Experimentally the highest temperatures and/or densities can be formed in relativistic heavy-ion collisions. Dilepton spectra were measured by CERES [1] and NA60 [2] at CERN-SPS energies. At lower energies (1-2 AGeV) the DLS [3] experiment and recently the HADES collaboration [4, 5, 6, 7] published their data for different collision systems.

Another site in the systematic studies of hadron properties is the so-called cold nuclear matter, i.e. a medium at ground state density and zero temperature. Experimentally it can be tested in heavy nuclei in proton -, pion - or photon - induced reactions. One has two different observables at hand: the spectral shape in the $e^+e^-$ invariant mass and/or the total yield for different nuclei. The measurement of the shape requires the decay taking place inside the nucleus. This is only true for short-lived vector mesons like the $\rho$ or longer lived mesons ($\omega$,$\phi$) at small momenta. The measurement of the yield or production cross section can be connected to the total width of the hadron via the so called transparency ratio [8, 9].

Journal of Physics: Conference Series **316** (2011) 012007 doi:10.1088/1742-6596/316/1/012007

So far, shape measurements in elementary reactions were done for the $\rho$ meson by the CLAS experiment at JLab[10] and the E325 experiment at KEK[11]. However, the results are not conclusive so far. For the $\omega$ and $\phi$ meson several experiments [12, 13, 14, 15] reported a sizable broadening of the total decay width inside the medium using the transparency ratio method. To contribute to this still unresolved issue the HADES collaboration measured the inclusive $e^+e^-$ pair production in proton induced reactions at $E_{kin}$ = 3.5 GeV. The reference spectrum was obtained in p + p reactions and the in-medium effects on the heavy nucleus Nb.

## 2. Experimental setup and measured $e^+e^-$ spectra

The High Acceptance Di-Electron Spectrometer HADES is installed at the GSI Helmholtzzentrum für Schwerionenforschung in Darmstadt and is provided with a proton or heavy-ion beam by the synchrotron SIS18. A detailed description of the spectrometer can be found in Ref. [4]. For the $e^\pm$ identification a hadron blind Ring Imaging CHerenkov detector (RICH) was used. Particle identification was supplemented by a time-of-flight measurement in a plastic scintillator wall (ToF) and an electromagnetic shower pattern in the Pre-Shower detector.
In our experiments a proton beam with a kinetic energy of $E_{kin} = 3.5\ GeV$ was incident on a liquid hydrogen target for the p + p run in 2007 [16] and on a 12-fold Nb target for the p + Nb run in 2008. The event selection was done in two steps. In the first trigger stage (LVL1), events with a charged particle multiplicity in the ToF wall of $M_{ch} \geq 3$ were selected. The second trigger stage (LVL2) selected events with at least one lepton candidate indicated by a ring in the RICH detector.
All possible combinations of identified $e^+/e^-$ tracks have been formed event by event and corrected for detector and reconstruction efficiencies. The latter ones were deduced using Monte-Carlo simulations embedded into real events. Invariant mass spectra of the unlike-sign pairs were constructed from single $e^+/e^-$ tracks. To increase the purity of the $e^+/e^-$ sample a cut on the single track momentum $0.08 < p_e/(GeV/c) < 2.00$ was applied.
The combinatorial background (CB) was extracted from all like-sign pair combinations inside the same event. Since the CB stems predominantly from external $\gamma$-conversion it could be reduced by cutting on the pair opening angle $\alpha_{ee} > 9^\circ$ and on the track fitting quality. By subtracting the CB from the unlike-sign pairs the signal spectrum was obtained.
The measured invariant mass distributions of $e^+e^-$ pairs are shown in the left panel of Fig. 1. The data points are normalized to the number of recorded LVL1 events. The signal to background ratio is $S/B \simeq 1$ at $M_{ee} \simeq 200\ MeV/c^2$ and increases up to $S/B \geq 10$ at $M_{ee} \simeq 750\ MeV/c^2$. The low-mass part of the spectrum is dominated by $\pi^0$ Dalitz decays ($M_{ee} < 150\ MeV/c^2$), while the intermediate part ($150\ MeV/c^2 < M_{ee} < 550\ MeV/c^2$) can be attributed to $\Delta(1232)$ and $\eta$ Dalitz decays[16]. The high mass part of the distribution is dominated by the direct decay of the light vector mesons $\rho$, $\omega$ and $\phi$. The peak around 780 $MeV/c^2$ corresponds to the decay of $\omega \rightarrow e^+e^-$ and the peak around 1000 $MeV/c^2$ reflects the decay of $\phi \rightarrow e^+e^-$. A mass resolution of $\frac{\Delta M}{M} \sim 2\%$ is extracted from a fit with a Gaussian distribution (see inset in Fig. 1). The remaining signal continuum in this mass region may originate from in-medium $\rho$-decays.

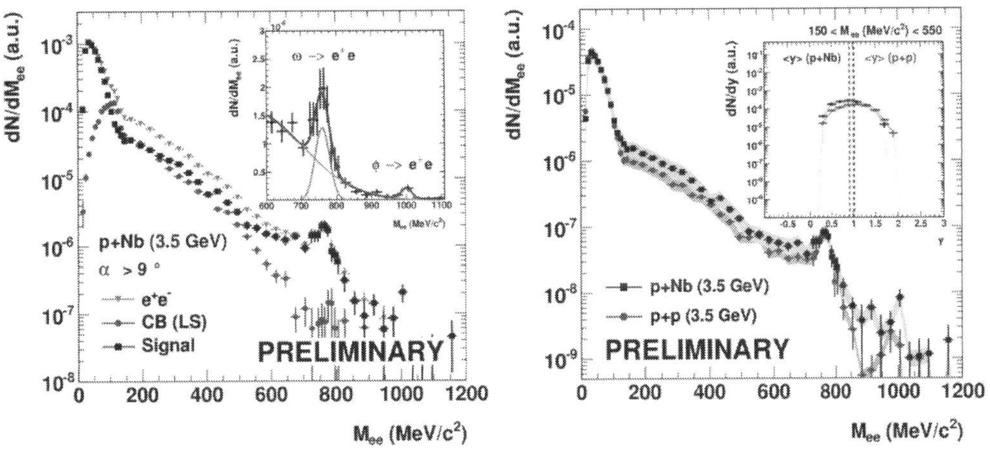


**Figure 1.** Left: Efficiency corrected invariant mass spectrum of all e$^+$e$^-$ pairs (red triangles), the combinatorial background (blue dots) and the signal pairs (black squares) for p+Nb collisions at $E_{kin} = 3.5\ GeV$. The inset shows the $\omega$ and $\phi$ meson yield. Right: Efficiency corrected e$^+$e$^-$ signal spectra for p+p (blue dots) and p+Nb (black squares) collisions. They are normalized to their yield $N_\pi$ in the $\pi^0$ mass region ($50\ MeV/c^2 < M_{ee} < 120\ MeV/c^2$). The systematic uncertainties are shown by the error bands. The inset shows the rapidity distributions, with the mean rapidity values given by the vertical lines.

## 3. Comparison of the p+p and p+Nb results

In a first step all spectra were scaled to the same yield $N_\pi$ in the $\pi^0$ mass region ($50\ MeV/c^2 < M_{ee} < 120\ MeV/c^2$). In the right panel of Fig. 1 the invariant mass distribution of the signal in p + Nb collisions is compared to the reference spectrum p+p. In the $\omega/\rho$ region the yield is comparable for both reaction systems relative to the $\pi^0$ yield. This suggests a similar vector meson production rate off the nucleus as for pions.
The yield in the intermediate mass region ($150\ MeV/c^2 < M_{ee} < 550\ MeV/c^2$) is enhanced by about 50%. To further investigate this enhancement different kinematic observables were studied. The inset of Fig. 1 shows the rapidity distributions of e$^+$e$^-$ pairs in this mass region for both reaction systems. The mean values for both reaction systems are indicated by vertical lines. For p + Nb, it is shifted towards target rapidity. This hints to the presence of an additional slow source, which has to be verified by detailed transport model calculations.
The spectral shape of the $\omega$ meson does not exhibit a clear signature of in-medium effects inside the error bars as visible in Fig. 1. The comparison of yields in different mass regions can lead to further insight about the in-medium widths of hadrons.

## 4. Cross sections

Total cross sections were obtained via the comparison of the measured yield $Y_H$ of a known physical process inside the HADES acceptance and its measured cross section $\sigma_{Meas}$. A global normalization factor

$$F_{NORM} = \frac{\sigma_{Meas}}{Y_H} \tag{1}$$

can then be calculated for $e^+e^-$ spectra.

In p+p collisions $F_{NORM}$ was obtained via the exclusive measurement of elastic p+p collisions and the known integrated cross section inside the HADES acceptance:

$$F_{NORM,pp} = \frac{\sigma_{ACC,elastic}}{Y_{H,elastic}}. \tag{2}$$

In p+Nb collisions such exclusive measurement was not possible. The physical source chosen there is the production of negative pions in proton induced reactions on nuclei. This was measured recently by the HARP-CDP collaboration [17] for a number of nuclei and projectile energies. Differential cross sections $\frac{d^2\sigma}{d\Omega dp}$ are given for bins in $0.1 < p_\perp$ (GeV/c) $< 0.9$ and $20° < \theta_{HARP-CDP} < 90°$ which covers fully the HADES polar angle range ($18° < \theta_{HADES} < 85°$). The system p + Nb was not measured but the systematics allows for an interpolation for our collision system.

In our experiment, negative pions were identified via their energy loss in the ToF detector. A detailed description of the analysis is given in [18]. A correction for efficiency was done in the same way as for $e^\pm$, additionally an acceptance correction in azimuthal angle was applied. To compare to differential cross sections measured by the HARP-CDP collaboration also the trigger bias was taken into account. This was achieved by a global factor $F_{TRIG}$ obtained from transport model (UrQMD [19]) calculations followed by full scale GEANT [20] simulations of the detector setup. The multiplicity of negative pions $M_{ACC,\pi}$ in the HADES acceptance for different cuts on the charged particle multiplicity $M_{ch}$ was then used to calculate $F_{TRIG}$:

$$F_{TRIG} = \frac{M_{ACC,\pi}(M_{ch} \geq 3)}{M_{ACC,\pi}(M_{ch} \geq 0)} = 1.42 \pm 0.14, \tag{3}$$

where $M_{ch} \geq 0$ means minimum bias events. The systematic error of 10 % was estimated from different scaling factors obtained from experiment and simulation, which was possible in the cases of $M_{ch} \geq 2$ and $M_{ch} \geq 3$. After correcting the pion spectra for trigger bias the normalization factor $F_{NORM,pNb} = 886 \pm 184$ mb was obtained by fitting the HADES data points bin-by-bin in $\theta$ and $p_\perp$ to the HARP-CDP results. Since the shape of the pt distribution turns out to be quite sensitive to both the bombarding energy and the target size, this was done for $0.3 < p_\perp$ (GeV/c) $< 0.9$ and $30° < \theta < 75°$ only. To interpolate the HARP-CDP data to our reaction system, a global scaling factor of 0.963 was applied to their 4.15 GeV p+Cu data set (interpolation of systematic HARP data to our collision system). The result is shown in the left panel of Fig. 2 for the polar angle region $50° < \theta < 60°$.

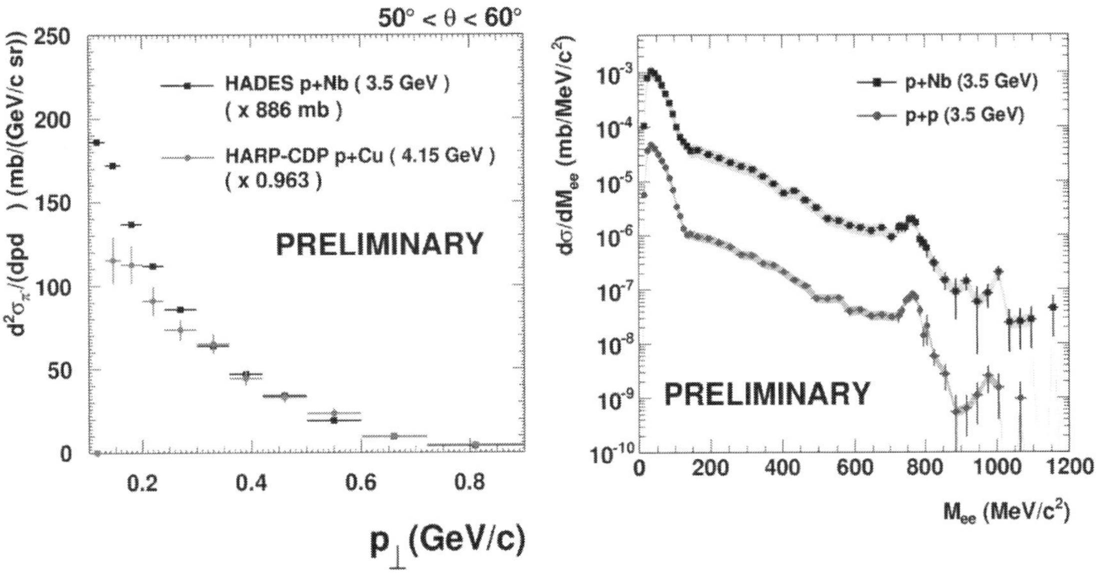


**Figure 2.** Left: Differential cross sections for the transverse momentum distribution of $\pi^-$: HADES data (black squares, only statistical errors) after scaling with the best fit value of $F_{NORM,pNb} = 886$ mb and HARP-CDP data (red circles, systematic and statistical errors) after system size scaling.
Right: Efficiency corrected $e^+e^-$ signal spectra for p+p (blue dots) and p+Nb (black squares) collisions. Given are differential cross sections after applying the normalization procedure. The systematic uncertainties are shown by the error bands.

## 5. Summary

We have measured inclusive $e^+e^-$ pair production in p+p and p+Nb collisions at $E_{kin} = 3.5\ GeV$. The comparison of invariant mass spectra scaled to their yield in the $\pi^0$ region shows the same signal strength in the $\omega$ pole mass region, whereas an excess of 50% over the elementary p+p processes is obtained in the intermediate mass region ($200\ MeV/c^2 < M_{ee} < 600\ MeV/c^2$). This excess is located at rapidities closer to target rapidity compared to the elementary processes in this mass interval, which hints to an additional slow source. The spectral shape of the $\omega$ meson does not show a significant difference for both reactions. For the analysis of total production cross sections a normalization of the dielectron spectra is needed, which was achieved via elastic collisions in the p+p reaction system and via the $\pi^-$ production in the p+Nb system. Final conclusions about possible in-medium changes of vector mesons in matter require a comparison to transport models. Work along this line is in progress.

## Acknowledgments

The HADES collaboration gratefully acknowledges the support by the BMBF through grants 06MT9156, 06GI146I, 06FY171, 06FY91001, and 06DR9059D (Germany), by GSI (TKrue1012, GI/ME3, OF/STR), by the Helmholtz Alliance HA216/EMMI, by the DFG Excellence Cluster 153 "Universe" (Germany), by the Helmholtz young researcher group VH-NG-330, by grants GA AS CR IAA100480803 and MSMTLC 07050 MSMT (Czech Republic), by grant KBN5P03B140 20, N N202 286038 and NN202198639 (Poland), by INFN (Italy), by CNRS/IN2P3(France), by grants MCYT FPA2000-2041-C02-02 and XUGA PGID FPA2009-12931 T02PXIC20605PN (Spain), by grant UCY- 10.3.11.12 (Cyprus), by grant PTDC/FIS/113339/2009 (Portugal) by INTAS grant 06-1000012-8861 and EU contract RII3-CT-506078.

## References

[1] D. Adamova et al. (CERES Collaboration), Phys. Lett. **B 666**, 425-429 (2008).
[2] S. Damjanovic et al. ( NA60 Collaboration), Nucl. Phys. **A 783**, 327 (2007).
[3] R. J. Porter et al., (DLS Collaboration), Phys. Rev. Lett. **79**, 1229–1232 (1997).
[4] G. Agakishiev et al. (HADES Collaboration), Eur. Phys. J. **A 41**, 243 (2009).
[5] G. Agakishiev et al. (HADES Collaboration), Phys. Rev. Lett. **98**, 052302 (2007).
[6] G. Agakishiev et al. (HADES Collaboration), Phys. Lett. **B 663**, 43 (2008).
[7] G. Agakishiev et al. (HADES Collaboration), subm. to Phys. Rev. **C**, (2011).
[8] M. Kaskulov, E. Hernandez and E. Oset, Eur. Phys. J. **A 31**, 245 (2007).
[9] P. Mühlich and U. Mosel, Nucl. Phys. **A 773**, 156 (2006).
[10] R. Nasseripour et al. (CLAS Collaboration), Phys. Rev. Lett. **99**, 262302 (2007).
[11] M. Naruki et al., Phys. Rev. Lett. **96**, 092301 (2006).
[12] M. Kotulla et al., Phys. Rev. Lett. **100**, 192302 (2008).
[13] T. Ishikawa et al. (SPring-8 Collaboration), Phys. Lett. **B 608**, 215-222 (2005).
[14] A. Polyanskiy et al. (ANKE Collaboration), Phys. Lett. **B 695**, 74 (2011).
[15] M. H. Wood et al. (CLAS Collaboration), Phys. Rev. Lett **105**, 112301 (2010).
[16] A. Rustamov et al. (HADES Collaboration), AIP Conf. Proc. **1257**, 736 (2010).
[17] A. Bolshakova et al. (HARP-CDP Collaboration), Eur. Phys. J. **C 62**, 697-754 (2009); Eur. Phys. J. **C 63**, 549-609 (2009); Eur. Phys. J. **C 64**, 181-241 (2009); Eur. Phys. J. **C 66**, 57-117 (2009).
[18] P. Tlustý, M. Weber, P. Salabura (HADES Collaboration), GSI Scientific Report 2010.
[19] M. Bleicher et al., J. Phys. **G 25**, 1859-189 (1999).
[20] Application Software Group, *GEANT – Detector Description and Simulation Tool, Program Library Long Writeups W5013*, CERN, 3.21 edition, 1993.

# Correlations of electrons from Heavy Flavor Decay with Hadrons in Au+Au and p+p Collisions

**Anne M. Sickles, for the PHENIX Collaboration**
Brookhaven National Laboratory, Upton, NY
E-mail: anne@bnl.gov

**Abstract.** Measurements of electrons from the decay of open-heavy-avor mesons have shown that the yields are suppressed in Au+Au collisions compared to expectations from binary-scaled p+p collisions. These measurements indicate that charm and bottom quarks interact with the hot-dense matter produced in heavy-ion collisions much more than expected. Here we extend these studies to two-particle correlations where one particle is an electron from the decay of a heavy-avor meson and the other is a charged hadron from either the decay of the heavy meson or from jet fragmentation. These measurements provide more detailed information about the interactions between heavy quarks and the matter, such as whether the modication of the away-side-jet shape seen in hadron-hadron correlations is present when the trigger particle is from heavy-meson decay and whether the overall level of away-side-jet suppression is consistent. We statistically subtract correlations of electrons arising from background sources from the inclusive electron-hadron correlations and obtain two-particle azimuthal correlations at sNN =200 GeV between electrons from heavy-avor decay with charged hadrons in p+p and also rst results in Au+Au collisions. We nd the away-side-jet shape and yield to be modied in Au+Au collisions compared to p+p collisions.

## 1. Introduction

One of the main goals of heavy ion physics is to understand the interactions between the produced matter (commonly understood as a strongly couple quark gluon plasma, sQGP) and fast partons. Measurements of the nuclear modification factor, $R_{AA}$, of $\pi^0$s [1] show a large suppression which is understood as resulting from parton-matter interactions throughout the time evolution of the matter. Further studies, in which a trigger $\pi^0$ is correlated with other hadrons in the event have shown a strong suppression of back-to-back jet-like correlations [2].

These results are qualitatively as expected from energy-loss models. However, heavy charm and bottom quarks have the ability, via their large mass, to provide additional constraints on the mechanisms by which the partons interact with the matter. Notably, heavy guarks are expected to suffer less energy loss via gluon radiation in the matter because of the dead cone effect [3].

Heavy quarks in heavy ion collisions have largely been studied via single electrons which come from the decays of $D$ and $B$ mesons which carry the original heavy quark. The $R_{AA}$ of these electrons has been shown to be significantly below 1 and comparable to that of $\pi^0$s [4]. Measurements of single electrons are sensitive to both charm and bottom quarks, with a mixture that depends on the electron $p_T$ [5, 6].

In order to further study these effects we have performed two-particle correlations of electrons from heavy flavor decay with other hadrons, in the same manner as has been done for $\pi^0$s. This can provide a more detailed view of the interactions between heavy quarks and the sQGP. However, there are additional complications when considering heavy quarks, thus baseline measurements in p+p collisions are vital. The results shown here have been published as Ref. [7].

## 2. Method

Experimentally, separating the correlations of heavy flavor electrons from those of non-heavy flavor electrons (such as those from photon conversions and Dalitz decays) is necessary. The procedure used is outlined in Ref. [7]. In general, a precise knowledge of the relative yields of heavy flavor and non-heavy flavor sources is needed. With that and a measurement of the correlations of non-heavy flavor electrons it is possible to statistically subtract the non-heavy flavor correlations from the total electron correlations (which include both the heavy flavor and non-heavy flavor correlations). This is analogous to the procedure used to separate direct photon triggered correlations from those triggered by decay photons in Ref. [8].

## 3. Results

The near side yields conditional yields are shown in Figure 1 and are consistent with those observed in p+p collisions. This is quantified by the ratio of the conditional yield in Au+Au collisions divided by the conditional yield in p+p collisions, $I_{AA}$, shown in Figure 2. This is a good cross check because those yields are expected to be dominated by the decay of the heavy meson ($D$ or $B$) which occurs far from the matter and thus should be unaffected by its presence. It is perfectly possible for the matter to enhance the near side correlations through some process, it is difficult to imagine a scenario where the yields are suppressed.

Figure 3 shows the away side conditional yields for both Au+Au and p+p collisions as a function of the associated hadron $p_T$ and for two different trigger $p_T$ selections and two ranges for the away side azimuthal angle integration. The correspondig $I_{AA}$ values are shown in Figure 4. In general at low $p_{T,hadron}$ the yields are enhanced in Au+Au collisions compared to p+p and at higher $p_{T,hadron}$ the yields are suppressed. Qualitatively, this is consistent to trends found in hadron-hadron correlations (see e.g. Ref. [9]). In order to make a more quantiative comparison we take into account that the electron $p_T$ is not the relevant $p_T$ with which to compare to the hadron-hadron $I_{AA}$ values. We use PYTHIA [10] simulations of the parent $B$ and $D$ (weighted according to Fixed Order Next to Leading Log (FONLL) [11] calculations of the mixture of charm and bottom) $p_T$s which give rise to the electrons in our trigger $p_T$ selections (see Table 1) and compare to hadron-hadron $I_{AA}$ at comparable $p_T$ values. This comparison is shown in Figure 4; good agreement is seen between the electron-hadron and hadron-hadron $I_{AA}$ values. It might also be reasonable to compare the hadron-hadron and electron-hadron $I_{AA}$ at similar parton $p_T$, accounting for the different fragmentation of heavy and light quarks. However, given the current uncertainties, the comparison at similar meson $p_T$ seems reasonable and gives good agreement.

| $p_{T,e}$ (GeV/$c$) | $\langle p_T \rangle_D$ (GeV/$c$) | $\langle p_T \rangle_B$ (GeV/$c$) | $\frac{b \to e}{(c \to e + b \to e)}$ | $\langle p_T \rangle_{meson}$ (GeV/$c$) |
|---|---|---|---|---|
| 1.5-2.0 | 3.4 | 4.4 | 0.15 | 3.6 |
| 2.0-3.0 | 4.1 | 4.7 | 0.26 | 4.3 |
| 3.0-4.0 | 5.6 | 5.6 | 0.42 | 5.6 |

**Table 1.** Mean transverse momentum of the parent $D$ and $B$ mesons contributing to the heavy-flavor electron $p_T$ bins used here. They are combined according to the fraction of heavy-flavor electrons from $b$ quarks, $\frac{b \to e}{(c \to e + b \to e)}$ according to the FONLL calculations [11] (as shown in Ref. [5]) to determine the mean heavy meson transverse momentum.

For many years, a strongly modified away side shape has been seen in hadron-hadron

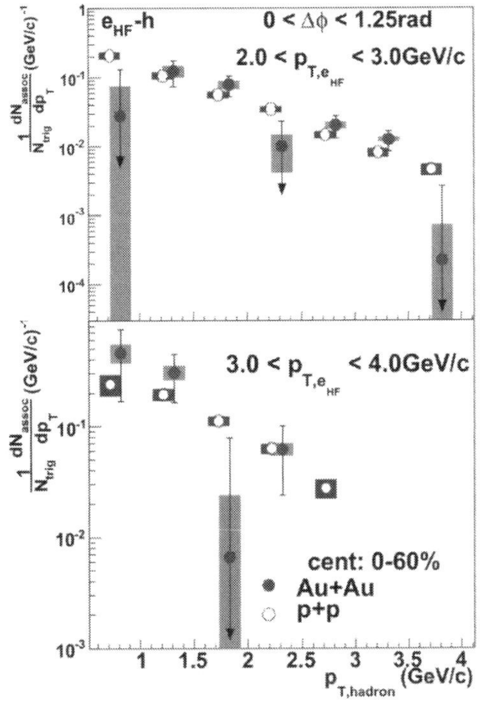


**Figure 1.** (color online) Near-side ($0< \Delta\phi <1.25$ rad) integrated yield for Au+Au (solid circles) and $p$+$p$ collisions (open circles) for $2.0< p_{T,e} <3.0$ GeV/$c$ (top panel) and $3.0< p_{T,e} <4.0$ GeV/$c$ (bottom panel) as a function of the associated hadron $p_T$. The overall normalization uncertainty of 9.4% in Au+Au and 7.9% in $p$+$p$ is not shown. Points are slightly shifted horizontally for clarity. Figure is from Ref. [7].

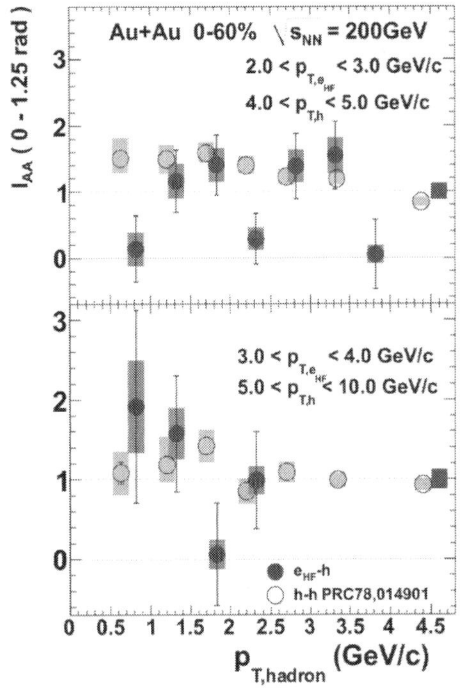


**Figure 2.** (color online) Near-side ($0< \Delta\phi <1.25$ rad) $I_{AA}$ for $2.0< p_{T,e} <3.0$ GeV/$c$ (top panel) and $3.0< p_{T,e} <4.0$ GeV/$c$ (bottom panel) as a function of the associated hadron $p_T$ for $e_{HF}$ (solid points) and hadron (open points) triggers (from Ref. [9]). The gray band around unity shows the overall normalization uncertainty (12.4%), which moves all points together. Points are slightly shifted horizontally for clarity. Figure is from Ref. [7].

correlations [12]. This has been attributed to various sources, among the most prevelant being the remains of a Mach cone [13]. With that motivation in mind we analyzed the away side shape for electron-hadron correlations. Heavy quarks, especially bottom, at moderate $p_T$ values travel at a velocity slower than the speed of light. Since the angle of the Mach cone opening depends on the velocity of the parton relative to the speed of sound, bottom quarks offer the opportunity to confirm a Mach cone by looking for a variation in the opening angle.

We measured the head-to-shoulder ratio, $R_{HS}$, which is the yield per radian measured into the head region, $2.51< \Delta\phi < \pi$ rad, divided by the same quantity in the shoulder region, $1.25< \Delta\phi <2.51$rad. The results for p+p and Au+Au collisions are shown in Figure 5. There is a significant increase in this ratio in Au+Au collisions compared to p+p collisions. No dependence on the associated particle $p_T$ is observed.

Since the publication of these data [7], there is evidence that the modified away side structure is due to fluctuations in the initial state of the Au+Au collisions which are then transported to the final state hydrodynamically [14, 15, 16]. The modification of the away side correlations in

Journal of Physics: Conference Series **316** (2011) 012008 doi:10.1088/1742-6596/316/1/012008

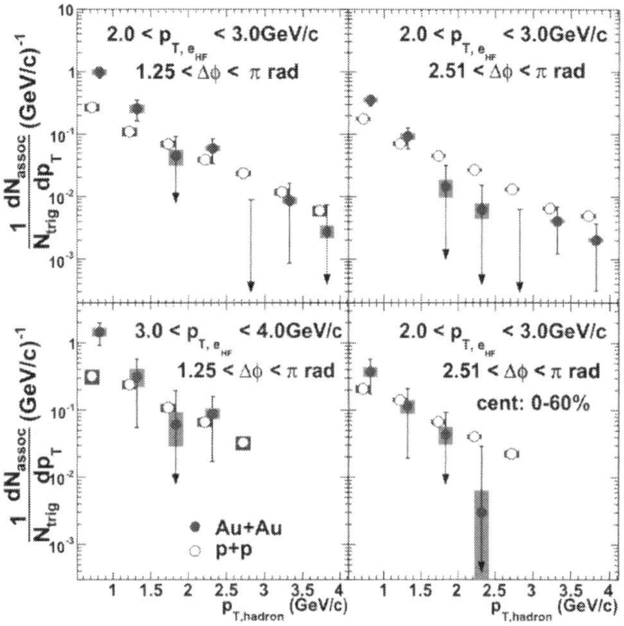


**Figure 3.** Away-side conditional yields for wide (left) and narrow (right) away-side $\Delta\phi$ integration ranges for Au+Au (solid points) and $p$+$p$ (open points). Top panels show 2.0< $p_{T,e}$ <3.0 GeV/$c$ and bottom panels shown 3.0< $p_{T,e}$ <4.0 GeV/$c$. Upper limits are for 90% confidence levels. The overall normalization uncertainty of 9.4% in Au+Au and 7.9% in $p$+$p$ are not shown. Points are slightly shifted horizontally for clarity. Figure is from Ref. [7].

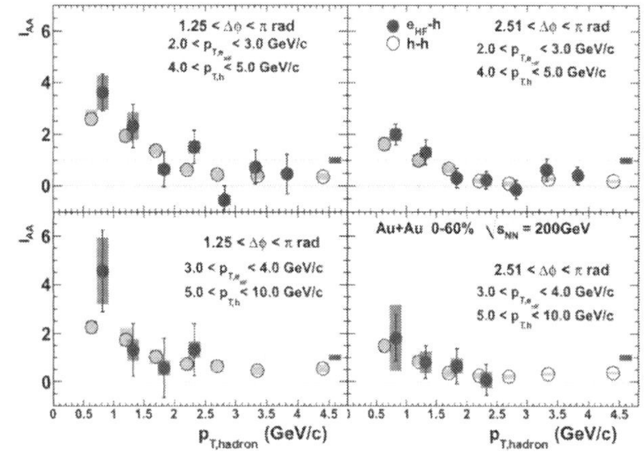


**Figure 4.** $I_{\rm AA}$ as determined from the away-side yields in Fig. 3. Two $\Delta\phi$ ranges are shown: 1.25< $\Delta\phi$ < $\pi$ rad (left panels) and 2.51< $\Delta\phi$ < $\pi$ rad (right panels). The gray band around unity shows the overall normalization uncertainty of 12.4%, which moves all points together. For comparison hadron-hadron $I_{\rm AA}$ values from Ref. [9] are also shown for trigger $p_T$ selections where the parent heavy meson has similar $p_T$ to the trigger light hadron (see Table 1). Points are slightly shifted horizontally for clarity. The solid horizontal line is at 0 and the dashed horizontal line is at 1. Figure is from Ref. [7].

electron-hadron correlations seen here is consistent with that understanding. In the trigger $p_T$ selection used in Figure 5, previous measurements already show a large $v_2$ value [4]. Thus, it would be expected that higher order Fourier coefficients (e.g. $v_3$), by which these fluctuations have been measured, would also be seen for these electrons (such a measurement has not been done, but would be very interesting).

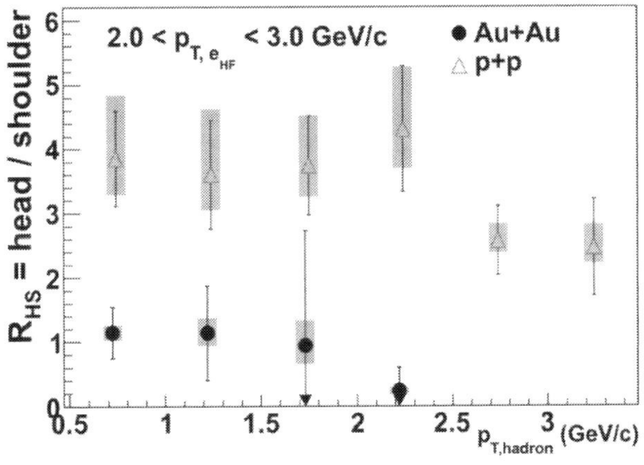


**Figure 5.** Ratio of the yield in the head region per radian to that in the shoulder region per radian for Au+Au (black) and *p*+*p* (red). Figure is from Ref. [7].

## 4. Conclusions

The mechanisms by which hard probes interact with the QGP are still not understood. Measurements involving heavy quarks have posed a particular problem to energy loss models based on radiative energy loss. However, there are many uncertainties in the theoretical calculations and the need for more and better measurements both at RHIC and the LHC. At PHENIX, the Silicon Vertex Detector was installed for the 2011 RHIC running period which will allow the separation of electrons from charm and bottom. This will be a major step forward to understand how the parton mass alters the effects of parton-matter interactions.

## References

[1] Adare A *et al.* (PHENIX) 2008 *Phys. Rev. Lett* **101** 232301
[2] Adare A *et al.* (PHENIX) 2010 *Phys. Rev. Lett* **104** 252301
[3] Dokshitzer Y and D K 2001 *Phys. Lett.* **B519** 199–206
[4] Adare A *et al.* 2006 *Phys. Rev. Lett.* **98** 172301
[5] Adare A *et al.* (PHENIX) 2009 *Phys. Rev. Lett.* **103** 082002
[6] Aggarwal M *et al.* (STAR) 2010 *Phys. Rev. Lett.* **105** 202301
[7] Adare A *et al.* (PHENIX) 2011 *Phys. Rev. C* **83** 044912
[8] Adare A *et al.* (PHENIX) 2009 *Phys. Rev. C* **80** 024908
[9] Adare A *et al.* 2008 *Phys. Rev* **C78** 014901
[10] Sjoestrand T *et al.* 2001 *Comp. Phys. Comm* **135** 238
[11] Cacciari M, Naso P and Vogt R 2005 *Phys. Rev. Lett.* **95** 122001
[12] Adler S S *et al.* (PHENIX) 2006 *Phys. Rev. Lett.* **97** 052301
[13] Casalderrey-Solana J, Shuryak E and Teaney D 2005 *J. Phys. Conf. Ser.* **27** 22–31
[14] Alver B and Roland G 2010 *Phys. Rev. C* **81** 054905
[15] Adare A *et al.* (PHENIX) (*Preprint* 1105.3928)
[16] Aamodt K *et al.* (ALICE) (*Preprint* 1105.3865)

# Jet-like correlations of heavy-flavor particles – from RHIC to LHC

**André Mischke**

ERC-Research Group *QGP-ALICE*, Institute for Subatomic Physics, Utrecht University, Princetonplein 5, 3584 CC Utrecht, the Netherlands

E-mail: a.mischke@uu.nl

**Abstract.** Measurements at the Relativistic Heavy Ion Collider (RHIC) at Brookhaven National Laboratory have revealed strong modification of the jet structure in high-energy heavy-ion collisions, which can be attributed to the interaction of hard scattered partons with the hot and dense QCD matter. The study of heavy-quark (charm and bottom) production in such collisions provides key tests of parton energy-loss models and, thus, yields profound insight into the properties of the produced matter. The high-$p_T$ yield of heavy-flavor decay electrons exhibits an unexpected large suppression. Since those single electrons have contributions from charm and bottom decays an experimental method is needed to investigate them separately. Heavy-flavor particle correlations provide information about the underlying production mechanism. In this contribution, a review on recent measurements on azimuthal correlations of single electrons and open charmed mesons at RHIC and perspectives of such measurements at the CERN-Large Hadron Collider (LHC) are presented. Moreover, it has been shown that next-to-leading-order (NLO) QCD processes, such as gluon splitting, become important at LHC energies. It will be demonstrated how this contribution can be determined through the measurement of the *charm content in jets*.

## 1. Introduction

The energy loss of partons is predicted to be a sensitive probe of the QCD matter created in high energy nucleus-nucleus collisions since its magnitude depends strongly on the color charge density of the matter traversed [1, 2]. In particular, the understanding of the flavor dependent coupling of quarks and the modification of their fragmentation function give essential information on the properties of the hot QCD matter produced in such collisions [3, 4]. Due to their large mass ($m > 1.2$ GeV/$c^2$), heavy quarks are believed to be produced predominantly in hard scattering processes in the early stage of the collision and, therefore, probe the complete space-time evolution of the expanding medium and their yields are sensitive to the initial gluon density [5]. Heavy-quark production by initial state gluon fusion also dominates in heavy-ion collisions where many, in part overlapping nucleon-nucleon collisions occur. It has been shown that charm production in the QCD medium might be significant at LHC energies [6]. The penetrating power of heavy quarks is much higher than for light quarks, providing a sensitive probe of the matter. Theoretical models based on perturbative QCD predict that the energy loss of heavy quarks in the medium is expected to be smaller compared to light-quarks and gluons due to the mass dependent suppression of the gluon radiation at small angles, the so-called dead-cone effect [7, 8].

Published under licence by IOP Publishing Ltd

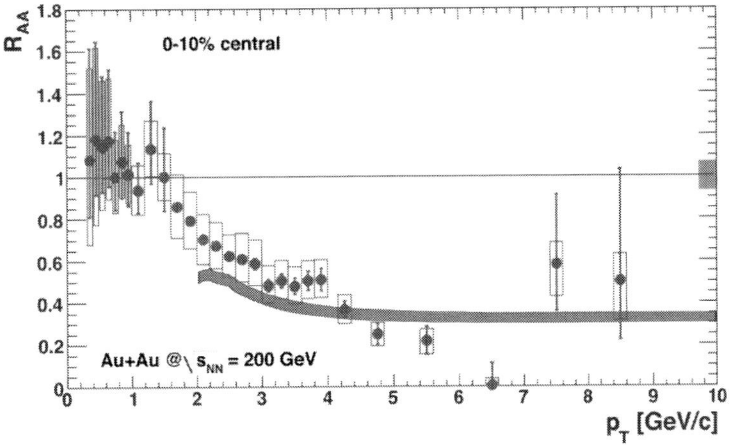


**Figure 1.** Nuclear modification factor $R_{\mathrm{AA}}$ of single electrons in the 10% most central Au+Au collisions at $\sqrt{s_{\mathrm{NN}}} = 200$ GeV measured by the PHENIX Collaboration. The data are compared with a collisional dissociation model (purple curve).

However, recent RHIC results on single electron production in nucleus-nucleus collisions have shown that the yield at high transverse momentum is suppressed to the same level as observed for light-quark hadrons. Azimuthal angular correlations of heavy-quark particles allow for the study of the fragmentation function of charm and bottom quarks separately, where high-$p_{\mathrm{T}}$ decay electrons are associated with open charmed mesons.

## 2. Single electron $R_{\mathrm{AA}}$

At RHIC charm and bottom quarks are identified by assuming that isolated electrons in the event originate from semileptonic decays of heavy-quark mesons. At high transverse momentum, this mechanism of electron production is dominant enough to reliably subtract other sources of electrons such as $\gamma$ conversions and $\pi^0$ Dalitz decays. STAR [9] and PHENIX measurements [10] in central Au+Au collisions have shown that the high-$p_{\mathrm{T}}$ yield of electrons from semileptonic charm and bottom decays is suppressed relative to properly scaled proton-proton collisions, quantified in the nuclear modification factor $R_{\mathrm{AA}}$ [9] (cf. Figure 1). This factor exhibits an unexpectedly similar amount of suppression as observed for light-quark hadrons, suggesting substantial energy loss of heavy quarks in the produced medium. This finding is not in line with expectations from the dead-cone effect. Energy-loss models incorporating contributions from charm and bottom do not describe the observed suppression sufficiently well (see discussions in [9, 10]). Although it has been realized that energy loss of heavy quarks by elastic parton scattering causing collisional energy loss is probably of comparable importance to energy loss by gluon radiation, yet the quantitative description of the suppression is still not satisfying. First attempts has been performed to distinguish between radiative and collisional energy loss [11]. Furthermore, it has been shown that the in-medium fragmentation of heavy quarks into heavy mesons which are collisionally dissociated, a process that can happen multiple times in the medium, may be significant in heavy-ion collisions. In addition, cold nuclear effects might explain part of the observed high-$p_{\mathrm{T}}$ yield suppression [12].

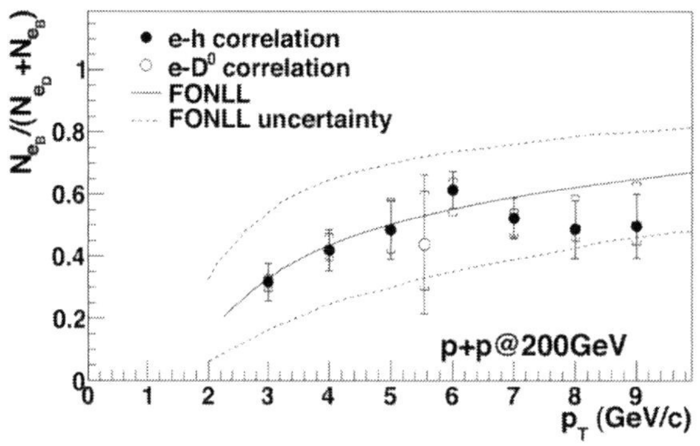


**Figure 2.** Relative bottom contribution to the single electron yields as a function of transverse momentum in $pp$ collisions at $\sqrt{s}$ = 200 GeV obtained from electron–hadron and electron–$D^0$ meson azimuthal correlations. Error bars are statistical and brackets are systematic uncertainties. The solid curve illustrates the FONLL calculation [13] with its theoretical uncertainties (dashed curve).

## 3. Azimuthal correlations of heavy-flavor particle

Perturbative QCD calculations at Fixed-Order plus Next-to-Leading Logarithm (FONLL) level have shown that the bottom contribution to the heavy-quark decay electrons is significant at intermediate $p_T$ (4-5 GeV/c) although the uncertainties are relatively large [13]. Therefore, measurements of the relative charm and bottom contribution to the single electrons is essential. Azimuthal angular correlations of single electrons and hadrons allow to exploit the different fragmentation of the associated jets [14]. Heavy quarks have, in general, a harder fragmentation function than gluons and light quarks, making the near-side correlation more sensitive to the decay kinematics. For the same electron transverse momentum the near-side electron–hadron angular correlation distribution from B decays is much broader than that from D decays.

Furthermore, azimuthal angular correlations of heavy-quark decay electrons and open charmed mesons yield important information about the charm and bottom contributions due to their different decay kinematics [15]. Flavor conservation implies that heavy quarks are produced in quark anti-quark pairs. A more detailed understanding of the underlying production process can be obtained from events in which both heavy-quark particles are detected. Due to momentum conservation, these heavy-quark pairs are correlated in relative azimuth ($\Delta\phi$) in the plane perpendicular to the colliding beams, leading to the characteristic back-to-back orientated sprays of particles. This correlation survives the fragmentation process to a large extent in $pp$ collisions. Charm quarks predominantly ($\approx$54%) hadronize directly and bottom quarks via B decays into $D^0$ mesons. The branching fraction for charm and bottom decays into electrons is $\approx$10% [16]. While triggering on the so-called leading electron (trigger side), the balancing heavy quark, which is identified by the $D^0$ meson, can be used to identify the underlying production mechanisms (probe side). Requiring like-sign electron–kaon pairs, where the kaons stem from the $D^0$ decays, the near-side correlation peak can be attributed to B decays only whereas the contribution to the away-side peak can mostly be attributed to charmed meson decays. Figure 2 depicts the relative bottom contribution to the single electrons from $pp$ collisions at $\sqrt{s}$ = 200 GeV measured by the STAR experiment [17]. The PHENIX Collaboration obtained similar results [18]. The data provide convincing evidence that the bottom contribution to the single electron yields is of the same magnitude than that of charm at and above $p_T$ = 5 GeV/c. The direct measurements of open charm and bottom mesons at RHIC will be possible with the high precision $\mu$ vertex detectors in the STAR and PHENIX experiments. Figure 3 illustrates

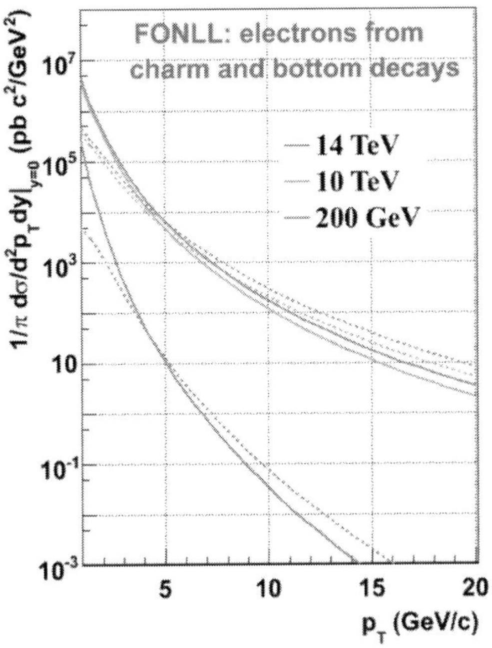


**Figure 3.** Electron production cross section from charm (full line) and bottom decays (dashed line) in *pp* collisions at RHIC and LHC energies obtained from FONLL calculations [19].

the charm and bottom contribution to the single electrons at 0.2, 10 and 14 TeV [19]. It is remarkable that the crossing point, where bottom starts to dominate over charm seems to be at around the same $p_T$ for all three collision energies. A possible explanation is that at the LHC energies more bottom is produced at small $p_T$ with respect to charm than at RHIC energies, which would bring the crossing point closer to zero. On the other hand the single spectra of bottom and charm are likely to be more similar at the LHC due to the fact that the quark mass is less relevant. This would push the crossing point away from zero. It seems that both mechanisms roughly compensate each other.

ALICE, A Large Ion Collider Experiment, is the dedicated detector to study hot QCD matter in lead-lead collisions at the TeV scale at the CERN-LHC. The initial energy density in the collision zone is expected to be about one order of magnitude higher than at the RHIC facility ($\approx$100 GeV/fm$^3$). The higher energy density allows thermal equilibrium to be reached more quickly and creates a relatively long-lived QGP phase. Therefore, it is expected that most of the in-medium effects will be enhanced. The first Pb+Pb data taken in November 2010 at $\sqrt{s_{\rm NN}} = 2.76$ TeV provided first exciting results [20]. Also first heavy-flavor results from *pp* collisions at $\sqrt{s} = 7$ TeV have been presented at this workshop [21]. The feasibility of heavy-flavor correlations in Pb+Pb collisions at LHC energies were studied using the PYQUEN event generator [22]. The azimuthal angular correlation of single electrons and $D^0$ mesons at mid-rapidity ($|\eta| < 0.9$) is studied in PYTHIA simulations of *pp* interactions and in PYQUEN simulations of the 5% most central Pb+Pb collisions at $\sqrt{s_{\rm NN}} = 5.52$ TeV. The left panel of Figure 4 shows both correlation distributions. For the *pp* collisions the correlation distribution exhibits a pronounced near-side peak, which can be attributed to B decays, and an away-side peak, which arises from charm decays. For Pb+Pb collisions, however, the near-side peak is expected to be suppressed whereas the away-side correlation peak disappear. A similar pattern have be found for di-hadron correlations with the first data from the RHIC facility (Figure 4,

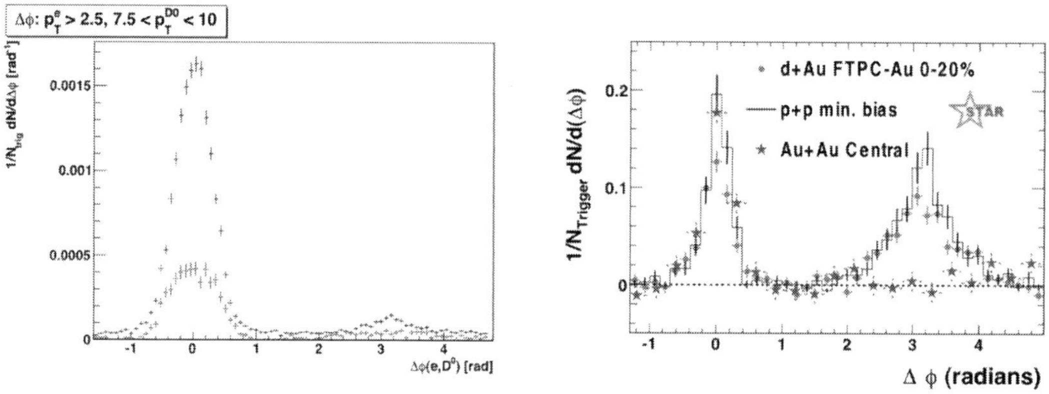


**Figure 4.** Left panel: Azimuthal angular correlation of single electrons and $D^0$ mesons from PYTHIA simulations of *pp* interactions (black histogram) and from PYQUEN simulations of the 5% most central Pb+Pb collisions at $\sqrt{s_{NN}} = 5.52$ TeV (red histogram). Right panel: di-hadron azimuthal correlation distribution from *pp*, *d*+Au and Au+Au collisions at $\sqrt{s_{NN}} = 0.2$ TeV measured by the STAR experiment.

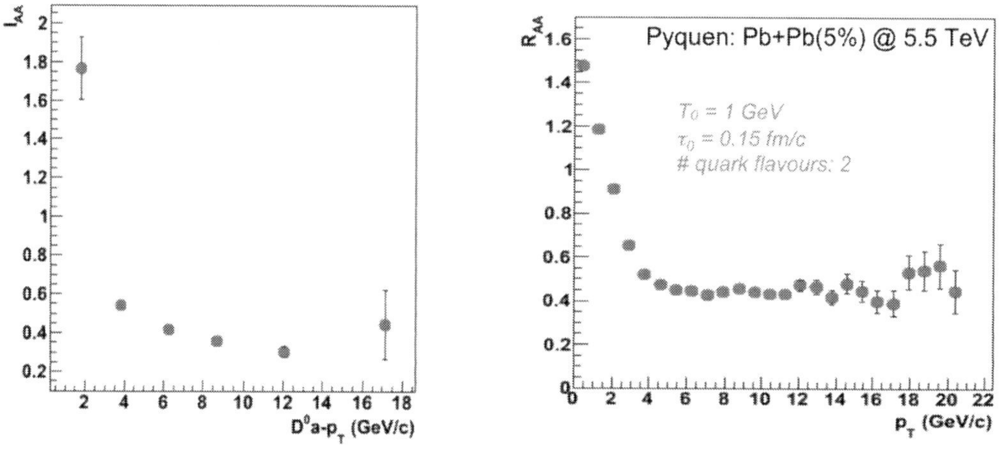


**Figure 5.** $I_{AA}$ (left panel) and $R_{AA}$ of single electrons (right panel) from PYQUEN simulations of the 5% most central Pb+Pb collisions at $\sqrt{s_{NN}} = 5.52$ TeV. The PYQUEN settings are specified in the right hand panel.

right panel) [23], indicating promising results to come from the first LHC data sample.

The yield of the near-side peak for an electron trigger $p_T$ of $2 < p_T^{trg-ele} < 4$ GeV/c was extracted from the integral of a Gaussian fit to the data ($\pm 3\sigma$ range around the peak position). The width of the near-side peak from PYQUEN (Pb+Pb case) is slightly broader compared to PYTHIA (*pp* case). The yield as a function of the associated $p_T$ of the $D^0$ mesons is shown in the left panel of Figure 5. The $I_{AA}$ decreases and stays constant at around $p_T = 8$ GeV/c. The magnitude of the yield suppression of the near-side peak in central Pb+Pb collisions at $\sqrt{s_{NN}} =$ 5.52 TeV is similar to the nuclear modification factor of the single electrons, as illustrated in the right panel of Figure 5. The single electron $R_{AA}$ agrees with the ALICE measurement [24] and previous theoretical model calculations [25].

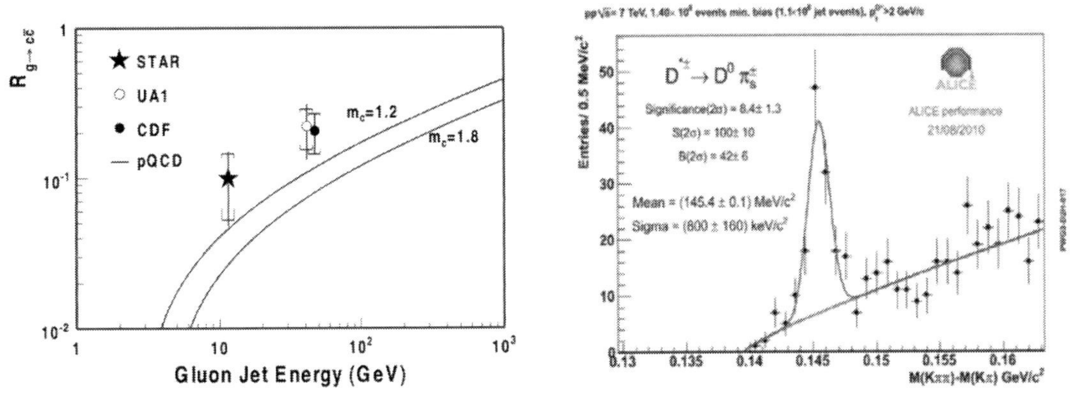


**Figure 6.** Left: Gluon splitting rate to charm pairs as a function of the gluon jet energy [32], compared with pQCD calculations. Right: Charged $D^*$ signal reconstructed with the ALICE experiment within the cone of jets in $pp$ collisions at $\sqrt{s} = 7$ TeV. See text for details.

## 4. Gluon splitting rate

In the previous sections is has been shown that the understanding of the heavy-flavor production mechanism is of considerable interest. Higher order sub-processes such as gluon splitting to $c\bar{c}$ pairs are expected to have a significant contribution to the total open charm yield at LHC energies [26]. The gluon splitting contribution is studied by identifying the *charm content in jets*. By utilizing jet-finder algorithms and isolating the near side of the jet-$D^*$ azimuthal angular correlation, the charm ($D^*$) content in a jet was measured. From the correlation strength on the near side, it is possible to estimate the amount of open charm production which results from gluon splitting.

The LEP experiments [27, 28, 29] have studied the $D^{*\pm}$-meson content in jets and demonstrated that the production from $Z^0$ decays in $e^+e^-$ collisions is dominated by $D^*$ mesons that carry large fractions of the jet momenta. This finding is consistent with the jets being produced from primary c (anti-)quarks. At higher collision energies ($\bar{p}p$ at $\sqrt{s} = 1.8$ TeV and 630 GeV) the CDF and UA1 experiments have observed $D^{*\pm}$ mesons in the cone of jets with transverse energies larger than 40 GeV [30, 31]. Their fractional momenta are found to be smaller, consistent with a different production mechanism in which the $D^*$ mesons originate from gluon splitting into $\bar{c}c$ pairs ($g \rightarrow \bar{c}c$ in the initial or final parton shower, with neither of the quarks from the $\bar{c}c$ pair participating in the hard QCD interaction) [26]. The STAR experiment measured a gluon splitting rate of about 10% in $pp$ at $\sqrt{s} = 200$ GeV [32]. The later three measurements are summarized in the left panel of Figure 6 compared to perturbative QCD [33, 34, 35]. Moreover, the STAR measurement is consistent with results from MC@NLO calculations [15], a QCD computation with a realistic parton shower model.

First performance studies have been conducted at LHC. The right panel of Figure 6 shows the charged $D^*$ signal reconstructed within the cone of jets with the ALICE experiment in $pp$ collisions at $\sqrt{s} = 7$ TeV [36, 37]. This is an important step towards the determination of the charm content in jets at the TeV scale.

## 5. Summary

Heavy quarks are particularly good probes to study the properties of hot QCD matter (especially its transport properties). RHIC data exhibit a relatively large energy loss of heavy quarks in the medium. Azimuthal angular correlations of single electron–hadron and single electron–$D^0$ mesons show that bottom decay yields are of comparable magnitude at and above $\approx$5 GeV/c, consistent with FONLL calculations, indicating that in particular bottom is more strongly suppressed than expected. At the LHC, heavy-flavor particle correlations will allow for the study of the possible modification of the fragmentation function in the medium. Next-to-leading-order QCD processes, such as gluon splitting, become important at LHC energies and their contribution can be studied through the measurement of the *charm content in jets*. First heavy-ion collisions in November 2010 mark the start of the study of hot quark matter in a new energy domain.

## Acknowledgments

The author would like to thank the organizers for the nice atmosphere and the stimulating discussions during the workshop.

The European Research Council has provided financial support under the European Community's Seventh Framework Programme (FP7/2007-2013) / ERC grant agreement no 210223. This work was supported in part by a Vidi grant from the Netherlands Organisation for Scientific Research (project number 680-47-232).

## References

[1] B. Müller and J.L. Nagle, Ann. Rev. Nucl. Part. Sci. **56**, 93 (2006).
[2] P.M. Jacobs and X.-N. Wang, Prog. Part. Nucl. Phys. **54**, 443 (2005).
[3] R. Rapp and H. van Hees, Quark-Gluon Plasma 4, World Scientific Publishing Co. Pte. Ltd., 2010 (arXiv:0903.1096).
[4] A.D. Frawley, T. Ullrich and R. Vogt, Phys. Rep. **462**, 125 (2008).
[5] Z. Lin and M. Gyulassy, Phys. Rev. **C51**, 2177 (1995).
[6] J. Uphoff, O. Fochler, Z. Xu and C. Greiner, Phys. Rev. **C82**, 044906 (2010).
[7] Y.L. Dokshitzer and D.E. Kharzeev, Phys. Lett. **B519**, 199 (2001).
[8] M. Djordjevic, M. Gyulassy and S. Wicks, Phys. Rev. Lett. **94**, 112301 (2005).
[9] B.I. Abelev *et al.* [STAR Collaboration], Phys. Rev. Lett. **106**, 159902 (2011); B.I. Abelev *et al.* [STAR Collaboration], Phys. Rev. Lett. **98**, 192301 (2007).
[10] A. Adare *et al.* [PHENIX Collaboration], arXiv:1005.1627 [nucl-ex], submitted to Phys. Rev. C, 2010; A. Adare *et al.* [PHENIX Collaboration], Phys. Rev. Lett. **98**, 172301 (2007).
[11] P.B. Gossiaux, J. Aichelin, T. Gousset and V. Guiho, J. Phys. **G37**, 094019 (2010).
[12] M. Durham for the PHENIX Collaboration, these proceedings.
[13] M. Cacciari, P. Nason and R. Vogt, Phys. Rev. Lett. **95**, 122001 (2005).
[14] A. Sickles for the PHENIX Collaboration, these proceedings.
[15] A. Mischke, Phys. Lett. **B671**, 361 (2009).
[16] K. Nakamura *et al.* (Particle Data Group), J. Phys. **G37**, 075021 (2010).
[17] M.M. Aggarwal *et al.* [STAR Collaboration], Phys. Rev. Lett. **105**, 202301 (2010).
[18] A. Adare *et al.* [PHENIX Collaboration], Phys. Rev. Lett. **103**, 082002 (2009).
[19] M. Cacciari and A. Mischke, private communication, 2009.
[20] J. Harris for the ALICE Collaboration, these proceedings.
[21] A. Grelli for the ALICE Collaboration, these proceedings.
[22] I.P. Lokhtin, A.M. Snigirev, Eur. Phys. J. **C46**, 211 (2006).
[23] J. Adams *et al.* [STAR Collaboration], Phys. Rev. Lett. **91**, 072304 (2003).
[24] S. Masciocchi for the ALICE Collaboration, Nucl. Phys. **A855**, 432 (2011).
[25] N. Armesto, Quark-Gluon Plasma 4, World Scientific Publishing Co. Pte. Ltd., 2010 (arXiv:0903.1330).
[26] E. Norrbin, T. Sjöstrand, Eur. Phys. J. **C17**, 137 (2000).
[27] R. Barate *et al.*, Eur. Phys. J. **C16**, 597 (2000);
[28] M. Acciarri *et al.*, Phys. Lett. **B476**, 243 (2000);
[29] G. Abbiendi *et al.*, Eur. Phys. J. **C13**, 1 (2000).
[30] F. Abe *et al.* [CDF Collaboration], Phys. Rev. Lett. **64**, 348 (1990).
[31] C. Albajar *et al.* [UA1 Collaboration], Phys. Lett. **B244**, 566 (1990).
[32] B. I. Abelev *et al.* [STAR Collaboration], Phys. Rev. **D79**, 112006 (2009).
[33] A. H. Mueller and P. Nason, Phys. Lett. **B157**, 226 (1985);
[34] M. L. Mangano and P. Nason, Phys. Lett. **B285**, 160 (1992);
[35] M. H. Seymour, Z. Phys. **C63**, 99 (1994).
[36] A. Grelli and A. Mischke, PoS (EPS-HEP 2009) 451.

[37] A. Grelli, J of Phys.: Conf. Ser. **270**, 012019 (2011).

27th Winter Workshop on Nuclear Dynamics (WWND 2011) IOP Publishing
Journal of Physics: Conference Series **316** (2011) 012010 doi:10.1088/1742-6596/316/1/012010

# $v_n$ measurements with forward rapidity $\Phi_n$ in 200GeV Au+Au collisions at RHIC-PHENIX

**Shinichi Esumi for the PHENIX collaboration**

Inst. of Physics, Univ. of Tsukuba
Tenno-dai 1-1-1, Tsukuba, Ibaraki 305-8571, Japan

esumi@sakura.cc.tsukuba.ac.jp

**Abstract**. Higher order event anisotropy $v_n$ have been measured at mid-rapidity with respect to the same order event plane $\Phi_n$ defined at the forward rapidity in 200GeV Au+Au collisions at RHIC-PHENIX experiment. Correlation between higher order event planes at different rapidity indicates the initial geometrical fluctuation is the dominant source of the higher order event anisotropy. Hydro-dynamic calculation seems to describe the measurements.

## 1. Higher order reaction plane correlation

The event anisotropy has been known to give specific rapidity dependence, symmetric and anti-symmetric for even and odd harmonics, respectively, when considering two black disk-like collisions with infinite number of participants in case of symmetric heavy ion collisions. It has also been suggested that the initial geometrical fluctuation of finite number of participant would give rapidity symmetric dependence even for odd harmonics [1]. In order to answer this question, whether which of the observed correlation is stronger between rapidity symmetric and anti-symmetric contributions, the higher order reaction planes are determined with various forward detectors, RXN: reaction plane detector ($|\eta|$ = 1.0~2.8), MPC: muon piston calorimeter ($|\eta|$ = 3.1~3.7), BBC: beam-beam counter ($|\eta|$ = 3.1~3.9) and ZDC: zero degree calorimeter ($|\eta| > 6.5$).

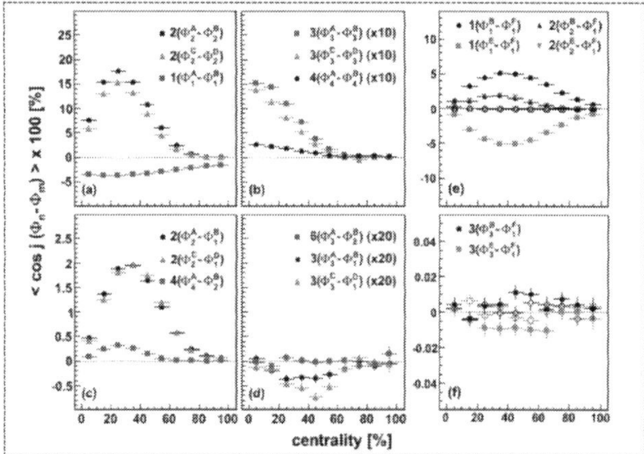


**Figure 1.** Various higher order event plane correlations between different rapidity are shown as a function of centrality. The choice of event planes are indicated as A:RXN$^{(+)}$, B:BBC$^{(-)}$, C:MPC$^{(+)}$, D:MPC$^{(-)}$, E:BBC$^{(+)}$ and F:ZDC$^{(+/-)}$, where (+) and (-) are for the positive and negative rapidity, (+/-) means combined event plane with flipping sign of (-). See details in the text.

Figure 1 shows the various higher order event plane correlations between different rapidity slices. The panel (a) and (b) show the same order event plane, where the 2nd order correlation is the largest, but the correlation between 3rd order harmonics is quite significant with a large rapidity gap. This is an indication of initial geometrical fluctuation is a dominant source of $\Phi_3$ correlation. The panel (c) and (d) show the cross harmonics correlation, where the correlation between 1st and 2nd as well as 2nd and 4th order have been known to show a significant signal from the $v_1$, $v_2$ and $v_4$ correlation. However, 1st and 3rd order shows a slight negative correlation, which could be an indication of the rapidity anti-symmetric $v_3$ contribution, although the magnitude of the correlation is about 50 times smaller than the 3rd order correlation. The 2nd and 3rd order shows invisible correlation within the current statistics. The panel (e) and (f) show the correlation of $\Phi_n$ at forward region with respect to the $\Phi_1$ at the spectator region. The rapidity anti-symmetric $v_1$ and symmetric $v_2$ signals are clearly seen, the rapidity anti-symmetric $v_3$ signal is weak but observed to have the same sign of $v_1$, which is known to be negative for pion dominating participant region. Figure 2 shows the higher order event plane resolutions and observed azimuthal distributions.

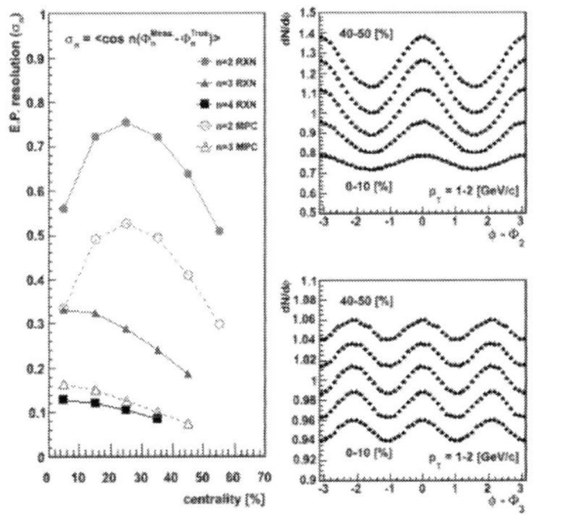


**Figure 2.** The extracted event plane resolution as a function of centrality is shown in the left panel. The solid symbols are resolutions with RXN, the open symbols are with MPC. The top- and bottom-right panels show measured azimuthal distribution with respect to the 2nd and 3rd order event planes, respectively, for 5 different centrality selections.

## 2. Two particle correlations with small and larger eta gaps

Two particle azimuthal correlation analysis methods have been used to extract the jet property, suppression and modification as well as event anisotropy. The recent hot topics on the jet correlation analysis have been the "ridge-like" near-side long-range eta correlation and "mach-cone-like" away-side double-peak correlation. It has been suggested that the both of them would have strong relation with $v_3$ signal, which seems to be dominant effect of initial geometrical fluctuation. The figure 3 and 4 show the comparison of two particle azimuthal correlation functions with large and small eta gaps. The 4 panels in figure 3 are inclusive photon and hadron azimuthal correlation functions measured within the central arm acceptance $|\eta|<0.35$ with varying the eta gap cuts. The 4 panels in figure 4 show the azimuthal correlation functions between central ($|\eta|<0.35$) and forward detectors, the left with RXN ($|\eta|$ = 1.0~2.8) and the right with MPC ($|\eta|$ = 3.1~3.7), while the top shows the fitting and subtraction including the first two harmonics to demonstrate and visualize the significant 3rd order harmonics in the correlation function. The bottom panels include the fitting and subtraction upto the 4th harmonics. This two particle correlation method also provides event anisotropy measurements that can be compared with event plane method.

Journal of Physics: Conference Series **316** (2011) 012010 doi:10.1088/1742-6596/316/1/012010

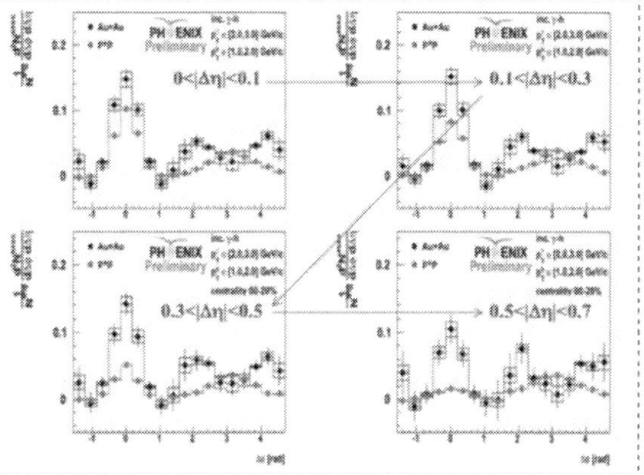


**Figure 3.** Correlation function between inclusive photon and hadron within the central arm (|η|<0.35), which is $v_2$ subtracted inclusive photon triggered jet associated hadron $\Delta\phi$ distribution.

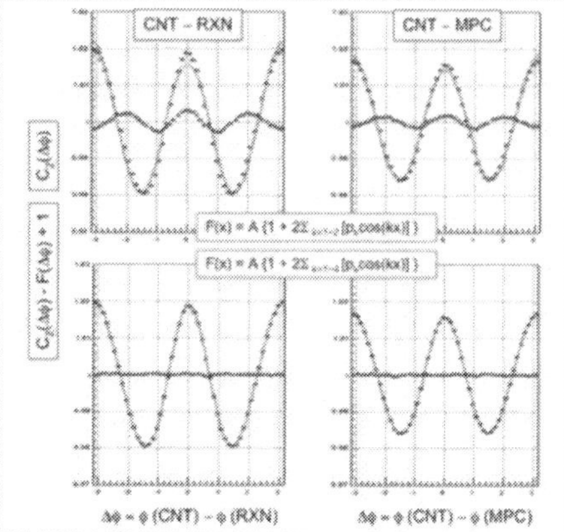


**Figure 4.** Correlation function between central (|η|<0.35) and forward RXN (|η| = 1.0~2.8) or MPC (|η| = 3.1~3.7).

## 3. Higher order event anisotropy

Figure 5 shows the measured higher order event anisotropy with respect to the higher order event plane defined at RXN (|η| = 1.0~2.8) as a function of $p_T$ for different centrality selections [2]. The different order harmonics show similar $p_T$ dependence. The magnitude of higher moments $v_3$ and $v_4$ are significant especially for the central collisions. Compared with the strong centrality dependence seen in $v_2$, a very weak centrality dependence of $v_3$ is observed. The $v_4\{\Phi_4\}$ is found to be about 2 times larger than the previously measured $v_4\{\Phi_2\}$, which is mainly from the overlapped geometry, while $v_4\{\Phi_4\}$ would have an additional contribution from the fluctuation.

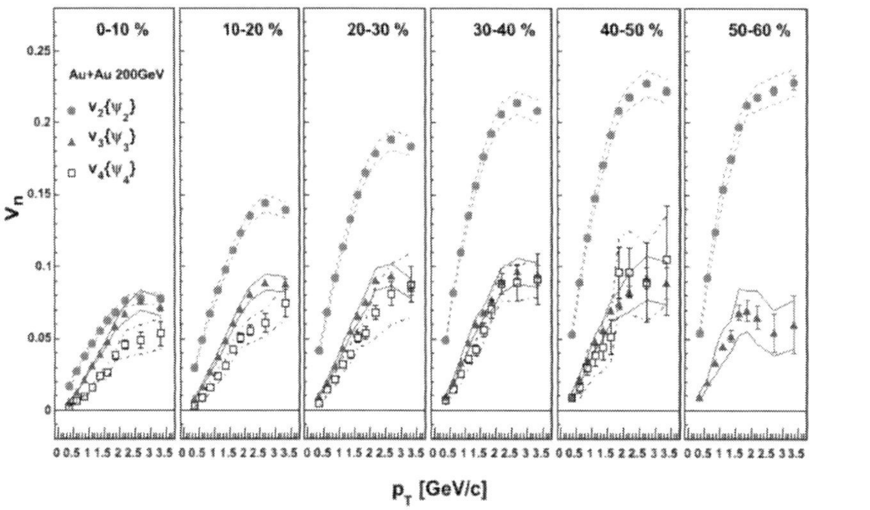


**Figure 5.** higher order event anisotropy as a function of $p_T$ .

The centrality dependence of higher order event anisotropy is shown in figure 6 as a function of number of participant for two $p_T$ slices, they are also compared with generalized eccentricity $\varepsilon_n$ from initial geometry with a glauber model calculation. Since one would expect there will be additional effect from hydro-dynamic expansion, one can not really compare the magnitude between $v_n$ and $\varepsilon_n$, however the relative difference between different harmonic orders could be taken in order to extract the effect from the expansion relative to the initial fluctuation.

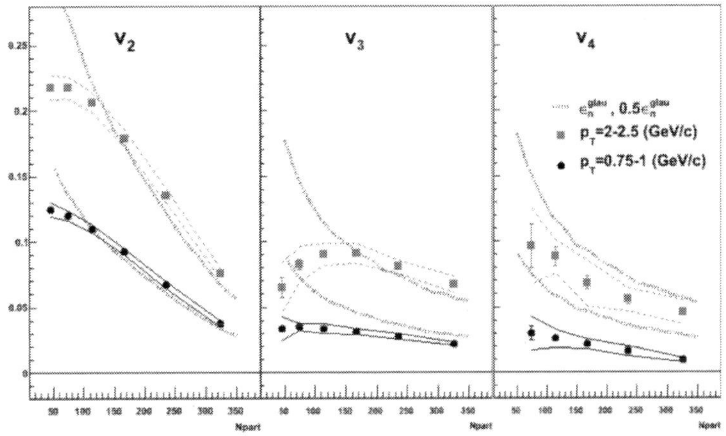


**Figure 6.** higher order event anisotropy as a function of number of participant

Figure 7 shows the $v_3$ measured with $\Phi_3$ from the forward rapidity compared with $v_3$ from the two particle correlation methods, which shows slightly larger $v_3$ because of the possible non-flow contribution from jets at higher $p_T$, however the agreement at lower $p_T$ between two experimental measurements as well as the agreement between hydro-dynamic calculation [3] and experimental data are remarkable.

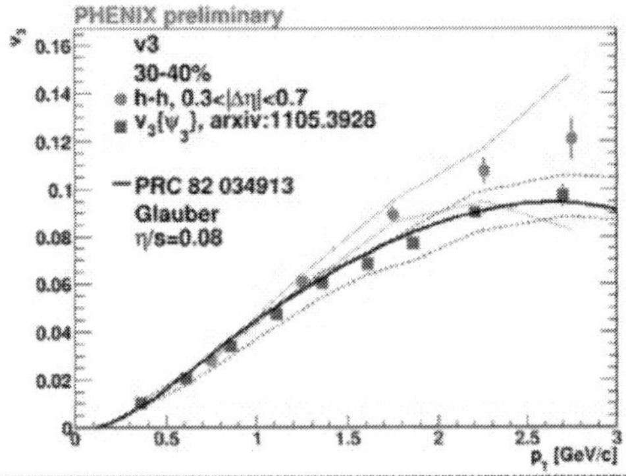


**Figure 7.** $v_3$ extracted with different methods compared with hydro-dynamic calculation

## 4. Summary

Higher order event anisotropy has been measured with higher order event plane defined at forward rapidity detectors and compared with two particle correlation method with various rapidity gap in 200GeV Au+Au colliaions at RHIC-PHENIX experiment. The results are compared with generalized initial geometrical eccentricity and with hydro-dynamic calculations. Higher order event anisotropy seems to be dominantly given by the initial geometrical fluctuation followed by the hydro-dynamic expansion.

## References

[1] B. Alver and G. Roland, Phys. Rev. C81, 054905 (2010)
[2] arXiv:1105.3928
[3] B. Alver et al., Phys. Rev. C82, 034913 (2010)

# CGC initial conditions at RHIC and LHC

**Javier L. Albacete[1], Adrian Dumitru[2,3], Yasushi Nara[4]**

[1] Institut de Physique Théorique - CEA/Saclay, 91191 Gif-sur-Yvette cedex, France.
[2] Department of Natural Sciences, Baruch College, CUNY, 17 Lexington Avenue, New York, NY 10010, USA
[3] RIKEN BNL Research Center, Brookhaven National Laboratory, Upton, NY 11973, USA
[4] Akita International University, Yuwa, Akita-city 010-1292, Japan

E-mail: nara@aiu.ac.jp

**Abstract.** Monte-Carlo implementations of $k_T$-factorization formula with both KLN and running-coupling BK unintegrated gluon distributions for nucleus-nucleus collisions are used to analyze recent experimental data on the particle multiplicities from RHIC(Au+Au@200GeV) and LHC(Pb+Pb@2.76TeV). We also compare the predicted transverse energy at midrapidity to new data from ALICE.

## 1. Introduction

Relativistic Heavy Ion Collider (RHIC) at BNL has started in 2000 and many interesting data have been obtained. Experiments at Large Hadron Collider (LHC) at CERN has just started. It is well known fact that nearly perfect fluid picture works to explain large elliptic flow discovered at RHIC and LHC. However, there is no detailed understanding on the particle production in high energy hadronic collisions. Especially subsequent non-equilibrium evolution of the system (Glasma) created in nuclear collisions is not known well. Without knowing the detailed glue dynamics in early stages of nucleus collision, correct initial conditions for hydrodynamics would not be obtained. First important issue is to understand the gluon production in the first moment of collision. In order to study such gluon production in nucleus-nucleus collision, we shall use the Monte-Carlo implementation of $k_t$-factorization formulation in which fluctuations of the position of nucleon inside a nucleus is taken into account, and we can study nucleus-nucleus collision event-by-event.

## 2. Theoretical Models

We will use the Monte-Carlo KLN (MC-KLN) model [1] and its extension to running coupling Balitsky-Kovchegov (MCrcBK) [2] for the computation of gluon production in heavy ion collisions. Both model apply the $k_t$-factorized formula [3] in the transverse plane perpendicular to the beam axis locally. The number distribution of produced gluons is given by

$$\frac{dN_g}{d^2r_\perp dy} \sim \kappa_g \frac{4N_c}{N_c^2-1} \int \frac{d^2p_\perp}{p_\perp^2} \int d^2k_\perp \, \alpha_s \, \phi_A(x_1, \boldsymbol{k}_\perp^2) \, \phi_B(x_2, (\boldsymbol{p}_\perp - \boldsymbol{k}_\perp)^2) \, , \tag{1}$$

with $N_c = 3$ the number of colors. Here, $p_\perp$ and $y$ denote the transverse momentum and the rapidity of the produced gluons, respectively. The light-cone momentum fractions of the colliding gluon ladders are then given by $x_{1,2} = p_\perp \exp(\pm y)/\sqrt{s_{NN}}$, where $\sqrt{s_{NN}}$ denotes the center of mass energy. A constant gluon multiplication factor $\kappa_g = 5$ is assumed in the MCrcBK model to obtain $p_\perp$-integrated hadron yields [2]. It does not appear in the corresponding formula for the transverse energy $dE_\perp/dy$ since gluon splitting and hadronization conserves energy.

At each grid point, we compute the thickness function $T_A(\boldsymbol{r}_\perp)$ to obtain the local saturation scale and then compute gluon production probability. Within a hard disk nucleon approximation, thinkness function is given by

$$T_A(\boldsymbol{r}_\perp) = \frac{\text{number of nucleon within } S}{S} \tag{2}$$

where we assume that the area $S$ has the same value as the inelastic proton-proton cross section at the incident energy of $\sqrt{s_{NN}} = 200$ GeV independent colliding energy.

In MC-KLN model, saturation momentum is defined as

$$Q^2_{s,A}(x;\boldsymbol{r}_\perp) = 2\ \text{GeV}^2 \frac{T_A(\boldsymbol{x}_\perp)}{1.53\ \text{fm}^{-2}} \left(\frac{0.01}{x}\right)^\lambda , \tag{3}$$

where $\lambda$ is a free parameter which is expected to have the range of $0.2 < \lambda < 0.3$ from HERA global analysis. In MC-KLN, we assume the gluon distribution function as

$$\phi_{A,B}(x, k_\perp^2; \boldsymbol{r}_\perp) \sim \frac{1}{\alpha_s(Q_s^2)} \frac{Q_s^2}{\max(Q_s^2, k_\perp^2)} , \tag{4}$$

On the other hand, in MCrcBK, $\phi$ is obtained from the Fourier transform of the numerical results of the running coupling BK (rcBK) evolution equation [7, 8, 9]:

$$\phi_{A,B}(k,x,b) = \frac{C_F}{\alpha_s(k)\,(2\pi)^3} \int d^2\mathbf{r}\; e^{-i\mathbf{k}\cdot\mathbf{r}}\, \nabla^2_{\mathbf{r}} \mathcal{N}_G(r, Y = \ln(x_0/x), b) . \tag{5}$$

where $C_F = (N_c^2 - 1)/2N_c$, $x_0 = 0.01$, and $\mathcal{N}_G$ is related to the quark dipole scattering amplitude that solves the rcBK equation, $\mathcal{N}$ as follows:

$$\mathcal{N}_G(r,x) = 2\mathcal{N}(r,x) - \mathcal{N}^2(r,x) . \tag{6}$$

In MC-KLN model, $x$ dependence is determined by eq. (3), but in rcBK, $x$ dependence can be obtained from the equation. Therefore, we expect that MCrcBK model has more predictive power than MC-KLN model.

## 3. Results

### *3.1. MC-KLN*

In the figure 1, the charged particle multiplicities at RHIC and LHC as a function of $N_{\text{part}}$ from KLN, fKLN, and MC-KLN model are compared to PHOBOS data.

In the KLN model, we plot the equation

$$\frac{1}{N_{\text{part}}} \frac{dN}{dy} = c \ln\left(\frac{Q_s^2}{\Lambda^2_{\text{QCD}}}\right) . \tag{7}$$

This equation is obtained by picking up the most important contribution of integration in $k_t$-factorized formula by using average nuclear saturation momentum. On the other hand, fKLN model [4] does not take into account the effects of fluctuations of nucleons inside a nucleus unlike the MC-KLN model. The main difference between KLN and fKLN model is that fKLN model

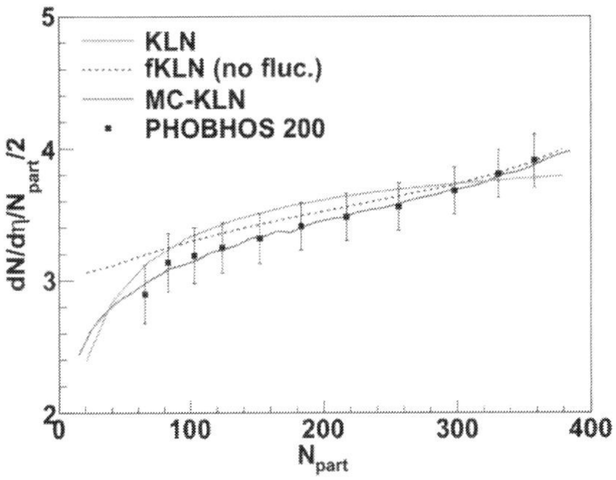


**Figure 1.** Centrality dependence of charged hadron multiplicity in Au+Au collision from the KLN, fKLN, and MC-KLN model are compared with PHOBOS data [5]

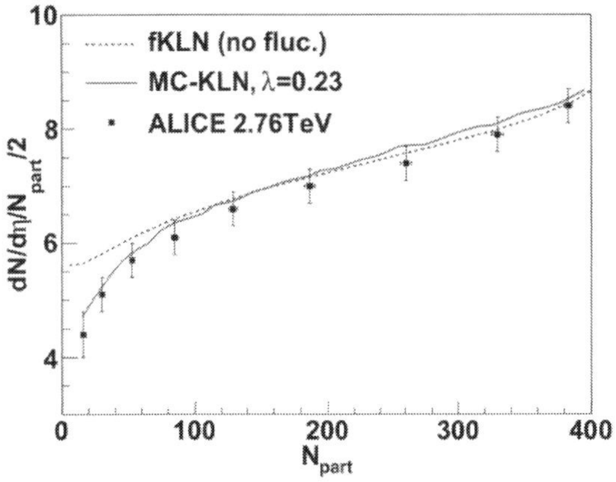


**Figure 2.** Centrality dependence of charged hadron multiplicity in Pb+Pb collision at $\sqrt{s_{NN}}$ = 2.76 TeV from the fKLN, and MC-KLN model are compared with ALICE data [6]

uses the local saturation scale instead of assuming a average nuclear saturation momentum as in KLN. Therefore, we see that computations with local saturation momentum improve results. A comparison between fKLN and MC-KLN model shows the effects of fluctuations of the position of nucleon inside a nucleus. One sees such effect in the peripheral collision which suggests that it is important to take into account fluctuation effect in the discussion of the centrality dependence of particle multiplicities.

In Fig. 2, the centrality dependence of charged hadron multiplicity in Pb+Pb collision at $\sqrt{s_{NN}}$ = 2.76 TeV from ALICE experiment is compared to the results from fKLN and MC-KLN. Where we use $\lambda = 0.23$ in Eq. (3) which controls the $x$-evolution speed. If $\lambda$ is small, multiplicity becomes small, and we may think additional particle production mechanism during the evolution of the system, especially before thermalization.

### *3.2. MCrcBK*

We now compare in Fig.3, the centrality dependence of charged hadron with the results obtained from the MCrcBK model in which unintegrated gluon function is taken from the numerical solution of the rcBK equation. The initial condition for the rcBK evolution is assumed to be MV model:

$$\mathcal{N}(r, Y\!=\!0) = 1 - \exp\left[-\frac{r^2 Q_{s0}^2}{4}\ln\left(\frac{1}{\Lambda r} + e\right)\right] , \tag{8}$$

with the initial saturation scale $Q_{s0}^2 = 0.2$ GeV$^2$ for nucleon and $\Lambda = 0.2$ GeV. For the running coupling, we use

$$\alpha_s(r^2) = \frac{12\pi}{5\ln\left(\frac{4C^2}{r^2\Lambda^2}\right)} \tag{9}$$

with $C = 1$ in solving rcBK equation. rcBK unintegrated gluon function with these parameter set describes the centrality dependence of charged hadron multiplicity for both RHIC and LHC which indicates that most of the gluon is produced at the first impact of the nuclear collision. However, we should check parameter dependence carefully. For example, if one changes the parameter $C$ in the running coupling, evolution speed changes.

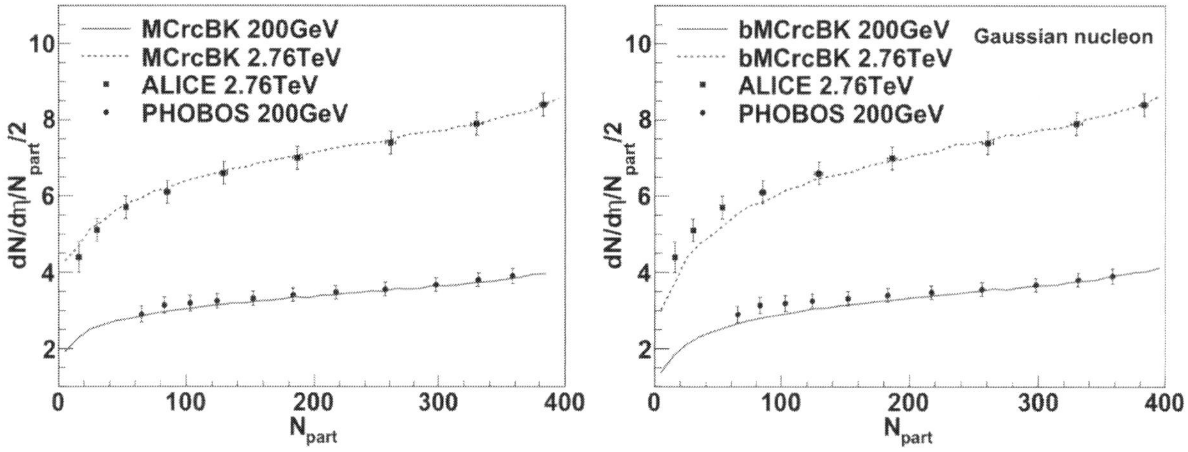


**Figure 3.** Centrality dependence of charged hadron multiplicity in Pb+Pb collision at $\sqrt{s_{NN}} = 2.76$ TeV and Au+Au collision at $\sqrt{s_{NN}} = 200$ GeV from the MCrcBK model are compared.

In the bottom up scenario [10], the number of gluon produced just after the collision will increase by a factor of $\alpha_s(Q_s^2)^{-2/5}$ during the subsequent evolution of the system. It would be interesting to include such effect into MCrcBK calculations.

*3.3. Gaussian shape*

So far we assumed that nucleus-nucleus collision is described by the incoherent sum of nucleon-nucleon collisions which will occur when transverse distance squared between two nucleons is smaller than the inelastic proton-proton cross section $\sigma_{NN}$ divided by $\pi$:

$$(x_i - x_j)^2 + (y_i - y_j)^2 \le \frac{\sigma_{NN}}{\pi} \tag{10}$$

This amounts to assume that nucleon is hard sphere (or disk). However, this approximation may not work in very high energy hadronic collisions. Let us think about the Gaussian shape nucleon in order to take into account the effect of extension of nucleon size as incident energy increases. In this case the thickness function becomes

$$T_p(r) = \frac{1}{2\pi B}\exp[-r^2/(2B)] \ , \tag{11}$$

and the probability of nucleon-nucleon collision $P(b)$ at impact parameter $b$ is

$$P(b) = 1 - \exp[-kT_{pp}(b)], \qquad T_{pp}(b) = \int d^2s \, T_p(s) \, T_p(s-b) \ . \tag{12}$$

where (perturbatively) $k$ corresponds to the product of gluon-gluon cross section and gluon density squared. We fix $k$ so that integral over impact parameter becomes the nucleon-nucleon inelastic cross section $\sigma_{NN}$ at the given energy:

$$\sigma_{NN} = \int d^2b \, (1 - \exp[-kT_{pp}(b)]) \, . \tag{13}$$

In this work, we use $\sigma_{NN} = 41.94$ mb for $\sqrt{s_{NN}} = 200$ GeV and $\sigma_{NN} = 61.36$ mb for $\sqrt{s_{NN}} = 2.76$ TeV. The result of this model is plotted in right hand side of Fig. 3. Model underpredicts the multiplicity at peripheral collisions. One possible interpretation may be the following: viscosity is large at large impact parameter, entropy production may be larger as impact parameter becomes large.

*3.4. Eccentricity and transverse energy*
Finally, we plot eccentricity defined by

$$\epsilon = \frac{\langle y^2 - x^2 \rangle}{\langle y^2 + x^2 \rangle} \tag{14}$$

in Fig. 4. Since the eccentricity is proportional to the magnitude of elliptic flow, it is important to know the initial value of this quantity. One sees that the incident energy dependence of the eccentricity is very small according to the results from MCrcBK.

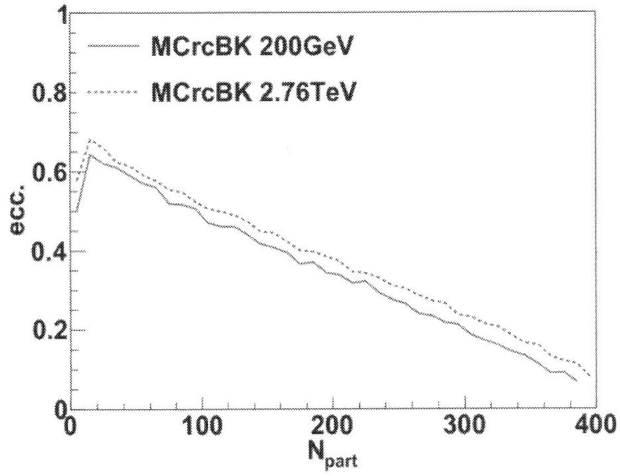


**Figure 4.** Centrality dependence of eccentricity with respect to reaction plane in Pb+Pb collisions at $\sqrt{s_{NN}} = 2.76$ TeV and in Au+Au collisions at $\sqrt{s_{NN}} = 200$ GeV.

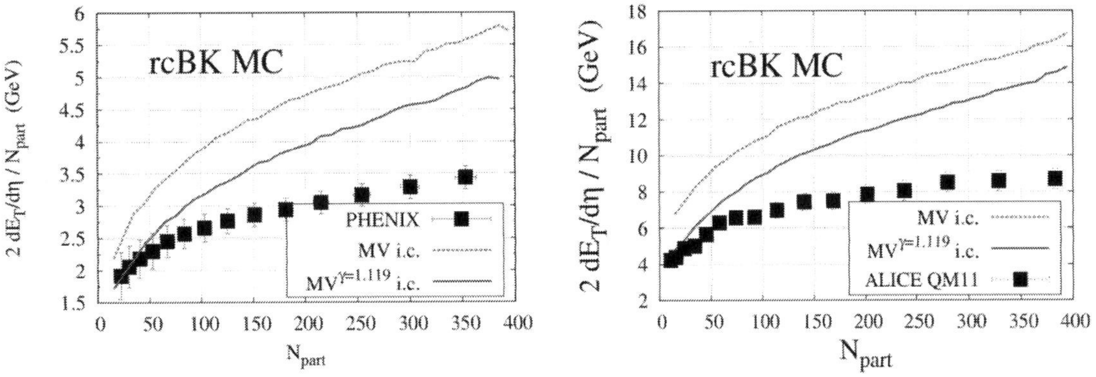


**Figure 5.** Centrality dependence of the transverse energy in Au+Au collision at $\sqrt{s_{NN}} = 200$ GeV (left) and Pb+Pb collision at $\sqrt{s_{NN}} = 2.76$ TeV (right). PHENIX data from [11], ALICE data from [12].

Fig. 5 shows the centrality dependence of the transverse energy at central rapidity for the MV initial condition, and for an initial condition featuring a more rapid fall-off at $k_\perp > Q_s$;

see ref. [2] for details. We note that $\sim 2.5\%$ ($0.5\%$) of the energy of the beams is predicted to be deposited initially into the central rapidity region in central collisions. Longitudinal hydrodynamic expansion ($-p\,\Delta V$ work) may reduce the transverse energy by up to a factor of 2 [13]. In all, the model based on rcBK evolution of the gluon distribution appears to be consistent with the multiplicity and transverse energy data over a good order of magnitude in both $N_{part}$ and $\sqrt{s_{NN}}$.

## 4. Summary

Within the MC-KLN and MCrcBK model, we compute the gluon production based on the picture of Color Glass Condensate. Both MC-KLN and MCrcBK model reproduce the centrality dependence of charged hadron multiplicity at RHIC and LHC energies which indicate that the entropy production during the thermalization process may be small as well as the small viscosity after thermalization. However, we need detail systematic study of particle production by looking at different observables together with the parameter dependence.

## Acknowledgments

The work of Y.N. was partly supported by Grant-in-Aid for Scientific Research No. 20540276. A.D. acknowledges support by the DOE Office of Nuclear Physics through Grant No. DE-FG02-09ER41620 and from The City University of New York through the PSC-CUNY Research Award Program, grant 63382-0041.

## References

[1] H. J. Drescher and Y. Nara, Phys. Rev. **C 75**, 034905 (2007); **76**, 041903(R) (2007).
[2] J. L. Albacete and A. Dumitru, 1011.5161[hep-ph], `http://physics.baruch.cuny.edu/node/people/adumitru/res_cgc`
[3] D. Kharzeev and M. Nardi, Phys. Lett. **B507**, 121 (2001); D. Kharzeev and E. Levin, ibid. **B523**, 79 (2001); D. Kharzeev, E. Levin, and M. Nardi, Phys. Rev. C **71**, 054903 (2005); Nucl. Phys. **A730**, 448 (2004).
[4] A. Adil, H. J. Drescher, A. Dumitru, A. Hayashigaki, and Y. Nara, Phys. Rev. C **74**, 044905 (2006).
[5] PHOBOS, B. B. Back *et al.*, Phys. Rev. **C65**, 061901 (2002), nucl-ex/0201005.
[6] K. Aamodt *et al.* [The ALICE Collaboration], Phys. Rev. Lett. **105** (2010) 252301; Phys. Rev. Lett. **106** (2011) 032301.
[7] J. L. Albacete and Y. V. Kovchegov, Phys. Rev. **D75**, 125021 (2007), arXiv:0704.0612 [hep-ph].
[8] J. L. Albacete, N. Armesto, J. G. Milhano, and C. A. Salgado, Phys. Rev. **D80**, 034031 (2009), 0902.1112.
[9] J. L. Albacete, N. Armesto, J. G. Milhano, P. Quiroga Arias and C. A. Salgado, arXiv:1012.4408 [hep-ph].
[10] R. Baier, A. H. Mueller, D. Schiff and D. T. Son, Phys. Lett. B **502**, 51 (2001); Phys. Lett. B **539**, 46 (2002); arXiv:1103.1259 [nucl-th].
[11] S. S. Adler *et al.* [ PHENIX Collaboration ], Phys. Rev. **C71**, 034908 (2005).
[12] A. Toia for the ALICE Collaboration, talk given at "Quark Matter 2011", 23 – 28 May 2011, Annecy, France
[13] M. Gyulassy, T. Matsui, Phys. Rev. **D29**, 419-425 (1984); M. Gyulassy, Y. Pang, B. Zhang, Nucl. Phys. **A626**, 999-1018 (1997); K. J. Eskola, K. Kajantie, P. V. Ruuskanen, K. Tuominen, Nucl. Phys. **B570**, 379-389 (2000); A. Dumitru, M. Gyulassy, Phys. Lett. **B494**, 215-220 (2000).

Journal of Physics: Conference Series **316** (2011) 012012 doi:10.1088/1742-6596/316/1/012012

# Ridges in heavy ion collisions and pp scatterings at the LHC

**K. Werner** [(a)], **Iu. Karpenko**[(b)], **T. Pierog** [(c)]

[(a)] SUBATECH, University of Nantes – IN2P3/CNRS– EMN, Nantes, France
[(b)] Bogolyubov Institute for Theoretical Physics, Kiev 143, 03680, Ukraine
[(c)] Karlsruhe Institute of Technology (KIT) - Campus North, Inst. f. Kernphysik, Germany

**Abstract.** We discuss recent developments in the theoretical desciption of ultrarelativistic collisions, concerning an event-by-event treatment of the hydrodynamical evolution in heavy ion collisions and also in high multiplicity pp scatterings at the LHC. The initial conditions are flux-tubes, which are formed following elementary (multiple) scatterings. We show that this picture leads in a natural way to the so-called ridge structures, observed in heavy ion and proton-proton collisions

Starting point of our consideration is a multiple scattering model in the spirit of the Gribov-Regge approach[1]. Here, one does not mean simply multiple hard scatterings, the elementary processes correspond to complete parton ladders, which means hard scatterings plus initial state radiation. In this case, this elementary process carries an important fraction of the available energy, and therefore we treat very carefully the question of energy sharing in the multiple scattering process. Open and closed ladders have to be considered in order to have a consistent quantum mechanical treatment. The corresponding graphs are squared, and we employ cutting rule techniques and Markov chains to obtain finally partial cross sections. The cut parton ladders are identified with longitudinal color fields or flux tubes, treated via relativistic string theory.

In case of very high energy *pp* collisions (at the LHC) or heavy ion scatterings already at RHIC, many flux tubes overlap and produce high energy densities. Let us consider the energy density at an early time in a Au-Au scattering at RHIC, as obtained from an EPOS simulation [1]. In fig. 1, we plot the energy density at different values of space-time rapidity $\eta_s$, as a function of the transverse coordinates $x$ and $y$. We observe a very bumpy structure concerning the $x-y$–dependence, whereas the variation with $\eta_s$ is small. There are in particular peaks in the $x-y$–plane, which show up at the same position at different values of $\eta_s$. So we have sub-flux-tubes which exhibit a long range structure in the longitudinal variable $\eta_s$. In fig. 1 , we clearly identify several sub-flux-tubes, with a typical width of the order of a fermi. This is exactly the width we obtain if we compute the initial energy density in proton-proton scattering at the LHC. This means, if a hydrodynamic treatment is justified for Au-Au collisions at RHIC, it is equally justified for pp scattering at the LHC, provided the energy densities are high enough, which seems to be the case.

We therefore employ a sophisticated hydrodynamical scenario (for details see [1]), for both heavy ions and proton-proton scatterings at the LHC [2, 3, 4], with initial conditions obtained from a flux tube approach (EPOS), compatible with the string model, used since many years for elementary collisions (electron-positron, proton proton), and the color glass condensate picture

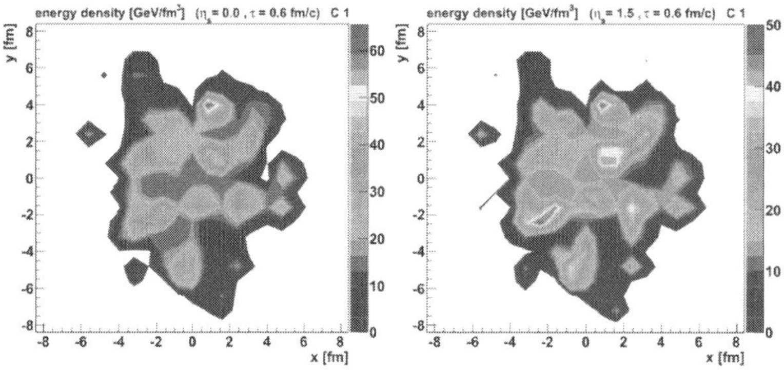


**Figure 1.** Energy density in central Au-Au scattering at 200 GeV.

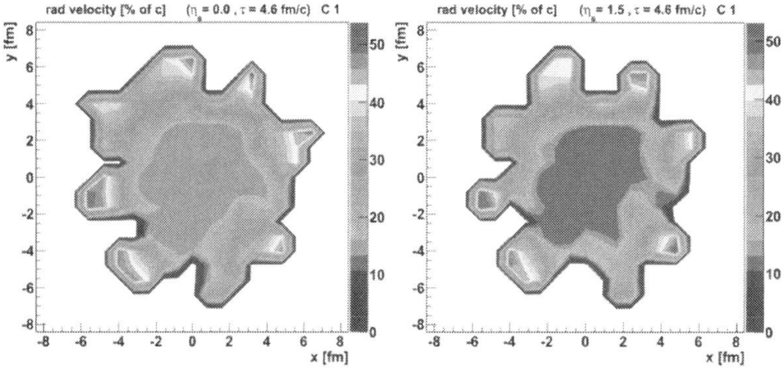


**Figure 2.** Radial flow velocity at a proper time $\tau = 4.6\,\mathrm{fm/c}$, at a space-time rapidities $\eta_s = 0$ and $\eta_s = 1.5$.

[5]. The equation-of-state is compatible with lattice gauge results of ref. [6]. We use a hadronic cascade procedure after hadronization from the thermal system at an early stage [7, 8].

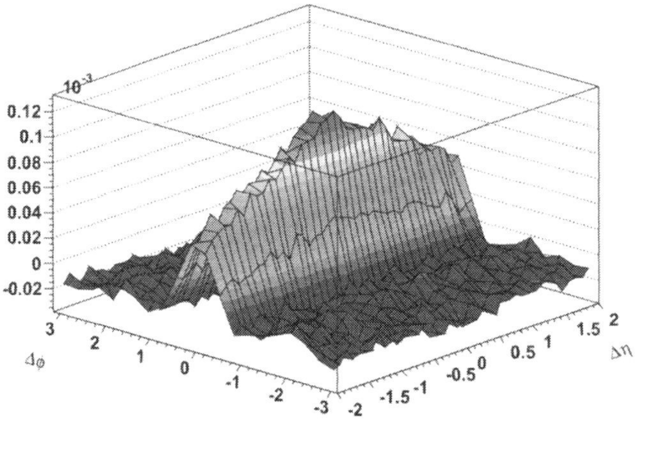


**Figure 3.** Dihadron $\Delta\eta - \Delta\phi$ correlation in a central Au-Au collision at 200 GeV.

In ref. [1], we test the approach by investigating all soft observables of heavy ion physics, in case of AuAu scattering at 200 GeV. Here, we are going to discuss some selected (and interesting) topics. In fig. 1, we see a complicated structure of the initial energy density in the transverse plane, but this structure is longitudinally translational invariant (same structure at different values of $\eta_s$). The equations of hydrodynamics preserve this translational invariance, and transport it to different quantities, as the radial flow, see fig. 2. As a consequence, particles emitted from different longitudinal positions get the same transverse boost, when their emission points correspond to the azimuthal angle of a common flow peak position. And since longitudinal coordinate and (pseudo)rapidity are correlated, one obtains finally a strong $\Delta\eta - \Delta\phi$ correlation, as seen in fig. 3, where we plot the dihadron correlation $dN/d\Delta\eta\, d\Delta\phi$, with $\Delta\eta$ and $\Delta\phi$ being respectively the difference in pseudorapidity and azimuthal angle of a pair of particles. Here, we consider trigger particles with transverse momenta between 3 and 4 GeV/c, and associated particles with transverse momenta between 2 GeV/c and the $p_t$ of the trigger, in central Au-Au collisions at 200 GeV. Our ridge is very similar to the structure observed by the STAR collaboration [9].

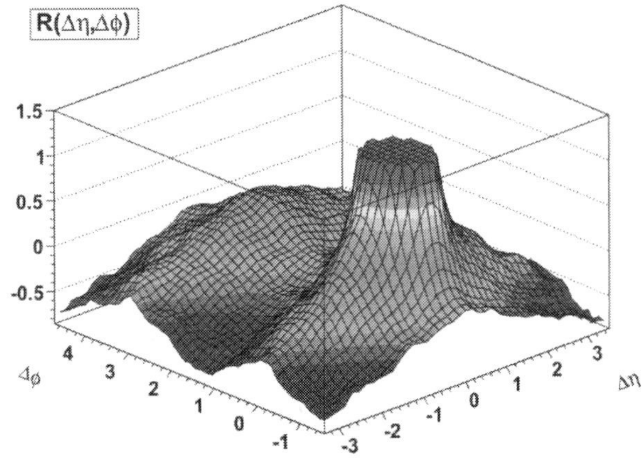


**Figure 4.** (Color online) Two particle correlation function $R$ versus $\Delta\eta$ and $\Delta\phi$ for high multiplicity events in $pp$ collisions at 7 TeV.

The CMS collaboration published recently results [10] on two particle correlations in $\Delta\eta$ and $\Delta\phi$, in $pp$ scattering at 7 TeV. Most remarkable is the discovery of a ridge-like structure around $\Delta\phi = 0$, extended over many units in $\Delta\eta$, referred to as "the ridge", in high multiplicity $pp$ events. A similar structure has been observed in heavy ion collisions at RHIC, as discussed earlier, and there is little doubt that the phenomenon is related to the hydrodynamical evolution of matter. This "fluid dynamical behavior" is actually considered to be the major discovery at RHIC.

So does $pp$ scattering provide as well a liquid, just ten times smaller than a heavy ion collision? It seems so! We showed recently [11] that if we take exactly the same hydrodynamic approach which has been so successful for heavy ion collisions at RHIC [1], and apply it to $pp$ scattering, we obtain already very encouraging results compared to $pp$ data at 0.9 TeV. In this paper, we apply this fluid approach, always the same procedure, to understand the 7 TeV results. In fig. 4, we show that our hydrodynamic picture indeed leads to a near-side ridge, around $\Delta\phi = 0$, extended over many units in $\Delta\eta$. One should note that the correlation functions are defined and normalized as in the CMS publication, so we can say that our "ridge" is quite close in

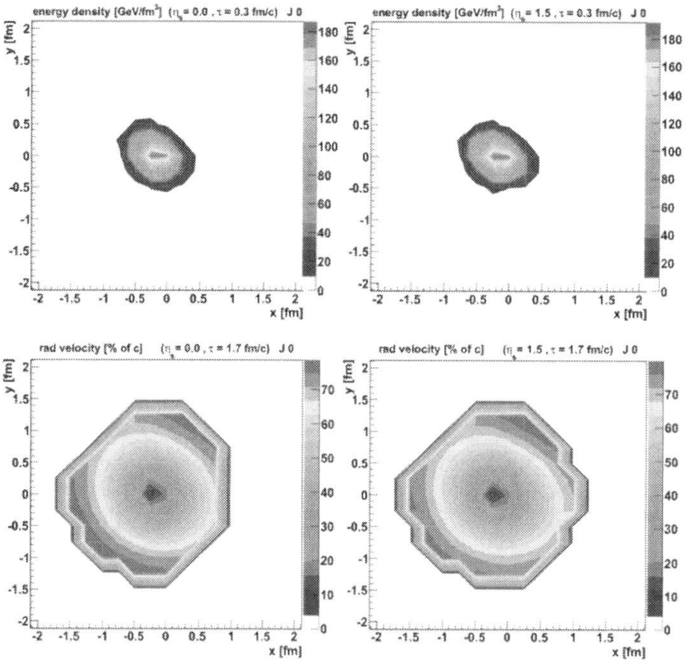


**Figure 5.** (Color online) Initial energy density (upper panel) and radial flow velocity at a later time (lower panel) for a high multiplicity $pp$ collision at 7 TeV at a space-time rapidity $\eta_s = 0$ (left) and $\eta_s = 1.5$ (right).

shape and in magnitude compared the experimental result. The experimental high multiplicity bin corresponds to about 7 times average, whereas in our calculation (extremely demanding concerning CPU power) "high multiplicity" refers to 5.3 times average (we actually trigger on events with 10 elementary scatterings). We cannot go beyond at the moment.

It is easy to understand the origin of the ridge, in a hydrodynamical approach based on flux tube initial conditions. Imagine many (say 20) flux tubes of small transverse size (radius $\approx 0.2$ fm), but very long (many units of space-time rapidity $\eta_s$). For a given event, their transverse positions are randomly distributed within the overlap area of the two protons. Even for zero impact parameter (which dominated for high multiplicity events), this randomness produces azimuthal asymmetries, as shown in fig. 5, upper panel. The energy density obtained from the overlapping flux tubes (details will be discussed later) shows an elliptical shape. And since the flux tubes are long, and only the transverse positions are random, we observe the same asymmetry at different longitudinal positions ($\eta = 0$ and $\eta = 1.5$ in the figure). So we observe a translational invariant azimuthal asymmetry!

If one takes this asymmetric but translational invariant energy density as initial condition for a hydrodynamical evolution, the translational invariance is conserved, and in particular translated into other quantities, like the flow. In fig. 5, lower panel, we show the radial flow velocity at a later time again at the two space-time rapidities $\eta_s = 0$ (left) and $\eta_s = 1.5$ (right). In both cases, the flow is more developed along the direction perpendicular to the principal axis of the initial energy density ellipse. This is a very typical fluid dynamical phenomenon, referred to as elliptical flow.

Finally, particles are produced from the flowing liquid, with a preference in the direction of large flow. This preferred direction is therefore the same at different values of $\eta_s$. And since $\eta_s$

Journal of Physics: Conference Series **316** (2011) 012012 doi:10.1088/1742-6596/316/1/012012

and pseudorapidity $\eta$ are highly correlated, one observes a $\Delta\eta\Delta\phi$ correlation, around $\Delta\eta = 0$, extended over many units in $\Delta\eta$: a particle emitted a some pseudorapidity $\eta$ has a large chance to see a second particle at any pseudorapidity to be emitted in the same azimuthal direction.

**Acknowledgments** This research has been carried out within the scope of the ERG (GDRE) "Heavy ions at ultra-relativistic energies. Iu.K. acknowledges partial support by the State Fund for Fundamental Researches of Ukraine (Agreement of 2011) and National Academy of Sciences of Ukraine (Agreement of 2011). T.P. and K.W. acknowledge partial support by a PICS (CNRS) with KIT (Karlsruhe).

[1] K. Werner, Iu. Karpenko, T. Pierog, M. Bleicher, K. Mikhailov, Phys. Rev. C 82, 044904 (2010)
[2] K. Werner, Iu. Karpenko, T. Pierog, Phys.Rev.Lett.106:122004,2011
[3] K. Werner, Iu. Karpenko, T. Pierog, M. Bleicher, K. Mikhailov, Phys. Rev. C 83, 044915 (2011)
[4] K. Werner, Y. Karpenko, T. Pierog. Il Nuovo Cimento Vol. 34 C, 85 (2011)
[5] L.D. McLerran, R. Venugopalan, Phys. Rev. D 49, 2233 (1994). ibid. D49, 3352 (1994); D 50, 2225 (1994)
[6] S. Borsanyi et al., arXiv:1007.2580
[7] M. Bleicher et al., J. Phys. G25 (1999) 1859
[8] H. Petersen, J. Steinheimer, G. Burau, M. Bleicher and H. Stocker, Phys. Rev. C78 (2008) 044901
[9] STAR Collaboration, B. Abelev et al., Phys. Rev. C 80 (2009) 64192
[10] CMS Collaboration, JHEP 1009:091,2010
[11] K. Werner, Iu. Karpenko, T. Pierog, M. Bleicher, K. Mikhailov, arXiv:1010.0400, Phys. Rev. C 83, 044915 (2011)

# Jet reconstruction and underlying event studies in p+p and d+Au collisions from STAR

**Jana Bielcikova for the STAR Collaboration**
Nuclear Physics Institute ASCR, Na Truhlarce 39/64, Prague 180 86, Czech Republic
E-mail: jana.bielcikova@ujf.cas.cz

**Abstract.** Description of parton interaction with the hot and dense nuclear matter created in heavy-ion collisions at high energies is a complex task, which requires a detailed knowledge of jet production in p+p and d+Au collisions. Measurements in these collision systems are therefore essential to disentangle initial state nuclear effects from cold nuclear matter effects, medium-induced $k_T$ broadening and jet quenching. To obtain complete description of the p+p (d+Au) collision it is also important to study particle production in the underlying event. The measured properties of underlying event can be used to tune the QCD based Monte-Carlo models.

In this paper some of the recent results on jet and underlying event properties in p+p and d+Au collisions at $\sqrt{s_{NN}}$ = 200 GeV measured by the STAR experiment are presented. In particular, the preliminary results on charged and strange particle fragmentation functions in p+p collisions are discussed and confronted with PYTHIA simulations. Next, the measurement of inclusive jet spectrum and di-jet correlations in d+Au collisions is presented and compared with the measurements in p+p collisions to estimate the size of cold nuclear matter effects. Finally, the study of underlying event properties in p+p collisions is shown and compared to PYTHIA simulation.

## 1. Introduction

Full jet reconstruction is considered as a promising tool for the quantitative study of properties of the hot and dense medium produced in heavy-ion collisions at RHIC energy. The complexity of heavy-ion collisions, where modifications of the fragmentation functions due to interaction of partons with the hot and dense medium are expected, requires inevitably a detailed understanding of jet production in p+p and d+Au collisions. Therefore measurements in these collision systems can be used as a baseline to disentangle initial state nuclear effects from cold nuclear matter effects, medium-induced $k_T$ broadening and jet quenching. In addition, it is also important to understand how the beam-beam remnants, multiple parton interactions and initial- and final-state radiation combine to produce particles in the underlying event (UE). The extracted underlying event properties serve as in important input for tuning of the QCD Monte-Carlo models such as PYTHIA.

The first results on inclusive jet spectra in p+p collisions at $\sqrt{s}$ = 200 GeV measured by the STAR experiment at RHIC in 2003-2004 were published in [1]. The measured inclusive jet cross-section is in a very good agreement with NLO pQCD calculations over seven orders of magnitude. During the data taking in 2006 and 2008, the STAR experiment collected large data samples of p+p and d+Au collisions at $\sqrt{s_{NN}}$ = 200 GeV. In this paper, the study of p+p collisions is extended by investigating fragmentation functions of charged and strange particles ($K^0_S$, $\Lambda$, $\bar{\Lambda}$)

as well as underlying event properties. Further, the paper discusses recent preliminary results on inclusive jet spectra and di-jet correlations in d+Au collisions.

## 2. Dataset and analysis

The preliminary results presented in this paper are based on analysis of p+p collisions at $\sqrt{s}$ =200 GeV measured in 2006 and p+p and d+Au collisions at $\sqrt{s_{NN}}$ =200 GeV recorded in 2008 by the STAR experiment. To increase yield of hard processes, two special triggers utilizing information from Barrel Electromagnetic Calorimeter (BEMC) were used in addition to minimum bias trigger. In 2006, a "jet-patch" trigger defined as a $\Delta\eta \times \Delta\phi = 1\times1$ patch of BEMC containing transverse energy $E_T > 8$ GeV was used. In 2008, an online BEMC high tower (HT) trigger was used. To satisfy the HT trigger condition, at least one tower of BEMC must contain $E_T$ above a predefined threshold, which in case of the data presented here was 4.3 GeV. In d+Au collisions, the Beam Beam Counter detector, located in the fragmentation region of Au nucleus, was used to select the 20% highest multiplicity events.

Jets were reconstructed using sequential recombination algorithms $k_t$ and anti-$k_t$ as well as seedless infrared safe cone algorithm SISCone, which are implemented in the FastJet package [2, 3, 4]. For the measurement of the neutral component of jets BEMC was used. The charged component of jets was measured by the Time Projection Chamber (TPC). Both TPC and BEMC have a full azimuthal coverage and their combined pseudorapidity acceptance is $|\eta| < 0.95$. This limits the fiducial acceptance of jets in pseudorapidity to 1-$R$, where $R$ is the resolution parameter used in jet reconstruction. To avoid double counting of energy (electrons, minimum ionizing particles and possible hadronic showers in BEMC), the so called 100% hadronic correction was applied: the momentum of a charged particle from TPC was subtracted off the neutral tower energy in BEMC, if the measured charged particle track pointed to a BEMC tower. In order to minimize biases due to dead areas and possible background in BEMC, the neutral energy fraction (NEF) of the total jet energy was required to be within $0.1 < \mathrm{NEF} < 0.9$. For charged particle tracks, an upper transverse momentum cut-off $p_T < 15$ GeV/$c$ was applied due to uncertainty in TPC tracking performance at large transverse momentum. Detailed investigation of this limitation is in progress.

In parallel to experimental data analysis, the analysis of jets and underlying event properties was also performed on simulated data. The simulations are based on PYTHIA 6.410, tuned to the CDF data at $\sqrt{s}$ =1.96 TeV (Tune A) and combined with a full STAR detector simulation based on GEANT. The simulated data were then analyzed using the same analysis chain as that for the measured data. We distinguish several levels of the simulated data: jet reconstruction at Monte Carlo particle level (PyMC) and at detector level including GEANT simulation (PyGe). In addition, to determine influence of background and its fluctuations on jet production in d+Au collisions, PYTHIA events were embedded in 0-20% highest multiplicity minimum bias d+Au events. This procedure was performed at detector level using reconstructed TPC tracks and signal in BEMC towers. This level of simulation is labeled as "PyBg".

## 3. Fragmentation functions in p+p collisions

Figure 1 shows the uncorrected fragmentation functions of charged particles for jets with transverse momentum $20 < p_T < 30$ GeV/$c$ measured in p+p collisions at $\sqrt{s} = 200$ GeV using the jet-patch triggered data [5, 6, 7]. As this trigger creates a neutral energy fragmentation bias for the triggered jet, the fragmentation functions are shown only for the di-jet partner on away side. The fragmentation functions are plotted as a function of $\xi = \ln(p_T^{\mathrm{jet}}/p_T^{\mathrm{hadron}})$. The jets were reconstructed by three different jet algorithms ($k_t$, anti-$k_t$ and SISCone) with a resolution parameter $R = 0.4$ and $R = 0.7$, respectively. All charged tracks measured by TPC and neutral towers from BEMC with $p_T(E_T) > 0.2$ GeV/$c$ were included in jet reconstruction. No significant dependence on jet algorithm is observed. The measured fragmentation functions are compared

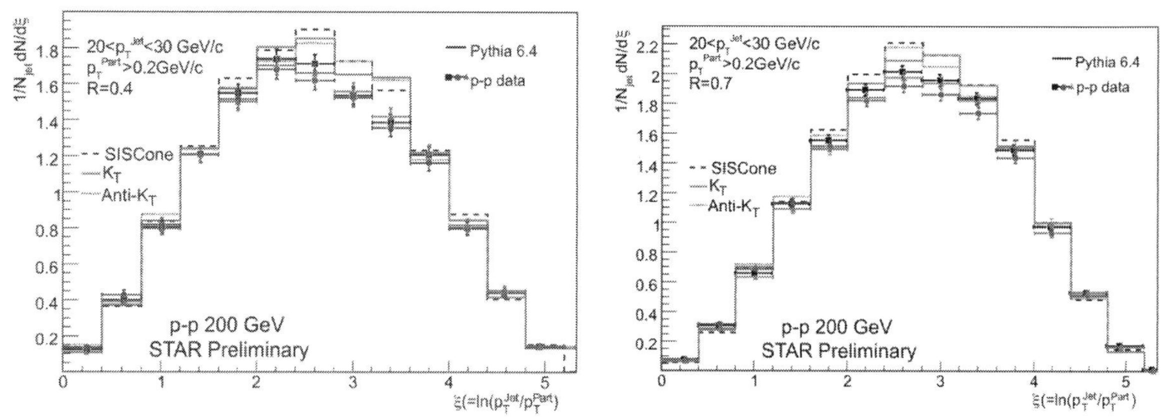


**Figure 1.** Charged particle fragmentation function as a function of $\xi$ at detector level for jets with 20$<$ $p_T$ $<$ 30 GeV/$c$ measured in p+p collisions at $\sqrt{s}$ = 200 GeV. The jets were reconstructed by three different jet algorithms (triangles: $k_t$, circles: anti-$k_t$, squares: SISCone) with a resolution parameter $R = 0.4$ (left) and $R = 0.7$ (right) [5, 6, 7]. The data are compared to PYTHIA simulations (histogram).

to PYTHIA simulations described above. For both resolution parameters studied there is a good agreement between the data and simulation. This suggests that the NLO contributions which are not included in PYTHIA are small at RHIC.

Further it is important to study in detail particle dependence of fragmentation functions as very little is known about identified fragmentation functions both experimentally and theoretically. While PYTHIA and NLO calculations agree reasonably well with the data on pion and proton production [8], theoretical description of strange particle production ($K$, $\Lambda$, $\Xi$) is a challenging task. The leading order calculations significantly underpredict the measured strange particle yields at intermediate and large transverse momenta [9]. Higher-order corrections to leading-order pQCD cross sections, which are in general large, can be accounted for by introducing a $K$-factor, which renormalizes the leading-order pQCD cross section. It was found that a reasonable agreement of strange baryon particle production with PYTHIA at RHIC energy can be achieved by introducing $K = 3$ [9]. However, such a large value of $K$ breaks the agreement of pion and proton data with PYTHIA.

The STAR experiment has excellent capability to reconstruct strange particles such as $K_0^S$ and $\Lambda$ via their typical V0 decay channels with a very good signal-to-noise ratio at $p_T > 1$ GeV/$c$. Figure 2 shows the measured fragmentation functions of $K_S^0$, $\Lambda$ and $\bar{\Lambda}$ particles in $p+p$ collisions at $\sqrt{s} = 200$ GeV at detector level [10]. The fragmentation functions were calculated for three different jet $p_T$ ranges: 10-15, 15-20, 20-40 GeV/$c$. The remaining small residual background under the mass peaks is not yet corrected for. The data are compared to polynomial fits to PYTHIA predictions. PYTHIA describes well the fragmentation functions of $K_S^0$, which is in agreement with measurements of $K^{\pm}$ fragmentation functions in $e^+ + e^-$ collisions for similar jet energy range. For $\Lambda$ and $\bar{\Lambda}$ fragmenation functions, PYTHIA predicts well the over-all yields, but it overestimates the yields at low $\xi$ and underestimates the yields at intermediate values of $\xi$. The integrated $\bar{\Lambda}/\Lambda$ and $\Lambda/K_S^0$ ratios fro $p_T > 1$ GeV/$c$ in the studied jet $p_T$ range are consistent with the values obtained from the minimum bias inclusive spectra when the same $p_T$ selection is applied [10]. This observation indicates that hard processes significantly contribute to $K_S^0$, $\Lambda$, and $\bar{\Lambda}$ production at $p_T > 1$ GeV/$c$. Further studies are needed to verify this. Corrections of the data to particle level as well as removal of remaining combinatorial background under mass

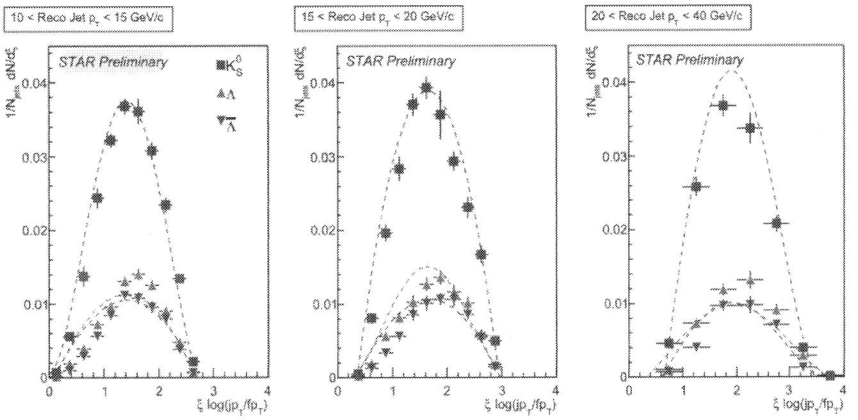


**Figure 2.** Fragmentation functions of $K_S^0$ (squares), $\Lambda$ (triangles) and $\bar{\Lambda}$ (upside-down triangles) as a function of the reconstructed jet $p_T$ in p+p collisions at $\sqrt{s} = 200$ GeV [10]. The curves show polynomial fits to PYTHIA 6.4 predictions. Both data and simulations are shown at detector level.

peaks are ongoing.

## 4. Inclusive jet spectrum in d+Au collisions

The inclusive transverse momentum spectrum of jets was measured in 10M 0-20% highest multiplicity d+Au collisions at $\sqrt{s_{NN}} = 200$ GeV. All charged tracks measured by TPC and neutral towers from BEMC with $p_T > 0.2$ GeV/$c$ were included in jet reconstruction. The jets were reconstructed with the anti-$k_t$ jet finding algorithm with a resolution parameter $R = 0.4$. For subtraction of d+Au underlying event background, a method based on active jet areas [11, 12] was applied on event-by-event basis. The inherent asymmetry of the d+Au collision system results in $\eta$ asymmetric underlying background. The $\eta$ dependence of underlying background was parametrized by a linear function and included in the background subtraction procedure.

To correct for detector effects in d+Au collisions, the PYTHIA simulation of p+p collisions was used. In addition, a correction accounting for the lower TPC reconstruction efficiency in d+Au relative to p+p collisions was applied to avoid bias due to different Jet Energy Scale (JES). The TPC reconstruction efficiency in d+Au was determined from a single charged pion embedding into measured minimum bias d+Au events on detector level. The higher tracking efficiency in the PYTHIA simulation was then artificially lowered to match the value in d+Au collisions. This procedure was applied prior to jet finding at PyGe and PyBg level of simulations.

To correct the jet spectrum to particle level, a bin-by-bin correction method was used. This correction procedure is based on the generalized efficiency determined as the ratio of PyMC to PyBg jet $p_T$ spectra, which is applied to the measured jet $p_T$ spectrum in d+Au collisions. The validity of this correction procedure relies on the fact that the shapes of the PyBg and the measured jet $p_T$ spectra are consistent. In [13] it was demonstrated that this condition is within statistical errors satisfied.

The fully corrected inclusive jet $p_T$ spectrum in d+Au collisions at $\sqrt{s_{NN}} = 200$ GeV is shown in Figure 3 [13, 14]. The systematic uncertainties are dominated by the JES uncertainty consisting of 10% systematic uncertainty in TPC charged particle reconstruction efficiency and 5% in BEMC calibration uncertainty, added in quadrature. Ongoing studies with a full jet embedding into measured d+Au events at raw detector level will enable to decrease the JES uncertainty.

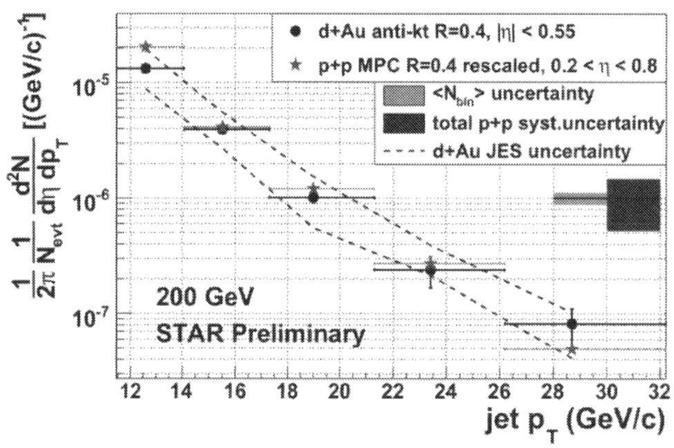


**Figure 3.** Inclusive jet $p_T$ spectrum in 0-20% highest multiplicity d+Au collisions at $\sqrt{s_{NN}}$ =200 GeV compared to $\langle N_{bin}\rangle$ scaled p+p spectrum from [1]. The systematic errors (grey boxes and a dashed band) are explained in the legend.

The jet spectrum in d+Au collisions is compared with the jet $p_T$ spectrum in p+p collisions from 2003-2004 for which a Mid Point Cone (MPC) jet algorithm with a cone radius $R = 0.4$ was used [1]. The effect of a slightly different $\eta$ acceptance on jet $p_T$ spectrum in case of p+p data was found to be less than 10% in the covered $p_T$ range. For comparison with the jet spectrum in d+Au collisions, the jet spectrum in p+p collisions was scaled by an average number of binary collisions $\langle N_{bin}\rangle$ = 14.6±1.7 calculated from the Monte-Carlo Glauber model for 0-20% highest multiplicity d+Au events. The jet spectra in both collision systems are consistent within current systematic errors dominated by the JES and MC Glauber model uncertainties. A future comparison of the inclusive jet spectrum in d+Au collisions with that in p+p collisions measured during the same data taking period in 2008 is expected to further decrease the systematic uncertainties because they will be largely correlated.

## 5. Nuclear $k_T$ broadening

Azimuthal correlations of di-jets can be used to quantify size of nuclear $k_T$ broadening in d+Au collisions relative to p+p collisions. The total transverse momentum of a di-jet ($k_{T,raw}$) can be determined from the di-jet opening angle $\Delta\phi$: $k_T = p_T \sin(\Delta\phi)$. To increase the di-jet yield, the HT triggered dataset with $E_T > 4.3$ GeV was used for both p+p and d+Au data measured in 2008. Jets were reconstructed with the anti-$k_t$ algorithm using the resolution parameter $R = 0.5$. In addition, a lower transverse momentum cut $p_T > 0.5$ GeV/$c$ was applied to charged tracks and neutral towers to reduce background. For di-jet analysis the two most energetic jets in a given event ($p_{\mathrm{T},1} > p_{\mathrm{T},2}$) satisfying $10 < p_{\mathrm{T},2} < 20$ GeV/$c$ were selected. To obtain a clean di-jet sample a cut on di-jet separation in azimuth $|\Delta\phi - \pi| < \pi/3$ was applied and measured $k_{T,raw}$ distributions were constructed as $k_{T,raw} = p_{\mathrm{T},1}\sin(\Delta\phi)$.

To quantify the size of detector and background effects on $k_{T,raw}$ distributions, a di-jet analysis was performed on PYTHIA simulated data which is demonstrated in Figure 4 together with a Gaussian fit of the simulated distributions [13, 14, 15]. Comparison of PyMC, PyGe and PyBg distributions shows that the extracted Gaussian widths of $k_{T,raw}$ distributions are within statistical errors consistent. This suggests that detector and d+Au background effects in data can in first order be neglected. It should be however noted, that in case of PYTHIA simulations without background (PyMC and PyGe), the Gaussian fit does not describe the data well.

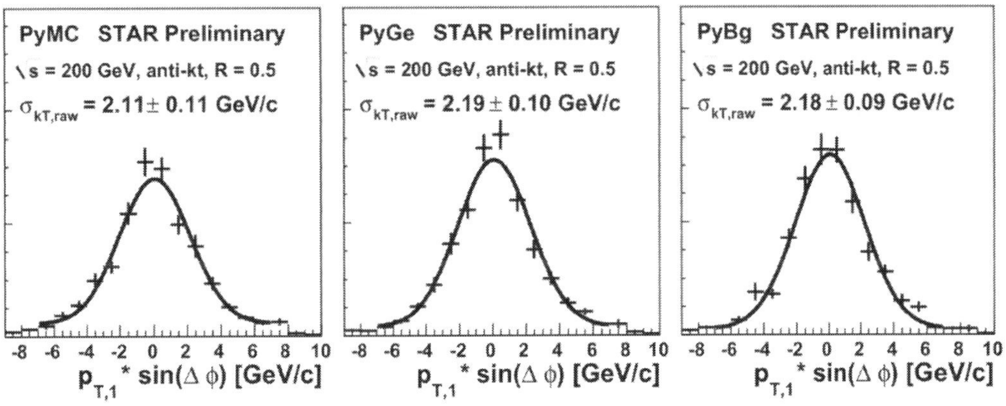


**Figure 4.** Distributions of $k_{T,raw} = p_{T,1}\sin(\Delta\phi)$ for $10 < p_{T,2} < 20$ GeV/$c$ in MC simulated data PyMC (left), PyGe (middle) and PyBG (right). Jets were reconstructed with the anti-$k_t$ algorithm with a resolution parameter $R = 0.5$. The distributions were fit with a Gaussian fit function.

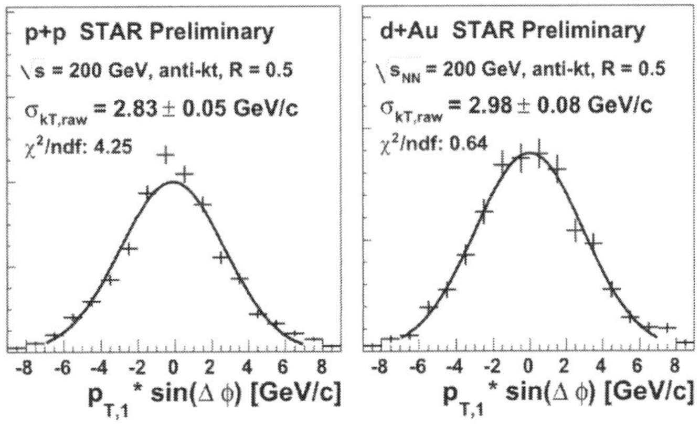


**Figure 5.** Measured distributions of $k_{T,raw} = p_{T,1}\sin(\Delta\phi)$ for $10 < p_{T,2} < 20$ GeV/$c$ in p+p (left) and 0-20% highest multiplicity d+Au collisions (right) at $\sqrt{s_{NN}} = 200$ GeV. Jets were reconstructed with the anti-$k_t$ algorithm with a resolution parameter $R = 0.5$. The distributions were fit with a Gaussian fit function.

The $k_{\mathrm{T,raw}}$ distributions for data from p+p and d+Au collisions are shown in Figure 5. The values of Gaussian widths extracted from the fits are $\sigma^{p+p}_{kT,raw} = 2.8\pm0.1$ (stat.) GeV/$c$ in p+p collisions and $\sigma^{d+Au}_{kT,raw} = 3.0\pm0.1$ (stat.) GeV/$c$ in d+Au collisions [13, 14, 15]. The widths in both collision systems are consistent within statistical errors indicating very small nuclear $k_T$ broadening. In case of p+p collisions, the Gaussian fit does not describe the shape of the $k_{\mathrm{T,raw}}$ distribution well similarly as observed for the PyMC and PyGe simulated data. The precise shape of the $k_{\mathrm{T,raw}}$ distribution is currently under investigation. Varying the size of $|\Delta\phi - \pi|$ cut for back-to-back di-jet selection between 0.5 and 1.0, the systematic uncertainty on the extracted Gaussian widths of $k_{\mathrm{T,raw}}$ distribution was estimated to be 0.2 GeV/$c$. The effects of $p_{\mathrm{T},2}$ range and jet finding algorithms were found to be negligible in both studied collision systems.

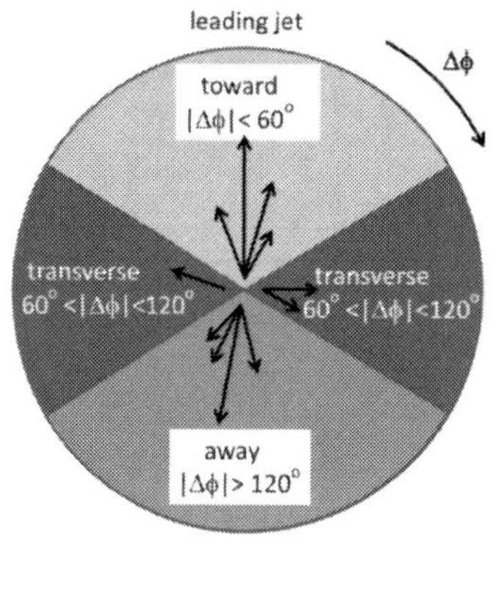


**Figure 6.** Cartoon of the event areas ("toward", "away", "transverse") in azimuthal angle difference relative to the leading jet ($\Delta\phi$).

## 6. Underlying event studies in p+p collisions

A full description of particle production in p+p collisions is a complex task which requires to incorporate both production of jets as well as production of particles in the so called underlying event within the same model. The underlying event (UE) is in general defined as everything but the hard scattering. It has contributions from soft and semi-hard multiple parton interactions (MPI), initial and final state radiation effects (ISR, FSR) and beam-beam remnants. As the study of UE presented below is focused on mid-rapidity, the contribution of beam-beam remnants to the UE properties can be neglected.

The UE study presented here follows the analysis technique developed by the CDF Collaboration at Tevatron [16, 17]. After reconstructing jets in a given event, the event is split into four sections schematically depicted in Figure 6. The sections are defined by their azimuthal angle difference with respect to the leading jet axis ($\Delta\phi$). The "toward" region is the region within $|\Delta\phi| < 60°$ around the leading jet axis and the "away" region is located at $|\Delta\phi| > 120°$. The remaining two regions labeled as "transverse" are placed at $60° < |\Delta\phi| < 120°$ relative to the leading jet axis. The underlying event contributes to all of the mentioned regions, but to access its properties the two "transverse" regions are used as there the contamination from jets is assumed to be small. According to the charged particle multiplicity ($N_{ch}$) the two transverse regions are further labeled as "TransMax" and "TransMin", where $N_{ch}(\text{TransMax}) \geq N_{ch}(\text{TransMin})$ in a given event. This selection enhances contribution of initial and final state radiation components of the hard scattered parton in TransMax region relative to TransMin region. The sensitivity to various physics processess in UE can be achieved by comparing properties of UE for a leading jet sample and a sample of events containing di-jets, which suppresses the probability of ISR and FSR effects. For this study events with two reconstructed jets with $p_T^{away}/p_T^{lead} > 0.7$ and $|\Delta\phi| > 150°$ were selected.

Figure 7 shows the uncorrected charged particle density and mean transverse momentum in TransMin, TransMax and away side jet regions as a function of leading jet $p_T$ in the di-jet sample. Jets were reconstructed with the $k_t$ algorithm with the resolution parameter $R = 0.7$ [5, 6, 7]. No significant dependence on jet algorithm was observed. The charged particle density as well as $\langle p_T \rangle$ are approximately independent of leading jet $p_T$, while the away-side jet charged particle density and $\langle p_T \rangle$ are rising with jet $p_T$, as expected. Compared to data measured at Tevatron ($\sqrt{s} = 1.96$ TeV), the values of chaged particle density and $\langle p_T \rangle$ in the underlying event at RHIC

Journal of Physics: Conference Series **316** (2011) 012013 doi:10.1088/1742-6596/316/1/012013

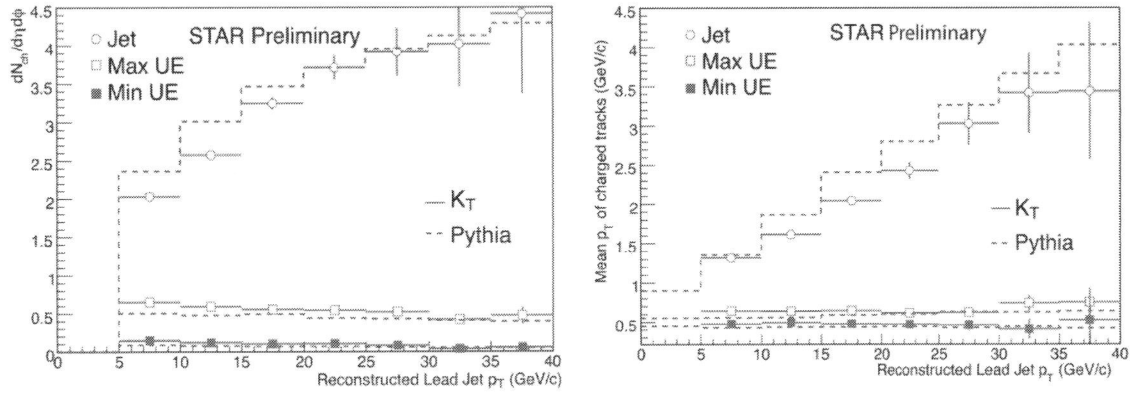


**Figure 7.** The uncorrected charged particle density (left) and charged particle mean $p_T$ (right) in the "di-jet" data set for the away side jet, TransMin and TransMax regions as a function of reconstructed leading jet $p_T$. The jets were reconstructed with the $k_t$ algorithm with a resolution parameter $R = 0.7$. The dashed histogram indicates the prediction from PYTHIA simulations (Tune A, $\epsilon = 0.25$). The data and simulation are presented at detector level.

energy are in general lower.

The measured data in Figure 7 are compared to PYTHIA simulation (Tune A) with MPI scaling factor $\epsilon = 0.25$ based on recent measurements at Tevatron energies ($\sqrt{s} = 630$, 1800 and 1960 GeV) [18]. PYTHIA describes the main features of the data well, however a closer investigation reveals that PYTHIA slightly underestimates the charged particle multiplicity and $\langle p_T \rangle$ in TransMax and TransMin regions while overestimating these quantities within the jet.

In Figure 8 the measured charged particle density in UE is investigated closer by comparing the results for the transverse regions between the leading and di-jet samples [5, 6, 7]. No significant difference between the jet samples is observed at RHIC energy which points to the fact that the hard scattered partons do not emit significant amount of large angle ISR and FSR. This observation is very different to that at Tevatron, where in $p+\bar{p}$ collisions at $\sqrt{s} = 1.96$ TeV the ratio of charged particle density in leading/di-jet samples was found to be about 0.65 [17]. The data in Figure 8 are compared with an expected charged particle density in case the multiplicity in the UE would follow a simple Poisson distribution with an average value of 0.36. This simple simulation suggests that at RHIC energy, the splitting of the TransMax and TransMin values is mostly due to the sampling of the total charged particle multiplicity measured.

## 7. Summary

In summary, various properties of jets in p+p and d+Au collisions at $\sqrt{s_{NN}} = 200$ GeV were discussed and compared to PYTHIA simulations. The measured fragmentation functions of charged particles and identified strange particles at detector level show a good agreement with PYTHIA simulations for charged and $K^0_S$ particles, while for $\Lambda$ and $\bar{\Lambda}$ PYTHIA fails to describe details of the fragmentation pattern.

The measured inclusive jet $p_T$ spectra and nuclear $k_T$ broadening in d+Au collisions from 2008 do not show any significant cold nuclear matter effects when compared to the measurements in $p+p$ collisions within current statistical and systematic uncertainties. These uncertainties will be further decreased by comparing the data to a new jet cross section measurement in p+p collisions from 2008 and improvements in the tracking efficiency determination from jet embedding in real d+Au data.

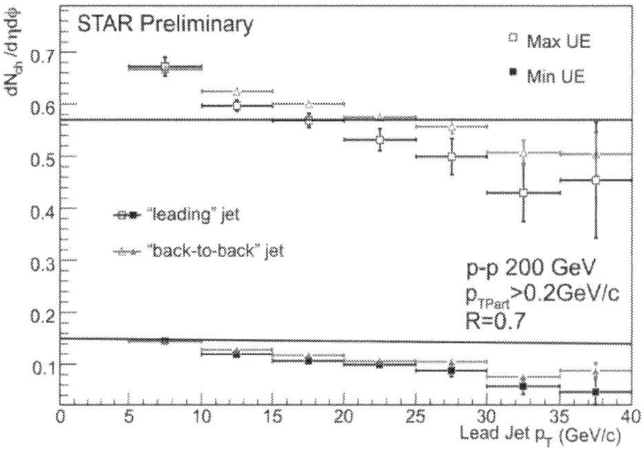


**Figure 8.** The uncorrected charged particle density in the TransMin and TransMax regions as a function of reconstructed leading jet $p_T$ using SISCone algorithm with a resolution parameter $R = 0.7$ shown separately for a leading jet and back-to-back jet samples. The data and simulation are presented at detector level. The solid line indicate expected charged particle density in case of simple Poisson distribution of charged particles in UE.

First studies of underlying event properties in p+p collisions show that its charged particle density and mean $p_T$ are largely independent of jet transverse momentum and the contributions from initial and final state radiation are small. Studies of UE in d+Au collisions are underway and are expected to bring further insights into details of particle production mechanism in cold nuclear matter.

## Acknowledgements

This work was supported in part by grants LC07048 and LA09013 of the Ministry of Education of the Czech Republic.

## References

[1] B. Abelev *et al.* (STAR Collaboration), *Phys. Rev. Lett.* **97** (2006) 252001.
[2] M. Cacciari, G. Salam, *Phys. Lett. B* **641** (2006) 57.
[3] G. P. Salam, G. Soyez, *JHEP* **0705** (2007) 086.
[4] M. Cacciari, G. Salam, G. Soyez, *JHEP* **0804** (2008) 063.
[5] H. Caines *et al.* (STAR Collaboration), *Nucl. Phys. A* **830** (2009) 263C.
[6] H. Caines *et al.* (STAR Collaboration), arXiv:0910.5203.
[7] H. Caines *et al.* (STAR Collaboration), *Nucl. Phys. A* **855** (2011) 376.
[8] J. Adams *et al.* (STAR Collaboration), *Phys. Lett. B* **637** (2006) 161.
[9] B. I. Abelev *et al.* (STAR Collaboration), *Phys. Rev. C* **75** (2007) 064901.
[10] A. Timmins *et al.* (STAR Collaboration), *J. Phys. G* **37** (2010) 094035.
[11] M. Cacciari, G. Salam, *Phys. Lett. B* **659** (2008) 119.
[12] M. Cacciari, G. Salam, G. Soyez, *JHEP* **0804** (2008) 005.
[13] J. Kapitan *et al.* (STAR Collaboration), *Nucl. Phys. A* **855** (2011) 412.
[14] J. Kapitan *et al.* (STAR Collaboration), *J. Phys. Conf. Ser.* **270** (2011) 012015.
[15] J. Kapitan *et al.* (STAR Collaboration), *Nucl. Phys. A* **830** (2009) 619C.
[16] R. Field *et al.* (CDF Collaboration), hep-ph/0510198.
[17] A. Cruz (CDF Collaboration), PhD Thesis, University of Florida, USA (2005).
[18] R. Field, AIP Conf. Proc. **828** (2006) 163.

# Jet reconstruction in heavy ion collisions

**A Dávila for the STAR Collaboration**

University of Texas at Austin, Physics Department, 1 University Station C1600, Austin,78712 TX, USA

E-mail: alandav@physics.utexas.edu

**Abstract.** The STAR experiment at RHIC detects Au+Au collisions at $\sqrt{S_{NN}} = 200\ GeV$. The very short lived ($\sim$ few fm/c) and highly dense medium created in these collisions has been probed with hadrons of high transverse momentum. Di-hadron azimuthal correlations and nuclear modification factor measurements have been used in the past to probe the medium. These measurements have known geometrical bias and limited kinematic reach. One way to improve on these limitations and get more insight into jet modification is to recover the hard scattered partons kinematics by jet reconstruction which supports the interesting possibility of transverse momentum redistribution with minimal energy loss. The $k_T$ and anti-$k_T$ jet finding algorithms are being used for jet reconstruction in STAR. One of the most significant challenges for jet reconstruction in the heavy ion environment comes from the correct characterization of the significant background fluctuations. The background is being subtracted by assigning areas to the jets and calculating per event background densities. The jet momentum irresolution due to background fluctuations has to be understood in order to recover the correct jet spectrum. Recent progress in jet reconstruction measurements and methodology in heavy ion collisions at STAR will be presented.

## 1. Heavy ion collisions' prior probes

Relativistic heavy ion collisions have been extensively studied via un-triggered particle correlations in ($\Delta\eta$ ,$\Delta\phi$)(all particles with transverse momentum($p_T$)above 150 MeV/c are accepted). STAR has measured the per final state particle 2D correlations at $\sqrt{S_{NN}} = 200, 62\ GeV$ Au+Au peripheral to central collisions [1]. The correlation signal can be fit by a function composed of a constant offset, a dipole and a quadruple in $\Delta\phi$, a $\Delta\phi$ independent Gaussian, and a 2D Gaussian [1]. The 2D Gaussian centered at $\Delta\eta$ $\Delta\phi$ can be attributed to minijets [1][2][3]. It is observed that the amplitude and $\Delta\eta$ width of the Gaussian increases with collision centrality deviating from glauber linear superposition [1]. The width on the $\Delta\phi$ variable has a small decrease in the most peripheral bins and stays constant as we move from mid central to central events. A jet signal structure survives even at the highest centrality in $\sqrt{S_{NN}} = 200\ GeV$ Au+Au events.

The transient character of the medium created in Heavy Ion Collisions (HIC) (lasting $\sim$ few fm/c) pushes physicists to look for probes created in the collisions themselves.Jets coming from hard scatterings constitute such probe. Distinguishing single jets in the HIC environment is challenging, therefore, high $p_T$ particles are used as jets proxies. The correlations in $\Delta\phi$ of high $p_T$ particles relative to different $p_T$ associated particles are studied. STAR measurements have established the suppression of high $p_T$ particles (3-6 GeV/c) recoiling from higher $p_T$ trigger hadrons (8-15 GeV/c) in central $\sqrt{S_{NN}} = 200\ GeV$ Au+Au collisions [5].Similarly, and enhancement of low $p_T$ (0.5-4 GeV/c) multiplicity and $p_T$ sum is found in the recoiling side of a trigger ($p_T$ = 4-10 GeV/c) [4]. A possible explanation of these observations is the softening of jet fragments. High $p_T$ hadron triggers have a surface bias. We preferentially select jets that have interacted the least with the medium. This bias can be used to study path length dependence of jet quenching (looking at the recoil side) but is undesirable if the goal is to measure the inclusive jet $p_T$ spectrum. A surface bias free measuremnt can be obtained via gamma-jets. The small mean free path of a direct gamma traversing the medium assures that it conveys its recoiling parton kinematics. The downside of this otherwise golden probe is its limited statistics at RHIC energies. An additional approach free of surface bias and with higher statistics is to perform full jet reconstruction.

## 2. Jet finding algorithms used at STAR

There are several jet finding algorithms in the literature (cone algorithms, sequential recombination, Gaussian filtering). A desirable characteristic of a jet finding algorithm (JFA) is that it would be infrared and collinear safe. This allows for meaningful comparison with pQCD calculations. The $k_T$ and anti-$k_T$ JFA from the FASTJET package are being used at STAR [6][7][8][9].This package includes a jet area calculation which is very useful for background estimation in heavy ion collision events [8]. A jet area is related to the likelihood of a jet to cluster background particles. It is the number of infinitesimal transverse momentum particles that would get clustered in a jet that is embedded in a dense event containing those particles. The input of any JFA is a set of particles to be clustered. STAR can combine charged hadron measurements with neutral particles due to its Time Projection Chamber (TPC) and Barrel Electromagnetic Calorimeter (BEMC). The TPC is embedded on a 0.5 Tesla magnetic field for transverse momentum determination. It can measure charged tracks of $p_T$ from 150 MeV/c up to $\sim$ 15 GeV/c before the $p_T$ determination becomes prohibitive. It has a pseudorapidty coverage of $|\eta| \leq 1.8$ [10]. The Barrel Electromagnetic Calorimeter (BEMC) consist of 4800 towers with granularity of 0.05x0.05 in $\delta\eta$, $\delta\phi$ size [11]. Gammas and $\pi^0$ decays are detected at the BEMC. STAR does not have $K_L^0$ nor neutron detection capabilities. Both TPC and BEMC have full azimuth coverage. The wide STAR detector acceptance makes it an excellent detector for jet reconstruction.

## 3. Preliminary jet measurements in central Au+Au Collisions at RHIC

Jets produced in p+p collisions at $\sqrt{S_{NN}} = 200\ GeV$ at STAR are well described by pQCD up to jet $p_T$ of 50 GeV [12]. The inclusive jet cross section was obtained with a midpoint cone algorithm with a radius = 0.4. A new measurement of the jet cross section in p+p was made with the anti-$k_T$ and $k_T$ JFAs and a R= 0.4 resolution parameter. The measurement with the two different JFAs is consistent with the published cross section and with pQCD calculations. We conclude that jets are a well calibrated probe to study any effects of the medium on them; furthermore, the choice of JFA does not affect the jets spectrum measurement in p+p collisions. Jet reconstruction gets more challenging in HICs.

A single central Au+Au collision at RHIC energies produces of the order of 5k particles with $\langle p_T \rangle \sim$ 0.5 GeV/c. Jets have to be reconstructed on top of this background. A given JFA cannot distinguish between particles coming from the fragmentation of a single hard scattered parton and any other particles present in a heavy ion collision on a jet by jet basis. The JFA recombination scheme chosen for these studies returns a 4 vector sum from the hadrons that fragmented from the hard scattered parton and possibly emitted gluons if the parton interacts with other colored objects. But that is not the end of the story; other hadrons in the vicinity of the jet-related particles get clustered to build the final jet as well. These "background" hadrons can come from another hard scattering, soft scattering, or elastic collision that just happen to overlap in $\eta - \phi$ space. The fundamental assumption in jet reconstruction studies in HIC is that the contribution of this background and the contribution coming from the jet can be factorized. We have chosen to use a jet-area based background correction scheme [8][9]. All particles in the event get clustered and the output clusters are all initially considered to be jets. A given event background is estimated by measuring the ratio of the jets's transverse momentum ($P_T^{cluster}$ ) over its area ($A_{jet}$), and extracting the median ($\rho$) of that ratio from the ensemble given by all the jets in the event. The expected background of any given jet is then $\rho A_{jet}$ and its measured transverse momentum: $P_T^{measured} = P_T^{cluster} - \rho A_{jet}$. Unfortunately, the event background is not uniform. These background fluctuations give irresolution in $P_T^{measured}$. The steeply falling $P_T$ cross section of high transverse momentum jets at $\sqrt{S_{NN}} = 200$ $GeV$ Au+Au collisions is severely affected by this irresolution which smears the measured spectrum. A $p_T$ cut in the particles used for jet reconstruction in Au+Au collisions can be applied to reduce the background contamination of the $p_T$ of the jet. This comes at the expense of possibly biasing your reconstructed jet signal towards jets that do not interact with the medium. It has been found that background fluctuations increase with the jet area ($\sim$ 6 GeV/c for R = 0.4 at $0-20\%$ most central Au+Au collisions ) [13]. A small resolution parameter is therefore desired with the caveat that you might miss more of the jet's energy in the presence of gluon radiation at angles bigger than the jet area. The bias and interplays of a $p_T$ cut and resolution parameter in jet reconstruction can be used to test jet-medium modification scenarios [14][15].

The $P_T^{jet}$ spectra of background corrected jets in Au+Au collisions has been measured at STAR via deconvolution(unfolding) of the background fluctuations which were initially assumed to have a Gaussian distribution. All particles in an event get clustered by the JFA. Some of those clusters of particles could lie on a region of the event with an upward background fluctuation and end up mocking a real high $p_T$ jet. Uncorrelated clusters of particles that contribute to the jet spectrum signal are called fake jets (uncorrelated in the sense of belonging to a single parton fragmentation, this cluster could have correlations among its constituents particles). The spectrum is corrected from fake jets contamination by subtracting the yield produced by running the JFA on central Au+Au events that had their particles randomized in azimuth. This effectively destroys any previous correlations and other less "destructive" schemes of fake jet estimations are under study.

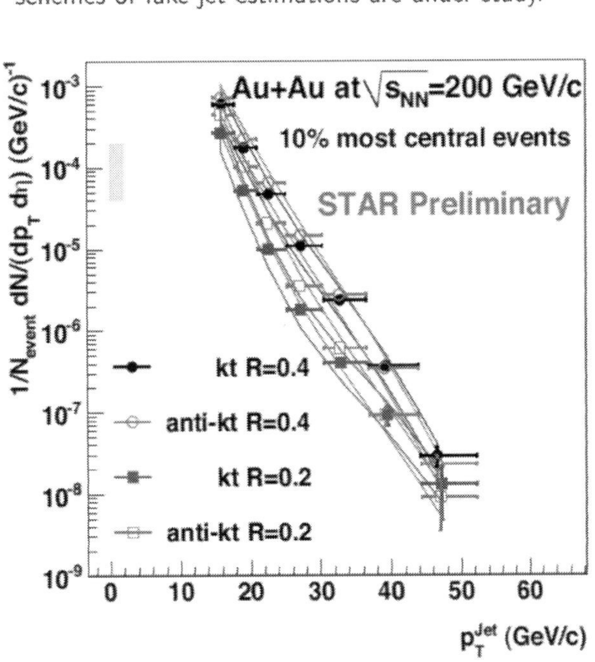


**Figure 1.** Inclusive jet $p_T$ spectrum in 10% most central $\sqrt{S_{NN}} = 200$ $GeV$ Au+Au collisions.$k_T$ and anti-$k_T$ JFA were used along with resolution parameters of R = 0.4 and R = 0.2.The yellow band is the jet energy scale uncertainty and the solid lines represend the systematic uncertainty due to event background fluctuations

STAR has preliminary results on the jets nuclear modification factor $R_{AA}$ at 10% most central $\sqrt{S_{NN}} = 200\ GeV$ Au+Au collisions [16]. $R_{AA}$ has been measured with two distinct JFA ($k_T$ and anti-$k_T$) and two resolution parameters (R=0.2, R=0.4). The nuclear modification factor of jets is consistent with one (Nbin scaling of jets production in Au+Au) for R = 0.4 but it is smaller than one for R = 0.2. This is a sign that not all jets are recovered since they should scale with the number of binary collisions due to their hard scattering origin. A decrease of $R_{AA}$ for smaller resolution R is to be expected since in the limit of $R \to 0$ we would recover the single hadron suppresion $R_{AA} =$ 0.2. In order to test if the smaller $R_{AA}$ in R=0.2 compared to R=0.4 is due to jet broadening, the ratio or spectra with the two resolution parameters was measured both in p+p and (0-10%) central Au +Au collisions at the same per nucleon pair center of mass energy. The jet yield ratio in the p+p system is smaller than zero and increases with jet's $p_T$. This result is consistent with vacuum Q-PYTHIA and PYTHIA + underlying event simulations. The increasing ratio with $p_T$ of the jet can be explained as a manifestation of the jet collimation with increasing energy measured in other experiments [17][18]. The yield ratio in Au+Au is suppressed. This is suggestive of a broadening of the jets in Au+Au collisions.

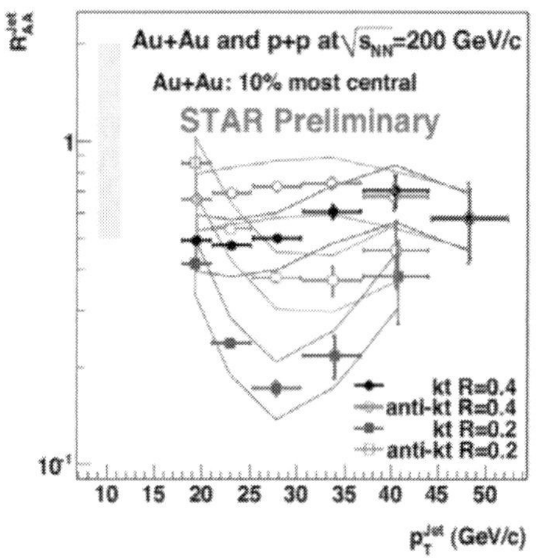


**Figure 2.** Nuclear modification factor in $\sqrt{S_{NN}} = 200\ GeV$ Au+Au collisions using two different JFA and two different resolution parameters. A clear suppression for R = 0.2 is seen.

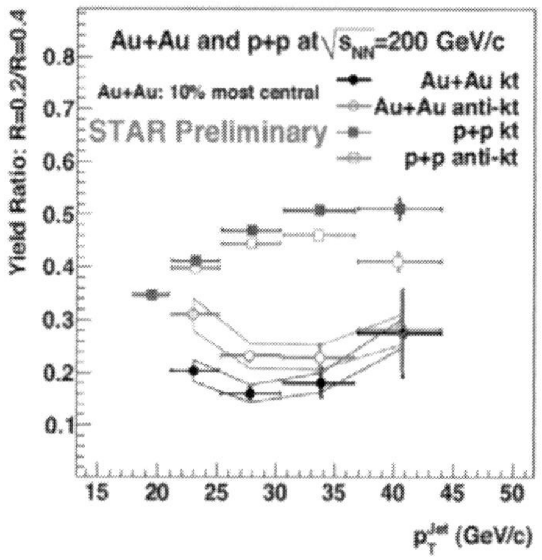


**Figure 3.** Yield ratio $(R = 0.2)/(R = 0.4)$ of jet $p_T$ spectra in $\sqrt{S_{NN}} = 200\ GeV$ p+p collisions (blue and magenta) and in $\sqrt{S_{NN}} = 200\ GeV$ Au+Au collisions (black and red). The Au+Au ratio gets suppressed.

One way to test that jets do get broaden is via jet+hadron correlations [19]. A reconstructed high $p_T$ jet ($P_T^{jet} >$ 20 GeV/c) is correlated with softer ($0.2 \le p_T \le 1.0$ GeV/c) associated particles. Triggering on a jet has the advantage of accessing higher kinematics and a clearer distinction of the correlation measure above background as compared to high $p_T$ hadron trigger correlations. The associated particle multiplicity in the direction of the jet ($\Delta\phi \sim 0$) and recoiling from the jet ($\Delta\phi \sim \pi$) increases in Au+Au collisions compared the the p+p reference. There is also a broadening in the correlation width more obvious in the recoiling side. Increasing the $p_T$ of the associated particles to 1 - 2.5 GeV/c results in an enhancement of the correlation amplitude on the jet direction's side and reduction in the away side amplitude (so that now it matches the amplitude in the away side in p+p collisions). The broadening on the away side correlation signal persists. These results are consistent with a broadening of the jet due to radiation of energy of the original parton.

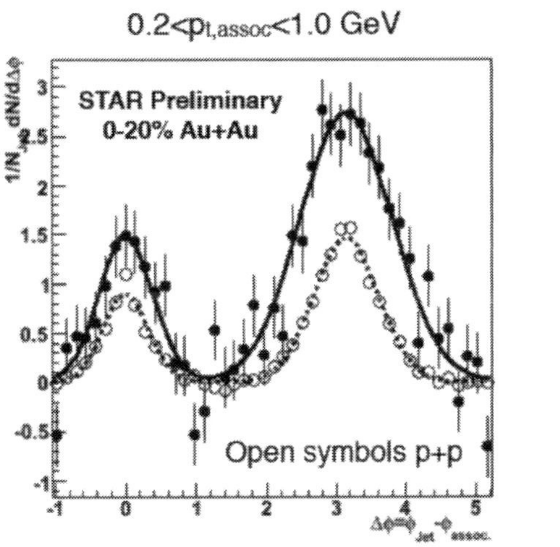


**Figure 4.** Jet-hadron correlation on $\Delta\phi$ for a triggered jet of $P_T^{jet} > 20\ GeV/c$ and associated hadrons of $0.2 < p_T^{assoc} < 1.0\ GeV/c$ for central Au+Au collisions (filled circles) and p+p collisions (hollow circles) at $\sqrt{S_{NN}} = 200\ GeV$

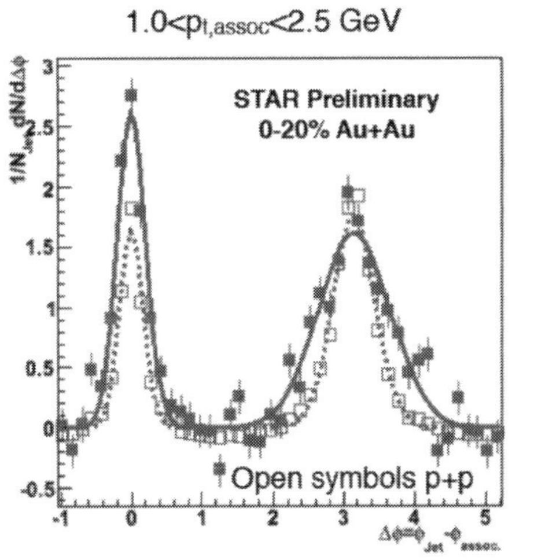


**Figure 5.** Jet-hadron correlation on $\Delta\phi$ for a triggered jet of $P_T^{jet} > 20\ GeV/c$ and associated hadrons of $1.0 < p_T^{assoc} < 2.5\ GeV/c$ for central Au+Au collisions (filled squares) and p+p collisions (hollow squares) at $\sqrt{S_{NN}} = 200\ GeV$

## 4. Recent background fluctuation studies

Much progress has been made in the study of the underlying HIC event background since the preliminary results on the jet spectra were first presented [16] [20]. The $P_T^{measured}$ irresolution can be quantified by embedding a known probe in a central event and measuring the excess (or deficit) of transverse momentum after background subtraction. The background fluctuation (jet irresolution) is therefore defined as: $\delta P_T = P_T^{cluster} - \rho A_{jet} - P_T^{embed}$. Where $\rho$ is calculated before the probe is embedded. We measure the probability of fluctuations $\delta P_T$ by embedding different $P_T^{embed}$ probes in several central Au+Au events. This probability is the response of a given signal to the heavy ion background. We find that the probability of fluctuations depends trivially on the area of the jet and also on its initial transverse momentum, but it does not depend strongly on the initial jet fragmentation pattern [20]. The measured jet spectrum is a convolution of the probability of fluctuations with the inclusive jet spectrum. The inclusive jet spectrum is what we would like to recover via deconvolution. It is worthwhile to notice that a simple statistical distribution describes the background fluctuations of jets reconstructed with random cones (R = 0.2), and the fast-jet anti $k_T$ algorithm on a simulated thermal background with particles randomly localized in $\eta - \phi$ space.

## 5. Jet Shapes Background

Jet shapes have been measured in $p - \overline{p}$ collisions at $\sqrt{S} = 1.96\ TeV$, in $p - e^+$ DIS at $\sqrt{S} = 300\ GeV$ and in $p - p$ collisions at $\sqrt{S} = 200\ GeV$ [17][21][22].

The following analysis is not concerned with the jet spectrum measurement itself but with the characterization of the fluctuations in the jets' differential shape measurement. The differential jet shape (DJS) is the rate of change of a jet's transverse momentum with distance in $\eta - \phi$ space from the jet axis normalized by the total jet's $P_T$. This definition only makes sense in practice as an average over many jets at the same $P_T^{jet}$. We have to measure the amount of $P_T$ at a distance r for each jet. The JFA response to the HIC background can have the following effects on the DJS. It could change the measured jets total transverse momentum. It could change the measured transverse

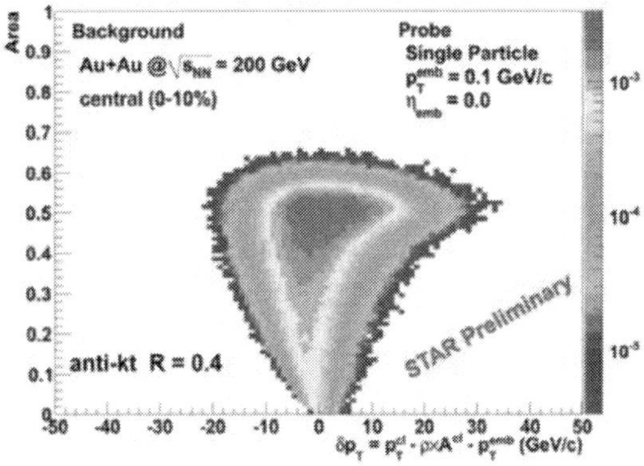


**Figure 6.** $\delta P_T$ and $A_{jet}$ distribution for a single particle $P_T^{embed} = 0.1$ GeV/c embedded in $\sim$ 8M 0-10% most central $\sqrt{S_{NN}} = 200\ GeV$ Au+Au events.An anti-$k_T$ with a resolution parameter of R = 0.4 JFA was used.

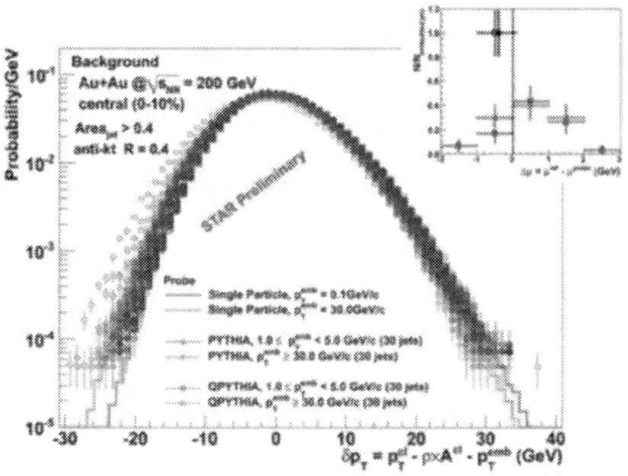


**Figure 7.** $\delta P_T$ distribution for different $P_T^{jet}$ embedded jets. Up to 30 different jets produced with PYTHIA and Q-PYTHIA were embedded. All the distributions are similar to within a factor of two, showing the independence of the distributions on the fragmentation pattern of the jets.

momentum at a distance r from the jet axis and it might change the original jet axis. The measured (background subtracted) jet shape will have a probability of fluctuations that we have to take into account in order to get the differential jet shape signal.

Following the jet area based background subtraction method we calculate the area of concentric rings of width $\delta r = 0.1$ and multiply it by $\rho$ to estimate the amount of background at a given distance from the jet axis. The total transverse momentum of the jet is still corrected using its total area. A point of warning is that this method will only probe the jets internal momentum distribution up to a distance smaller than your R parameter, missing the region where very interesting phenomena are encountered (i.e. $\Delta\eta$ ridge).

It has been found that the background fluctuations grow with area [13]. Therefore, we expect smaller fluctuations in the shape measurement closer to the jet axis (smaller ring) but those fluctuations are correlated with the total $\delta P_T$ of the jet. The measured DJS at a distance r is:

$$\frac{1}{N_{jet}} \sum_{jets} \frac{1}{\delta r} \frac{P_T^{jet}(r) - 2\pi r \rho \delta r}{P_T^{jet} - \rho A_{jet}} \tag{1}$$

The probability of fluctuations is obtained by embedding a probe with known differential jet shape signal (a single high $p_T$ pion, whose jet shape is zero at any distance r) and subtracting that signal after having applied the background subtraction. The obtained distribution is normalized to one. For the case of a single pion the fluctuation is

$$\delta\psi(r) = \frac{1}{\delta r} \frac{p_T^{cluster}(r) - 2\pi r \rho \delta r}{p_T^{cluster} - \rho A_{jet}} \tag{2}$$

We embedded a single $p_T = 30$ GeV/c pion at $\eta = \phi = 0$ in $\sim$ 150k central Au+Au events at $\sqrt{S_{NN}} = 200\ GeV$ from Run 2007 in order to extract the probability of fluctuation of the DJS as a function of distance from the jet axis (r). We reconstruct the jet containing the embedded pion with the anti-$k_T$ JFA using a resolution parameter R = 0.6. We calculated the background using the $k_T$ JFA with the same resolution parameter. The $k_T$ algorithm is preferred to the anti-$k_T$ algorithm as a $\rho$ estimator due to the tendency of clustering jets of very small areas of the later. The presented probability of fluctuations is for charged particles only and the background at a given ring was calculated with charged particles accordingly.

Both charged and neutral particles are used to obtain the total jet transverse momentum. An area cut was imposed in the studied jets in order to avoid a trivial edge deficit problem. Only jets with area, $\pi R^2 \leq A_{jet} \leq \pi R^2 + 0.05$, were accepted. Only particles with $p_T > 0.2$ GeV/c were used. An $|\eta| < 1 - R$ cut was imposed on jets to assure that the complete jet area is within the detector acceptance. The central Au+Au events to embed the pion where chosen to have a z vertex cut of $< 30$ cm and reference multiplicity $> 400$. The transverse momentum of charged particles matching a signal from a calorimeter tower is subtracted to avoid double counting. The probability of fluctuations show the expected area dependence being wider for bigger r (i.e. bigger ring area). A value of zero $\delta\psi$ means that we have accomplished perfect background subtraction. We have chosen an r bin size of 0.1. Expected typical values of $\psi(r)$ at the considered jet energy vary between 0-10 [23]. Therefore a fluctuation of 1 can represent a change in the DJS 10% or more. The probability of fluctuations for a relative distance of r/R = 0.249 has 93.5% (74.6 %) of its distribution in the $\delta\psi(r/R)$ interval of [-1, 1] ([-0.5, 0.5]). This indicates that there is a $\sim 25\%$ of probability that the jet shape measurement after background subtraction is off by an amount bigger than 0.5. The probability of fluctuations for a relative distance from the jet axis of r/R = 0.849 has 74.2% (46.3 %) of its distribution in the $\delta\psi(r/R)$ interval of [-1, 1] ([-0.5, 0.5]). The probability of fluctuations of the DJS for a 30 GeV/c jet are narrow enough that we can attempt a deconvolution of the measured signal in central Au+Au collisions.

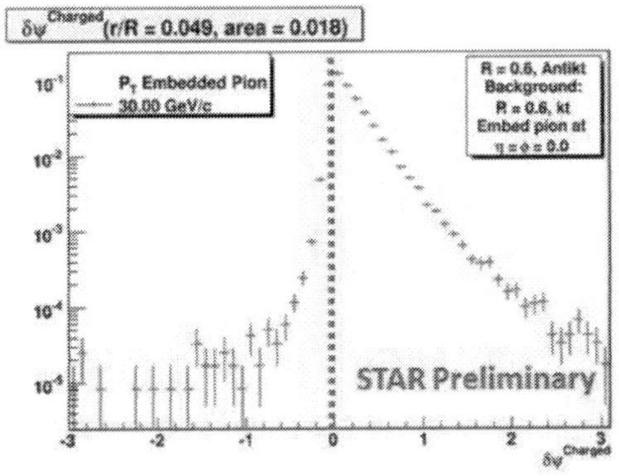


**Figure 8.** $\delta r$ probability distribution for a $P_T = 30\ GeV/c$ embedded pion at a relative distance from the jet axis of r/R = 0.049. Jet reconstruction was done with anti-$k_T$ ($k_T$ for background determination), R = 0.6 JFA. The ring area is 0.018.

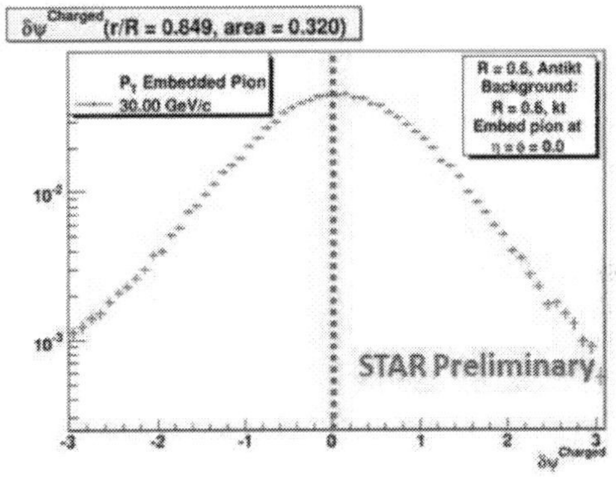


**Figure 9.** $\delta r$ probability distribution for a $P_T = 30\ GeV/c$ embedded pion at a relative distance from the jet axios of r/R = 0.849. The ring area is 0.320. The fluctuations in the DJS are wider for a greater area as expected.

## 6. Conclusions

Jets are a calibrated probe that can be used to expand current studies of the medium created in HIC at RHIC. Preliminary results from jet reconstruction at STAR in $\sqrt{S_{NN}} = 200\ GeV$ 0-10% most central Au+Au collisions indicate possible jet profile broadening. Jet $R_{AA}$ is found to be consistent with one for resolution parameter R = 0.4 but is suppressed at R = 0.2. The yield ratio of reconstructed jets with these two parameters show a suppression compared to the same ratio in p+p. This is consistent with a possible broadening of the jets. Jet+hadron correlations confirm these broadening. Progress has been made on the understanding of the distributions of background fluctuations in central Au+Au events. The distribution of fluctuations of the measured jet transverse momentum is found to be independent of fragmentation patterns. This is important to be able to reconstruct the spectrum in the presence of quenched jets that are expected to have different fragmentation patterns compared to vacuum fragmenting jets. Studies to characterize the effect of background fluctuations for differential jet shapes look promising for Jets of $P_T = 30$ GeV/c.

## References

[1] Daugherity M (STAR Collaboration) 2008 *J. Phys.*G **35** 104090
[2] Trainor T A, Porter R J and Prindle D J 2005 *J. Phys.*G **31** 809
[3] Porter R J and Trainor T A 2004 eprint hep-ph/0406330v1
[4] Adams J *et al.* STAR Collaboration 2005 *Phys. Rev. Lett.* **95** 152301
[5] Adams J *et al.* STAR Collaboration 2006 *Phys. Rev. Lett.* **97** 162301
[6] Salam G P, Soyez G 2007 JHEP 0705 **86**
[7] Cacciari M, Salam G P and Soyez G 2008 JHEP 0804 **063**
[8] Cacciari M, Salam G P and Soyez G 2008 JHEP 0804 **005**
[9] Cacciari M, Salam G P 2008 *Phys.Lett.* B **659** 119-126
[10] Ackermann K H *et al.* 2003 *Nucl. Instr. and Meth* A **499** 624-632
[11] Beddo M *et al.* 2003 *Nucl. Instr. and Meth* A **499** 721
[12] Abelev B I *et al.* STAR Collaboration 2006 *Phys. Rev. Lett.* **97** 252001
[13] Putschke J 2009 *Eur. Phys. J.* C **61** 629-35
[14] Salur S 2009 *Eur. Phys. J.* C **61** 761-67
[15] Vitev I, Zhang B W and Wicks S 2009 *Eur. Phys. J.* C **61** 139-44
[16] Ploskon M 2009 *Nucl. Phys.* A **830** 255c-58c
[17] Acosta D 2005 *Phys. Rev.* D **71** 112002
[18] Aad G 2011 *Phys. Rev.* D **83** 052003
[19] Ohlson A 2011 These proceedings
[20] Jacobs P 2011 *Nucl. Phys.* A textbf855 299-302
[21] Breitweg J 1998 *Eur. Phys. J.* C **8** 367-80
[22] Sakuma T 2009 *STAR Collaboration, PhD thesis*
[23] Vitev I and Zhang B W S 2010 *Phys. Rev. Lett.* **104** 132001

# Jet-hadron correlations in STAR

**Alice Ohlson for the STAR Collaboration**

Physics Department, Yale University, 272 Whitney Ave., New Haven, 06520-8120 CT, USA.

E-mail: `alice.ohlson@yale.edu`

**Abstract.** In recent years, the study of dihadron correlations has been one of the primary methods used to investigate the propagation and modification of hard-scattered partons through the QGP. Due to recent advances in jet-finding algorithms, it is now possible to use reconstructed jets in these correlation studies, extending the kinematic reach compared to dihadron analyses. The results of the jet-hadron correlation analysis indicate a broadening and softening of jets that interact with the medium. Jet-hadron correlations can also be used to assess the systematics of other jet-like correlation analyses, such as 2+1 correlations. It is shown that the jets selected in 2+1 correlations are relatively unmodified. Future work will include an analysis of jet-hadron correlations with respect to the event plane to measure the pathlength dependence of parton energy loss. The first steps in this analysis indicate that complications arise when calculating the event plane in the presence of a jet as well as in calculating jet $v_2$. The data analyzed were collected by the STAR detector in $\sqrt{s_{NN}} = 200$ GeV Au–Au collisions at the Relativistic Heavy Ion Collider (RHIC).

## 1. Introduction

Jets are used to study the energy loss of partons which traverse the hot and dense medium produced in ultrarelativistic heavy-ion collisions. Models of partonic energy loss fall into two general categories: radiative/collisional energy loss models[1] (where radiative energy loss is expected to dominate for light partons) and "black-and-white" models (for example, the core-corona model[2]). In the former, partons lose energy and are scattered as they traverse the medium. In angular correlation studies this would manifest itself as a difference between the widths and yields of the jet peaks in Au–Au compared to p–p. In the latter, jets either exit the medium entirely unmodified or are completely thermalized and indistinguishable from the bulk. While the overall rate of jets would be suppressed in inclusive studies, in angular correlation studies (which are normalized per trigger jet) there would be no observed difference in jet shapes between Au–Au and p–p. These two classes of models can be distinguished by analyzing jet shapes and yields with jet-hadron angular correlations[3].

## 2. Data sets and Analysis

In jet-hadron correlations the azimuthal distribution of associated particles (all charged hadrons in the event) is studied with respect to the axis of a reconstructed jet. Utilizing a reconstructed jet as a trigger increases the range of accessible parton kinematics compared to dihadron correlations[4].

The data analyzed here were collected by the STAR detector at the Relativistic Heavy Ion Collider (RHIC) in Run 6 p–p and Run 7 Au–Au collisions at $\sqrt{s_{NN}} = 200$ GeV. The main detectors used in this analysis are the Time Projection Chamber (TPC) and the Barrel Electromagnetic Calorimeter (BEMC). Both detectors have full $2\pi$ azimuthal coverage and span a pseudorapidity range of $|\eta| < 1$. Charged tracks are reconstructed in the TPC and the BEMC measures the neutral energy component. Corrections for double-counting of electrons and charged hadronic energy deposition in the BEMC are applied.

Jets are reconstructed using the anti-$k_T$ algorithm from the FastJet package[5]. Only tracks and neutral towers with $p_T$>2 GeV/$c$ are used in the jet reconstruction in order to minimize the effect of background fluctuations. Furthermore, the reconstructed jets used in this analysis must include a BEMC tower that fired the high tower (HT) trigger. The online HT trigger requires that more than 5.4 GeV of transverse energy is deposited in one BEMC tower ($\Delta\varphi \times \Delta\eta = 0.05 \times 0.05$). It is expected that the $p_T$ cut and HT trigger requirements bias the trigger selection towards harder fragmenting jets which interact less with the medium, potentially due to surface production[6].

The raw jet-hadron correlations are $\Delta\varphi$ distributions, where $\Delta\varphi = \varphi_{jet} - \varphi_{assoc}$. In Au–Au the raw correlations include a large contribution from the heavy-ion background. It is necessary to subtract this background in order to analyze the jet signal (the nearside and awayside jet peaks at $\Delta\varphi = 0$ and $\Delta\varphi = \pi$, respectively). There are two complications associated with this background subtraction: First, using the ZYAM ("Zero Yield at Minimum") technique is unfavorable because it overestimates the background level in the presence of broad jet peaks. Second, the $v_2$ of the trigger jets is *a priori* unknown. In this analysis background levels are estimated by fitting. The signal is assumed to have the functional form in Equation 1.

$$\text{Nearside Gaussian} + \text{Awayside Gaussian} + B\left(1 + 2v_2^{assoc}v_2^{jet}\cos(2\Delta\varphi)\right) \quad (1)$$

In Equation 1, $v_2^{assoc}$ is given by the standard STAR parameterization ($^1/_2(v_2\{2\} + v_2\{4\})$ as a function of $p_T$) and $v_2^{jet}$ is estimated to be $v_2\{2\}$ at $p_T = 6$ GeV/$c$. Because $v_2^{jet}$ is unknown, the lower and upper bounds on the uncertainty of $v_2^{assoc}v_2^{jet}$ are conservatively estimated to be 0 and 150% of $v_2\{2\}(6\ \text{GeV}/c) \cdot v_2\{2\}(p_T^{assoc})$. The correlations with a flat background subtracted are shown in Figure 1 for three $p_T^{assoc}$ bins. The awayside peak appears enhanced in Au–Au compared to p–p in the lowest $p_T^{assoc}$ bin and suppressed in the highest $p_T^{assoc}$ bin. In the intermediate $p_T^{assoc}$ bin the Au–Au awayside peak appears broader than in p–p.

## 3. Jet-Hadron Correlations - The Nearside

The highly biased trigger jet sample allows easier comparison between nearside jets in Au–Au and p–p. On average, background particles with $p_T > 2$ GeV/$c$ create an excess of $p_T \sim 1$ GeV/$c$ within the jet cone. The energy of the p–p reference jets is consequently shifted down by 1 GeV/$c$ with respect to the Au–Au jet energies to more accurately compare similar partonic energies. When comparing the Gaussian widths of the nearside jet peaks in p–p and Au–Au, the Au–Au peaks appear broader (even after the -1 GeV/$c$ shift). Additionally, there seems to be slight enhancement in the ratio of the nearside associated yields, $I_{AA}$ (Equation 2), below $p_T^{assoc} = 2$ GeV/$c$.

$$I_{AA}(p_T^{assoc}) = \frac{Y_{AA}(p_T^{assoc})}{Y_{pp}(p_T^{assoc})} \quad (2)$$

There are two possibilities for this enhancement: (1) The enhancement in the low-$p_T^{assoc}$ yield could be due to the ridge, $v_3$, or other bulk effects that have not been accounted for (i.e. the enhancement is related to the bulk), or (2) it could be due to energy loss on the nearside (i.e. the enhancement is related to the jet). Under the assumption that the latter possibility is the cause for the enhancement, then the energy loss can be estimated by summing up the excess $p_T$ below $p_T^{assoc} = 2$ GeV/$c$, yielding a charged energy difference $\Delta E \sim$2 GeV/$c$. This can be

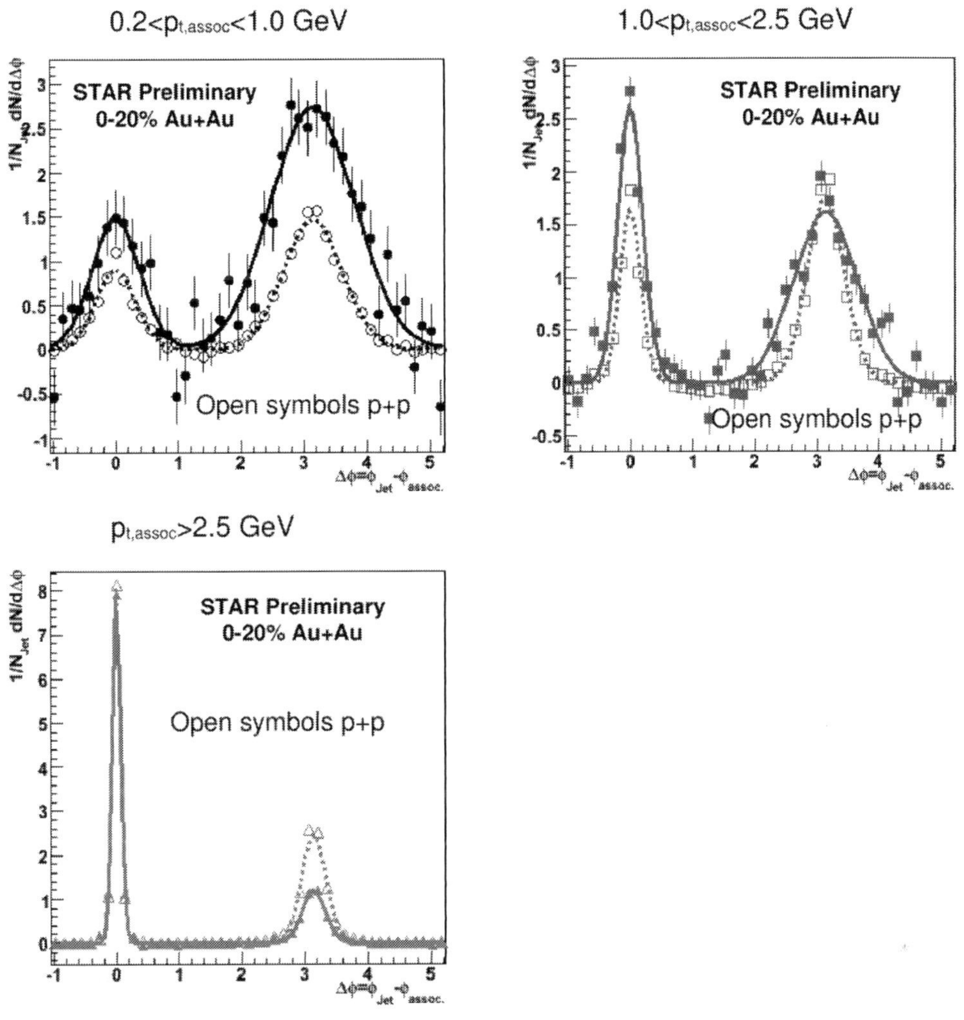


**Figure 1.** Jet-hadron correlations with a flat background (no $v_2$ modulation) subtracted for jets with reconstructed $p_T^{jet}$ >20 GeV/$c$. Three $p_T^{assoc}$ bins are shown: top left 0.2 <$p_T^{assoc}$ <1.0 GeV/$c$, top right 1.0 <$p_T^{assoc}$ <2.5 GeV/$c$, bottom left $p_T^{assoc}$ >2.5 GeV/$c$. Closed symbols are Au–Au (0-20% centrality) and open symbols are p–p.

accounted for by adjusting the p–p reference up by $1.5 \cdot \Delta E$ (where the factor of 1.5 is due to the neutral energy fraction).

The possible modification of the nearside jet peak can also be quantified by the difference in energy between Au–Au and p–p as a function of $p_T^{assoc}$, $D_{AA}$ (Equation 3), and its integral, $\Delta B$ (Equation 4).

$$D_{AA}(p_T^{assoc}) = Y_{AA}(p_T^{assoc}) \cdot p_{T,AA}^{assoc} - Y_{pp}(p_T^{assoc}) \cdot p_{T,pp}^{assoc} \quad (3)$$

$$\Delta B = \int dp_T^{assoc} D_{AA}(p_T^{assoc}) \quad (4)$$

When $\Delta B \sim 0$ the suppression at high-$p_T$ is fully balanced by enhancement at low-$p_T$. On the nearside it is observed that $\Delta B \sim 1 - 2$ GeV/$c$ before the p–p energy shift, indicating that the energy is in large part balanced. After the p–p reference is shifted by $1.5 \cdot \Delta E$, the nearside $\Delta B$ is even smaller. This indicates the possibility that there is energy loss on the nearside,

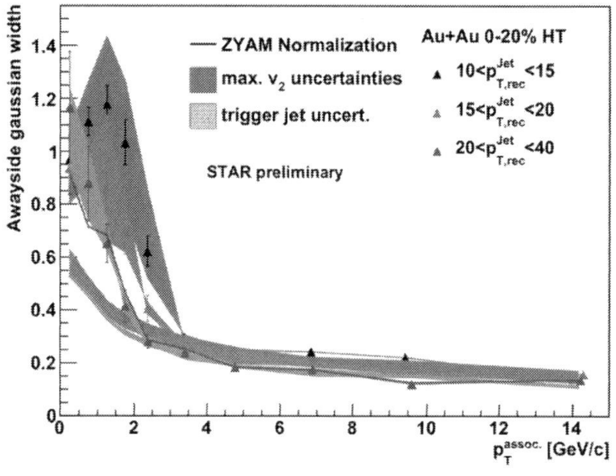


**Figure 2.** The Gaussian widths of the awayside jet peaks in Au–Au (upper shaded bands with symbols) are compared to p–p (lower shaded bands without symbols) for three reconstructed jet energy ranges: 10 $<p_T^{jet}<$15 GeV/$c$ in black, 15 $<p_T^{jet}$ $<$20 GeV/$c$ in red, 20 $<p_T^{jet}$ $<$40 GeV/$c$ in blue. The awayside jet peaks in Au–Au are significantly broadened compared to p–p for jet energies between 10 and 40 GeV/$c$.

even in high tower trigger jets with stringent jet-finding criteria, although this certainly needs to be studied more thoroughly. In the following analysis of the awayside jets, Au–Au will be compared to p–p with both energy shifts applied (+3 GeV/$c$ and -1 GeV/$c$). The uncertainty on the trigger jet energy scale is estimated to be $\pm$2 GeV/$c$.

## 4. Jet-Hadron Correlations - The Awayside

It is expected that the surface bias of the nearside jet will cause the recoil parton to travel through a significant amount of the medium, therefore making the awayside the ideal place to study any parton energy loss and jet modification. The awayside Gaussian widths (Figure 2) indicate significant broadening of the jet peak in Au–Au compared to p–p. The awayside $I_{AA}$ (Figure 3) indicates significant softening of the awayside jets in Au–Au. This modification can also be clearly seen in $D_{AA}$ (Figure 4). Since $\Delta B \sim 1 - 2$ GeV/$c$, it is clear that most (but not all) of the high-$p_T^{assoc}$ suppression is compensated for by low-$p_T^{assoc}$ enhancement. All of these observations are indicative of significant jet quenching on the awayside. The observed modifications in the jet shapes and yields between Au–Au and p–p are in qualitative agreement with the radiative energy loss picture while the black-and-white models are disfavored.

## 5. "2+1" and Jet-Hadron Correlations

In 2+1 correlations, a pair of back-to-back high-$p_T$ hadrons is used as a dijet proxy. Typical dihadron correlations are then performed with respect to both trigger particles[7]. It is suspected that the dijet proxy requirement in 2+1 correlations selects jets which are unmodified, possibly due to surface bias. The systematics of 2+1 correlations can be explored through jet-hadron correlations, by performing the jet-hadron analysis when there is a high-$p_T$ hadron opposite the reconstructed trigger jet. Furthermore, the 2+1 correlation analysis used a $p_T^{assoc}$ cut-off at 1.5 GeV/$c$; in jet-hadron correlations the analysis can be extended below this cut. In the jet-hadron correlations shown in Figure 5 a dijet trigger is required in the range $|\varphi_{jet} - \varphi_{trig} - \pi| < 0.2$. As the $p_T$ cut on the dijet trigger is raised from 0 to 2 to 4 GeV/$c$, the awayside jet peak in Au–Au becomes more and more similar to p–p. Additionally, a comparison of the jet and event populations selected by the various $p_T^{trig}$ thresholds indicates that 2+1 correlations are biased towards harder jets as well as more peripheral events. This supports the idea that the jet population selected in 2+1 correlations is biased towards unmodified jets.

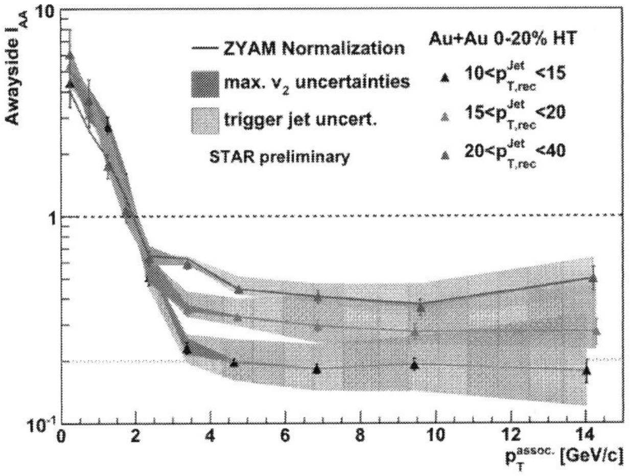


**Figure 3.** The awayside $I_{AA}$ (defined in Equation 2) is shown for three reconstructed jet energy ranges: 10 $<p_T^{jet}$ <15 GeV/$c$ in black, 15 $<p_T^{jet}$ <20 GeV/$c$ in red, 20 $<p_T^{jet}$ <40 GeV/$c$ in blue. $I_{AA}(p_T^{assoc}) = 1$ would indicate no difference in the awayside jet peak charged hadron yield between Au–Au and p–p. The awayside $I_{AA}$ indicates significant softening of the awayside jet for jet energies between 10 and 40 GeV/$c$.

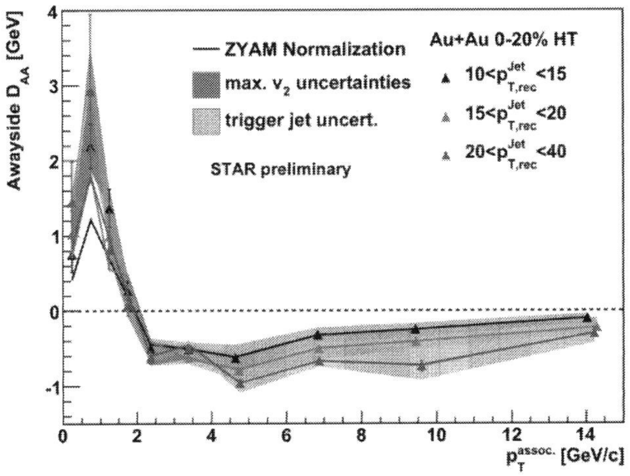


**Figure 4.** The awayside $D_{AA}$ (defined in Equation 3) is shown for three reconstructed jet energy ranges: 10 $<p_T^{jet}$ <15 GeV/$c$ in black, 15 $<p_T^{jet}$ <20 GeV/$c$ in red, 20 $<p_T^{jet}$ <40 GeV/$c$ in blue. $D_{AA}(p_T^{assoc}) = 0$ would indicate no difference in the energy of the awayside jet fragments between Au–Au and p–p. The awayside $D_{AA}$ illustrates how high-$p_T^{assoc}$ suppression is compensated for by low-$p_T^{assoc}$ enhancement.

## 6. Jets with respect to the Event Plane

Future work on jet-hadron correlations will include an investigation of the pathlength dependence of partonic energy loss via an analysis of the awayside jet modification as a function of the relative angle between the trigger jet and the event plane. Before performing the jet-hadron analysis with respect to the event plane, however, it is necessary to first investigate the relation between the trigger jet and the reaction plane.

Figure 6 shows that there is a significant correlation between the axis of a HT trigger jet and the event plane (reconstructed using Equation 5).

$$\tan(2\Psi_{EP}) = \frac{\sum_n \sin(\varphi_n)}{\sum_n \cos(\varphi_n)} \tag{5}$$

Although this result appears to clearly show a non-zero jet $v_2$, there is another potential competing effect: the jet may be biasing the event plane calculation. When tracks within a cone of $R = 0.4$ in $(\eta, \varphi)$-space around the reconstructed jet axis are excluded from the event plane calculation, the correlation between the reconstructed jet and the event plane is reduced.

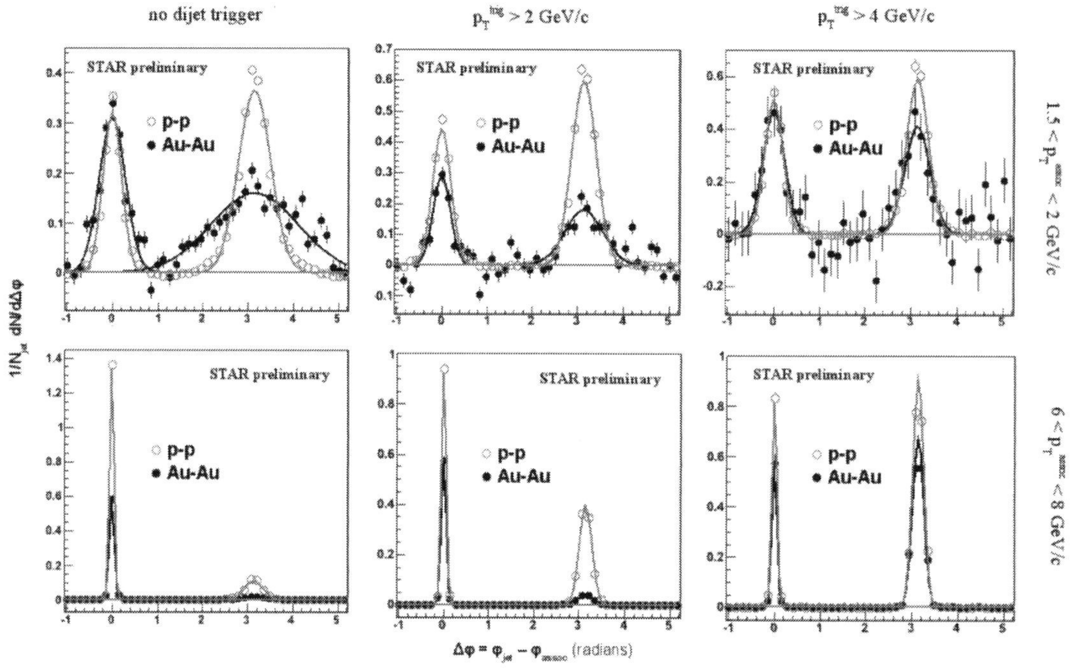


**Figure 5.** Jet-hadron correlations with the requirement of a dijet trigger (high-$p_T$ hadron opposite the reconstructed jet axis, $|\varphi_{jet} - \varphi_{trig} - \pi| < 0.2$): left $p_T^{trig} > 0$ GeV/$c$, center $p_T^{trig} > 2$ GeV/$c$, right $p_T^{trig} > 4$ GeV/$c$. The reconstructed jet energy is $10 < p_T^{jet} < 20$ GeV/$c$ and two $p_T^{assoc}$ bins are shown: top $1.5 < p_T^{assoc} < 2$ GeV/$c$ and bottom $6 < p_T^{assoc} < 8$ GeV/$c$. As the dijet trigger $p_T$ threshold rises the awayside jet peak in Au–Au (black closed symbols) becomes narrower and taller to match the awayside peak in p–p (red open symbols).

Therefore the presence of a jet affects the calculation of the event plane, but the question remains: is there a jet $v_2$ beneath the artificial jet-event plane bias? The jet-event plane bias can be further investigated through simulation: PYTHIA[8][9] jets are embedded isotropically in a thermal background. The jets are randomly distributed with respect to the reaction planes of the background event and therefore entirely uncorrelated ($v_2^{jet} = 0$ by construction). The event plane is calculated in the standard way (Equation 5) and the distribution of the relative angle between the jet and the event plane is shown in Figure 7. For PYTHIA jets with $E_T^{PYTHIA} = 15$ GeV the jet has an artificial $v_2^{obs}$ of $\sim 7\%$ (before dividing by the event plane resolution, meaning the artificial jet $v_2$ is actually higher), and for jets with $E_T^{PYTHIA} = 30$ GeV the artificial jet $v_2^{obs}$ is $\sim 9\%$. Therefore the jet pulls the event plane significantly. Attempts to remove the jet-event plane bias in simulation (by removing particles in the jet cone or in a $\eta$ wedge, or by using the Lee-Yang Zeros method to calculate the event plane[10]) have been so far unsuccessful. Future work will investigate the possibility of using forward detectors (FTPC, BBC, ZDC-SMD) to calculate the event plane in the presence of a jet, but long-range correlations associated with the jet may have an effect on event plane reconstruction at forward rapidity as well.

## 7. Conclusions

Jet-hadron correlations can be used to investigate the modification of partons that traverse the medium created in heavy ion collisions. The results indicate that in high tower trigger events

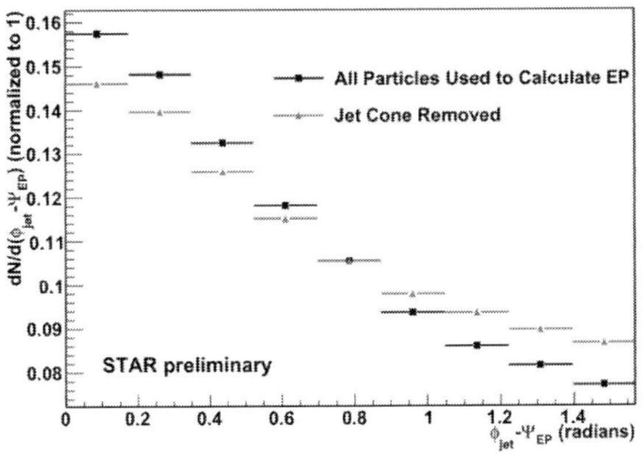


**Figure 6.** The relative angle between a HT trigger jet ($p_T^{jet} >$ 10 GeV/$c$) and the event plane. The event plane is reconstructed from all charged tracks in the TPC with $p_T < 2$ GeV/$c$ and without $p_T$ weighting using Equation 5. The result (black squares) is compared to the correlation between the jet axis and event plane when tracks within a cone of $R = 0.4$ in $(\eta, \varphi)$-space around the reconstructed jet axis are excluded from the event plane calculation (pink triangles).

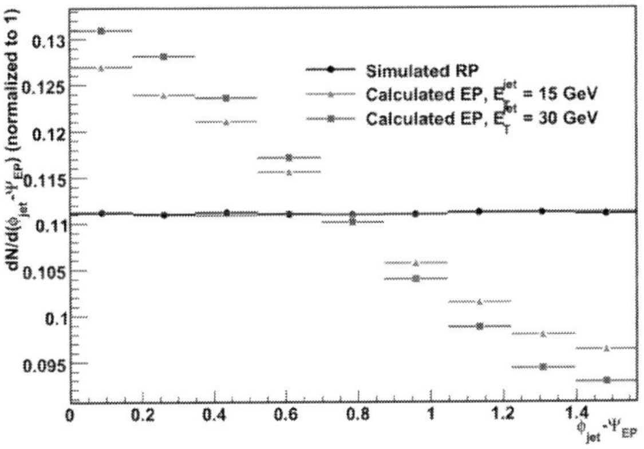


**Figure 7.** A jet is embedded in a thermal background ($T$ = 0.291 GeV) with $p_T$- and centrality-dependent $v_2$ (10-20% centrality). By construction $v_2^{jet} = 0$ (black circles). The correlation between the jet axis and the reconstructed event plane is shown for PYTHIA jets with $E_T^{PYTHIA} = 15$ GeV (red triangles) and $E_T^{PYTHIA} = 30$ GeV (blue squares).

the recoil jet is significantly broadened and softened in Au–Au compared to p–p, in qualitative agreement with the radiative energy loss picture. Furthermore, most of the high-$p_T$ suppression is balanced by low-$p_T$ enhancement. Jet-hadron correlations also indicate that the dijet trigger requirement in 2+1 correlations biases the jet sample towards unmodified jets. A study of jets with respect to the event plane shows that the presence of a jet significantly biases the event plane calculation. An analysis of jet $v_2$ and jet-hadron correlations with respect to the event plane is in progress.

## Acknowledgments

This research was supported in part by an award from the Department of Energy (DOE) Office of Science Graduate Fellowship Program administered by the Oak Ridge Institute for Science and Education for the DOE. This work was also supported in part by the facilities and staff of the Yale University Faculty of Arts and Sciences High Performance Computing Center.

## References

[1] A. Majumder and M. van Leeuwen, Prog. Part. Nucl. Phys. **66** (2011) 41.
[2] K. Werner, Phys. Rev. Lett. **98** (2007) 152301.
[3] J. Putschke for the STAR Collaboration, Hard Probes 2010 Proceedings, Nucl. Phys. A **855** (2011) 83.
[4] M.M. Aggarwal et al., Phys. Rev. C **82** (2010) 024912.
[5] M. Cacciari and G. Salam, Phys. Lett. B **641** (2006) 57.

Journal of Physics: Conference Series **316** (2011) 012015 doi:10.1088/1742-6596/316/1/012015

[6] T. Renk and K. Eskola, Phys. Rev. C **75** (2007) 054910.
[7] H. Pei for the STAR Collaboration, these proceedings.
[8] T. Sjostrand, S. Mrenna and P. Skands, JHEP **05** (2006) 026.
[9] T. Sjostrand, S. Mrenna and P. Skands, Comput. Phys. Commun. **178** (2008) 852.
[10] A. Bilandzic, N. van der Kolk, J.-Y. Ollitrault and R. Snellings, Phys. Rev. C **83** (2011) 014909.

Journal of Physics: Conference Series **316** (2011) 012016 doi:10.1088/1742-6596/316/1/012016

# Jet studies in STAR via 2+1 correlations

**Hua Pei, for the STAR collaboration**

Post-doc research associate, 845 W Taylor St, MC 273, Dept. of Physics, University of Illinois at Chicago, Chicago, IL, 60607, U.S.A.

E-mail: hpei@uic.edu

**Abstract.**
This paper report results from an analysis based on a new multi-hadron correlation technique. Jet-medium interactions and di-jet surface emission bias are explored at RHIC, via events tagged by back-to-back high transverse momentum hadrons pairs as proxy of di-jets, then studying the associated hadron distributions. In contrast with traditional two- and three-particle correlations with a single trigger particle, the associated hadron distribution at both trigger sides reveals no evident modification in either relative pseudo-rapidity, $\Delta\eta$, or relative azimuthal angle, $\Delta\phi$, from d+Au to central Au+Au collisions. This observation stands for either back-to-back triggers of similar kinematic selections (symmetric, of $5 < p$ $< 10$GeV/$c$ and $4 < p$ $< 10$GeV/$c$) or of highly different kinematic selections (asymmetric, of $10 < E$ $< 15$GeV/$c$ and $4 < p$ $< 10$GeV/$c$). The associated hadron yields and spectra are also measured on each trigger side, to gain additional insights of medium properties.

## 1. Introduction

Di-hadron correlation measurements in heavy ion collisions have provide deeper view of the hot and strongly interacting matter at RHIC in additional to inclusive single particle spectrum. The effects of jet-medium interactions are observed in the broad away-side distributions ($\Delta\phi \sim \pi$) of associated hadrons in di-hadron azimuthal correlations and their softened transverse momentum ($p_T$) spectra [1, 2, 3, 4]. For associated hadrons with transverse momentum between $1.0 < p_T < 2.0$ GeV/$c$, in addition to the overall broadening of the away-side, a dip at $|\Delta\phi|$ $\pi$ is reported in [4, 5, 6]. Various scenarios and different mechanisms of jet-medium interactions have been proposed to explain the experimental observations [7, 8, 9, 10]. On the near-side ($\Delta\phi \sim 0$) modification of the correlation structure is also observed [2, 11, 12, 13]. In [12] the same side peak is described by two components: the ridge, a long-range $\Delta\eta$ plateau, and the small-$\Delta\eta$ jet-like peak. It is argued that the ridge might arise from jet-medium interactions. However, recent three-particle analyses in $\Delta\eta - \Delta\eta$ space has reported possible decoupling of the jet and ridge features [14]. On the other hand, the existing correlation analyses, even those three-particle ones, use a single high$-p_T$ trigger and do not specifically identify the away side jet, hence the pseudo-rapidity correlation relative to the away-side jet cannot be studied.

Di-hadron azimuthal correlations of two high-$p_T$ hadrons have been observed to exhibit jet-like peaks in both near- and away-sides [15, 16]. Little modification is apparent in the shape of the peaks, but the amplitude of the away-side peak shows a strong suppression in central Au+Au collisions with respect to d+Au collisions. The data may be interpreted in two scenarios: in-medium energy loss followed by in-vacuum fragmentation, and finite probability to escape the medium without interactions. As jet-medium interactions manifest themselves via shape changes of correlation structures predominantly in the softer momentum range (below 2 GeV/$c$ at RHIC), it is important to investigate the modification dynamics in the same events where back-to-back jets are already present. The analysis in this paper is using a new 3-particle (2 + 1) correlation technique, which measures the degree of correlation of a softer particle with a di-hadron trigger pair as proxy of di-jets. The technique utilizes a second high-$p_T$ trigger on the away-side of the highest-$p_T$ particle in the event as an estimate of the away-side jet axis direction. The asymmetry between energies of two triggers are varied as an attempt to control the path length each parton traveled in the medium, thus the difference between their final energy.

## 2. Analysis

### *2.1. Data selection*

The analysis presented in this paper used data collected by the STAR (Solenoidal Tracker at RHIC) experiment [17] in the years 2003-2004, and 2007-2008. Data samples consist of events from collisions with the center-of-mass per nucleon pair energy of 200 GeV, including 5.5M d+Au of year 2003, 14M minimum bias Au+Au and 18M central triggered Au+Au of year 2004, 1.1M high-energy BEMC (Barrel Electro-Magnetic Calorimeter) triggered Au+Au and 74M minimum bias Au+Au of year 2007, 6M high-energy BEMC triggered d+Au and 46M minimum bias d+Au of year 2008. Based on the charged track multiplicity at mid-rapidity ($|\eta| < 0.5$) recorded by the STAR Time Projection Chamber (TPC) [18] the minimum bias Au+Au data are divided into multiple centrality bins. They correspond to 0-20%, 20-40%, 40-60%, and 60-80% of the total geometric cross-section. The central-trigger data represent approximately 12% of the total cross-section. To express centrality dependences, the number of participating nucleons ($N_{part}$) and the number of binary collisions ($N_{coll}$) for each centrality bin used in the analysis are calculated via Monte-Carlo Glauber model. The charged tracks from TPC for the analysis have identical quality cuts as in [5].

For the main (first) trigger, the highest-$p_T$ particle in an event is selected, and is required to be $p_T^{trig1}$ between 5 to 10 GeV/$c$ as a charged track in TPC for year 2003-2004, or $E_T^{trig1}$ between 10 to 15 GeV as a BEMC cluster for year 2007-2008. The second trigger is always a charged track in TPC, and its $p_T^{trig2}$ is selected to be above 4 GeV/$c$. Two high-$p_T$ particles are considered to be a di-jet-like pair if they conform to the requirement of relative azimuthal difference of $|\phi_{trig1} - \phi_{trig2} - \pi| < \alpha$. The $\alpha$ is chosen to be 0.2 corresponding approximately to $1\sigma$ of the away-side peak width in relative azimuth for two high-$p_T$ hadrons [15], such that the contribution from random combinatorics is relatively low. The associated hadron distributions in $\Delta\phi$ and $\Delta\eta$ coordinates are reconstructed with respect to each trigger direction. The associated soft particle yield is measured relative to both triggers using the following expression

$$\frac{d^2N}{d\Delta\eta d\Delta\phi} = \frac{1}{N_{trig}\epsilon_{\text{pair}}}\left(\frac{d^2N_{raw}}{d\Delta\eta d\Delta\phi} - a_{zyam}\frac{d^2N_{Bg}}{d\Delta\eta d\Delta\phi}\right)$$

where $N_{trig}$ is the number of trigger pairs, corrected for a dilution effect due to randomly associated pairs, and $d^2N_{raw}/d\Delta\eta\, d\Delta\phi$ is the associated hadron distribution for all hadrons in triggered events. The associated hadrons $p_T$ (1.5 GeV/$c < p_T^{assoc} < p_T^{trig1}$) is selected to coincide with the range where the away-side dip and the near-side ridge in di-hadron correlations are reported. The "raw" distribution is corrected for single-track efficiency and acceptance effects. This correction factor ($\epsilon_{\text{pair}}$) includes single track reconstruction efficiency, which depends on $\eta$, $\phi$, $p_T$ and multiplicity. It also corrects for the pair acceptance obtained by the mixed-event technique. The item $\frac{d^2N_{Bg}}{d\Delta\eta d\Delta\phi}$ represents the total background, consisting of multiple terms. The dominant background contribution is from random combinatorics between trigger and associates, and is accounted for by mixing trigger pairs and associated hadrons from different events, selected to have similar multiplicity and primary vertex position to minimize geometry variance. The initially uniform distribution (apart from efficiency and acceptance effects) from mixing-events is then modulated by the flow term

$$f(\Delta\phi) = 1 + \frac{2v_2^{trig1}v_2^{assoc} + 2v_2^{trig2}v_2^{assoc}\frac{\sin(2\alpha)}{2\alpha}}{1 + 2v_2^{trig1}v_2^{trig2}\frac{\sin(2\alpha)}{2\alpha}}\cos(2\Delta\phi)$$

where $v_2^{trig1}$, $v_2^{trig2}$, $v_2^{assoc}$ are sample-averaged values of the elliptical flow coefficient for primary and secondary triggers, and associated hadrons, respectively, to account for elliptical flow effects, which are independent of $\eta$ in the range studied [19, 20]. The multiplicity and $p_T$ dependence of the elliptical flow coefficients ($v_2$) for triggers and associated particles are obtained from averaging the results from the Event Plane and Four-particle Cumulant methods [19]. The background due to randomly associated triggers in the initial selection of the trigger pairs is also considered [6]. The di-hadron correlations for the appropriate kinematic selections are used to determine the per-trigger correlated background for such pairs, and trigger-trigger correlation is used to estimate the ratio of such pairs to total trigger pairs. The overall background level $a_{zyam}$ is estimated with the Zero-Yield at Minimum (ZYAM) method [21, 6, 22]. For the 2+1 correlation the zero-yield region is chosen to be more than $3\sigma$ (corresponding to 1.3 radians) away from both jet-like peaks. A fit of double Gaussian plus a $v_2$-modulated background is applied to estimate the systematic error due to the ZYAM method. The transverse momentum spectra for associated hadrons is measured by selecting the hadrons within 0.5 radians in relative azimuth and 0.5 in relative pseudorapidity of the respective trigger direction, with same procedure of removing background.

### *2.2. Systematic errors*

Various sources of systematic uncertainties on measured yields are evaluated. Single track reconstruction efficiency contributes 5% overall normalization uncertainty for the associated hadron yields in year 2003-2004. As for year 2007-2008, the study is still going on and a very conservative 15% uncertainty is applied. The systematic uncertainty in the pair-acceptance correction due to finite statistics of the mixed-events is estimated to be less than 5%. The uncertainty due to the elliptical flow correction is estimated from the difference in the $v_2$ results from the Event-Plane and Cumulant methods. The uncertainty is found to be 5% in most central, less than 1% in most peripheral and 9% in mid-central Au+Au events. This uncertainty is not applicable to d+Au events. Systematic uncertainty due to ZYAM normalization of the background level was estimated by varying the $\Delta\phi$ range for minimum yield region, and is found to be ~10% in the Au+Au data and less than 5% in d+Au events. The related uncertainty is correlated for comparison between same- and away-sides. Systematic uncertainty due to correlated background subtraction from the di-jet sample is determined to be less than 3% for both d+Au and Au+Au events. This error is correlated between the same- and away-sides, and was estimated by varying background normalization for the trigger-trigger correlation in a similar manner as for 2 + 1 correlation. Finally, the uncertainty in the di-hadron distributions used for estimating the correlated background term arises mainly from the uncertainty in $v_2$ and ZYAM normalization. This is evaluated in a similar manner and estimated to vary the final result from less than 1% in d+Au events to about 5% in central Au+Au data. The errors above are strongly correlated between same- and away-side distributions, and mostly cancel out for such comparisons.

## 3. Result

### *3.1. Symmetric triggers*

Figure 1 shows the work from most recent STAR 2+1 correlation publication [25]. The analysis was based on data from year 2003-2004. The $\Delta\phi$ and $\Delta\eta$ projections (symmetrized about 0) of

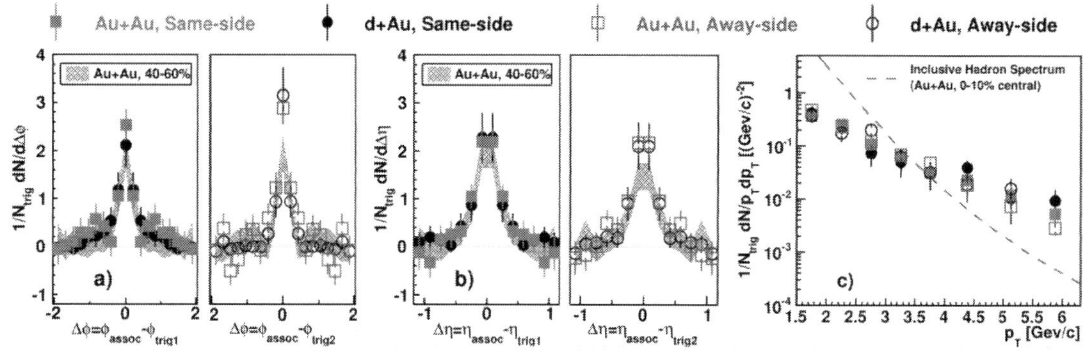

**Figure 1.** Projections of 2+1 correlation on $\Delta\phi$ ($a$) and $\Delta\eta$ ($b$) for 200 GeV top 12% central-triggered (squares) and mid-central (band) Au+Au and minimum bias d+Au (circles) data [25]. Errors shown are statistical. The kinematic selection is as follows: $5 < p_T^{trig1} < 10$ GeV/$c$, 4 GeV/$c$ $< p_T^{trig2} < p_T^{trig1}$, $1.5 < p_T^{assoc} < p_T^{trig1}$. $c$) Transverse momentum distributions per trigger pair for the same- and away-side hadrons associated with di-jet triggers ($|\Delta\phi| < 0.5$, $|\Delta\eta| < 0.5$). Errors are the quadrature sum of the statistical and systematic uncertainties. Inclusive charged hadron distribution 10% most central Au+Au data [24] is shown for comparison.

the correlation function are plotted, measured for central and mid-central Au+Au collisions and compared with the corresponding measurement from minimum bias d+Au data. The $\Delta\phi$ and $\Delta\eta$ projections in figure 1$a$ and 1$b$ reveal jet-like peaks on both the same- and away-sides. More important is that the observed structures from Au+Au data of all centralities are consistent within errors with the ones in d+Au collisions. This constitutes the first observation of not only the near-side, but also the away-side correlation structure in central Au+Au reproducing those of d+Au as a reference (without hot quark matter), in both $\Delta\phi$ and $\Delta\eta$, for the associated hadrons in this kinematic range. These results differ significantly from earlier di-hadron correlation observations, where substantial differences in both yields and shapes on both the same and away-sides were observed between Au+Au and d+Au data [5]. First, no evidence of the Mach-cone effect in the three particle (2+1) correlations studied is present in figure 1$a$. Second, the ridge properties is explored by looking at $\Delta\eta$ projections for di-jet triggered correlations for same-side as well as away-side triggers in figure 1$b$. The same-side integrated yields at $0.5 < \Delta\eta < 1.5$ are consistent with zero; however, statistical limitations ban exclusion of the ridge magnitudes reported in [12]. As a conclusion, the $2+1$ results provide no evidence of a "ridge" or "cone" associated with the direction of either of the high-$p_T$ triggers, consistent with lack of medium-induced effects on those di-jets selected by this analysis. The possible effects of radial flow on Mach-cone-like and ridge signals in di-hadron correlations have been studied in [23]. No such radial flow-related effects are expected for tangentially emitted trigger pairs.

The associated hadron spectra is further explored to complement the correlation shape analysis. Figure 1$c$ shows same- and away-side associated hadron yields plotted vs transverse momentum in d+Au and 12% central Au+Au collisions. Within the uncertainties no difference is observed between associated hadron spectra on same-side or away-side, Au+Au or d+Au data samples. This differs from earlier measurements from di-hadron correlations in a similar kinematic range, where significant softening of the away-side associated hadron spectra are shown. [2].

To check for the possible away-side softening which is indicative of energy deposition in the medium, the jet energy is estimated by summing the transverse momentum of trigger particle and the charged hadrons used for the spectra in the fiducial range, then compared between the same-side and away-side for d+Au and Au+Au data. Such difference of energy estimates between the same-side and away-side results in a major improvement of systematic uncertainties as no background re-normalization/subtraction is needed and uncertainties on elliptical flow cancel. The result in the 12% central Au+Au data $\Delta(Au + Au) = \Sigma(p_T)^{same} - \Sigma(p_T)^{away} = 1.59 \pm 0.19$ GeV/$c$, similar to the minimum bias d+Au data value of $\Delta(d + Au) = 1.65 \pm 0.39$ GeV/$c$. The initial state kinematic effects are expected to cause similar differences [26, 27], and calculated theoretically to be 1.6 GeV/$c$. Note this 1.6 GeV/$c$ is before partons lose energy into medium.

The absence of a jet quenching signal or a suppression of associated hadrons suggests that the $2 + 1$ particle triplets considered in this analysis are biased towards surface jet emission, i.e. di-jet events where both jets are emitted nearly tangentially to the medium surface. In such a scenario, one would expect that di-jet production rates are determined by the surface area of the fireball. This possibility is investigated by studying the centrality dependence of jet and di-jet production rates in 200 GeV Au+Au data.

### *3.2. Comparison with surface-bias model simulation*

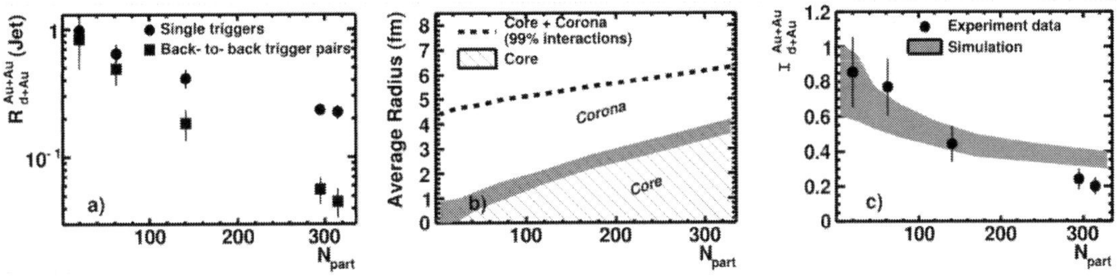


**Figure 2.** Comparison between data and model [25]. *a*) Relative production rates for jets and di-jets in year 2004 Au+Au collisions of different centralities with respect to year 2003 d+Au data. *b*) Calculations of the Glauber-based core/corona model that accommodates inclusive hadron suppression and single jet production rate results. *c*) Conditional di-jet survival probability in Au+Au data compared to d+Au reference. The band shows the expectations from Glauber-based core/corona model described in the text. Error bars in *a*) and *c*) are the quadrature sum of the statistical and systematic uncertainties.

Figure 2 is also from most recent STAR 2+1 correlation publication [25]. In figure 2$a$ shows the centrality dependence of the nuclear modification factors $R_{d+Au}^{Au+Au}$ (ratio of binary-scaled per-event trigger counts in Au+Au and d+Au data) for the primary (single) triggers and di-jet triggers. The observed $R_{d+Au}^{Au+Au}$ for single triggers is consistent with the suppression factors observed in inclusive charged hadron measurements [24], while correlated trigger pair production rates are suppressed even further.

To examine if these results are consistent with the purely tangential jet production scenario, a Monte Carlo Glauber calculations is applied to find the spatial distribution of hard scattering were performed for each centrality bin. Here a simplistic scenario is assumed where the interaction zone in heavy-ion collisions consists of a completely opaque core (full jet attenuation) surrounded by a permeable corona (no jet-medium interactions). Participant eccentricity for the medium core shape has been used. This has been found most suitable for elliptical flow results [29]. Energy loss fluctuations as discussed in [28] have not been taken into account. The size of the core is estimated directly from the $R_{AA}$ suppression measurements [30, 24] and the calculated core eccentricity. It is important to note that this $R_{AA}$ is the only input of data to tune this MC-model. The model then calculates the di-jet rate and $R_{d+Au}^{Au+Au}$ on its own as predictions, and compare with the experimental data to test its validity. In the tuning of core/corona size, the total size of the interaction area is calculated by requiring the integrated collision density inside this area to be 99% of the total $N_{coll}$. It is further assumed that all observed trigger hadrons come from the corona. Core/corona sizes that accommodate the inclusive hadron suppression results are shown in figure 2*b*. The gray band shows propagation of uncertainties in the published $R_{AA}$ values into our calculations.

In figure 2*c* displays the conditional di-jet survival rates. The expected rates from core/corona model are shown as a band, where the width reflects the uncertainty in the published $R_{AA}$ data. The conditional di-jet survival probability in the data are shown as points, which indicate the double ratio $I^{Au+Au}_{d+Au} = \frac{R_{AA}(triggerpairs)}{R_{AA}(singletriggers)}$. Both data and MC-model calculations reflect any changes in probability to find an away-side trigger for each primary trigger in Au+Au data relative to d+Au, i.e., the conditional di-jet survival probability. For example, at 12% most central Au+Au data such probability is approximately 20 % $\pm 2(stat) \pm 4.5(syst)$ compared to d+Au, as is shown the symbols in figure 2*c* for comparison with the above model. However, it must be pointed out that core emission where neither of the di-jets interacted with the medium cannot be ruled out by this analysis.

### *3.3. Asymmetric triggers*

While surface emission model works well with the experimental data, it is important to understand how path-length dependent energy-loss models adapt to the 2+1 correlation measurement. In a scenario where in-medium energy loss followed by in-vacuum fragmentation, selecting trigger pairs of similar $p_T$ ($> 5GeV/c$ vs. $> 4GeV/c$) can possibly bias towards partons losing similar energy, assuming both near-/away-side triggers from not fully surface biased partons. Since both partons travel similar path length of medium then fragment in the medium, they reach similar energy loss or survival probability. This argument can be tested by picking up those trigger pairs of high and asymmetric $p_T$ cuts, as proxy of di-jets which have big asymmetry in energy loss, and measure the relative difference between Au+Au and d+Au for comparison of near-/away-sides.

Such analysis is done from STAR year 2007 Au+Au and 2008 d+Au data. Benefited from the much improved coverage of BEMC since year 2007, the first trigger is now selected as those high-energy BEMC tower clusters, whose transverse energy $E_T$ can be pushed to very high region of $10 < E_T^{trig1} < 15$ GeV, while still maintaining a good energy resolution. The triggered events are mostly within 0-20% centrality in Au+Au. They are mainly $\pi^0$s because of the big coverage of single tower 0.5x0.5 in $\Delta\eta$ by $\Delta\phi$ space that include both decay photons, while there can be up to 30% contribution from direct photons of same $E_T$ [31] in most central Au+Au. On the other hand, such direct gamma contamination won't affect the validity of asymmetric trigger test by comparing the summed energy on both trigger sides, since the direct gamma is expected to not losing energy in the medium. Thus away-side parton has to lose more energy and make away-side summed energy (trigger and associates) smaller if path-length dependence stands. The second trigger is still selected to be charged tracks of 4 GeV/$c < p_T^{trig2} < 10$ GeV/$c$, so as to satisfy large $\Delta p_T$ between two triggers.

As is shown in the figure 3 and 4, even with such asymmetric trigger selections, the correlation functions shown evident jet-like structures, and they are still of similar shape and magnitude from d+Au to central Au+Au collisions, at both same-/away-sides. The systematic errors previously discussed are not shown on the plots. While they strongly correlated between near-/away-sides, these errors could affect the relative difference of both trigger sides between Au+Au and d+Au. No evident ridge or dip structure is observed at near-/away-side of Au+Au data. The away-side magnitude is higher than near-side in both Au+Au and d+Au, and this difference is possibly due to the direct-gamma contamination. On the other hand, if the surviving di-jets with their jet-like structure sustained mainly come from partons losing energy or escaping from deep within the medium, picking up asymmetric trigger pairs would sample asymmetric parton (jet) energies,

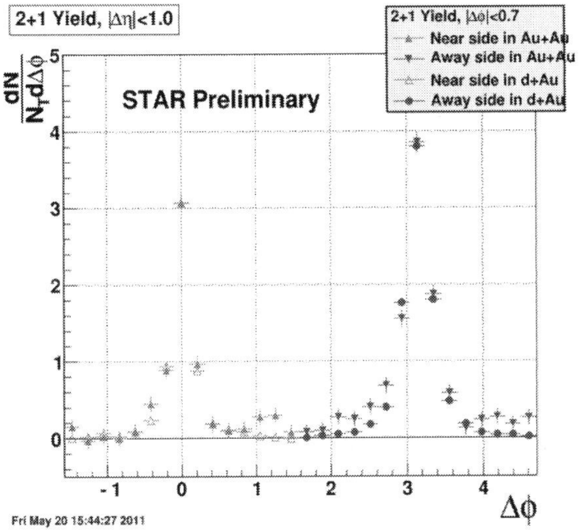


**Figure 3.** Projections of 2+1 correlation on $\Delta\phi$ for 200 GeV high-energy BEMC tower triggered Au+Au (solid) and d+Au (open) data. Errors shown are statistical. The kinematic selection is as follows: $10 < E_T^{trig1} < 15$ GeV, 4 GeV/$c$ < $p_T^{trig2} < 10$ GeV/$c$, $1.5 < p_T^{assoc} < 10$ GeV/$c$.

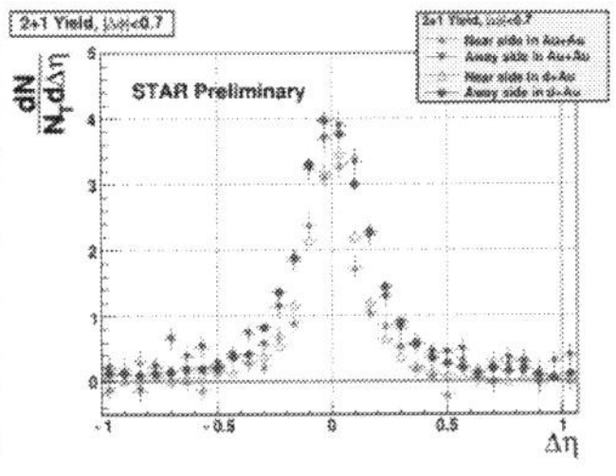


**Figure 4.** Projections of 2+1 correlation on $\Delta\eta$.

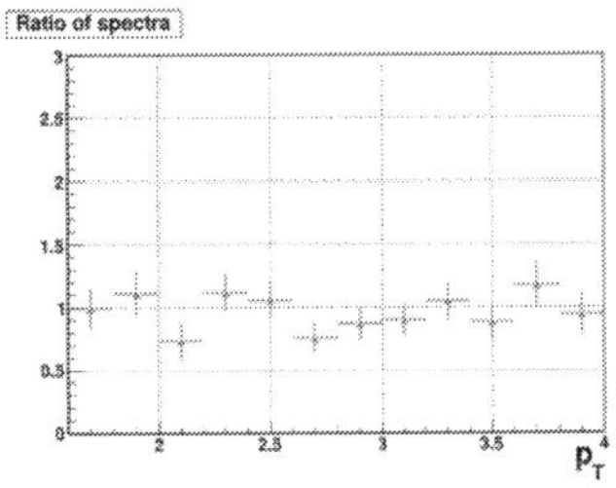


**Figure 5.** Near-side spectra of Au+Au divided by d+Au.

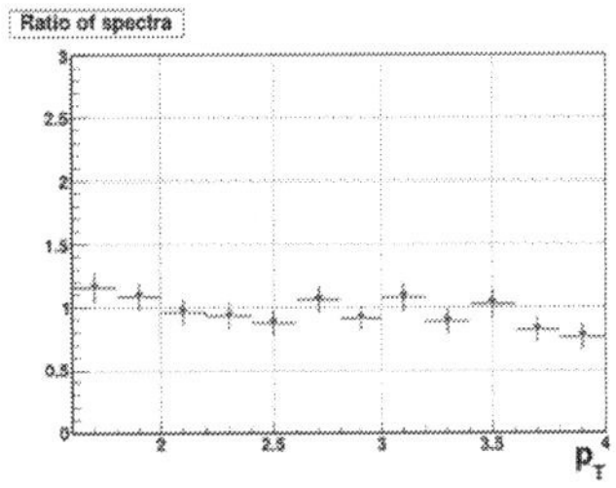


**Figure 6.** Away-side spectra of Au+Au divided by d+Au.

i.e., different levels of surface-bias. By comparing such difference between Au+Au and d+Au, the measurement doesn't indicate the asymmetry in energy loss of parton pairs expected from path-length dependence model [26].

Also noted is that even with the conservative systematic errors we assigned, most of them are strongly correlated between near-/away-sides. That is, even if we did shift either Au+Au or d+Au points by 20%, any strong asymmetry between near-/away-sides of Au+Au relative to d+Au won't be hidden.

The spectrum of associates at either trigger side is further analyzed to explore the jet-like structures. Same as Figure 1c, associated hadrons are selected within 0.5 radians in $\Delta\phi$ and 0.5 in $\Delta\eta$ of the respective trigger. Figure 5 and 6 shows the ratios of such spectrum of Au+Au divided by d+Au. The ratio is consistent with flat and close to unity at either trigger side, showing no evidence of softening or strong suppression. This measurement further supports the assumption of strong surface-bias whenever a jet-like structure is observed.

## 4. Summary

To summarize: the mechanisms of jet-medium interactions in ultra-relativistic heavy ion collisions was investigated using a novel technique involving three particle $(2+1)$ correlations of hadrons associated with a correlated pair of back-to-back high $p_T$ particles. Both same-side and away-side peaks of di-jet triggered correlations from central Au+Au data are found similar to the structures observed in d+Au collisions at the same energy, and same in multiple centrality bins of minimum bias Au+Au data The associated hadron spectra on each side of a di-jet primary trigger are also found to be of similar, in shapes and integrated yields, in central Au+Au data between same- and away -sides, and are consistent with those measured in d+Au data. Also no modification is observed in correlations of associated hadrons in the momentum range of $p_T^{assoc} > 1.5$ GeV/$c$ with the kinematic selection studied. All these are in contrast with earlier measurements with a single trigger. Systematic assessment of di-jet production rates supports tangential emission bias in a simplistic core/corona scenario, showing no evidence for di-jets interacting with the core. For the purpose of complement, the measurement was done in both cases of symmetric to very asymmetric trigger pairs, to test the possible asymmetry of energy loss predicted by path-length dependence models. No such asymmetry is observed at this stage within the errors.

We thank the RHIC Operations Group and RCF at BNL, the NERSC Center at LBNL and the Open Science Grid consortium for providing resources and support. This work was supported in part by the Offices of NP and HEP within the U.S. DOE Office of Science, the U.S. NSF, the Sloan Foundation, the DFG cluster of excellence 'Origin and Structure of the Universe'of Germany, CNRS/IN2P3, FAPESP CNPq of Brazil, Ministry of Ed. and Sci. of the Russian Federation, NNSFC, CAS, MoST, and MoE of China, GA and MSMT of the Czech Republic, FOM and NWO of the Netherlands, DAE, DST, and CSIR of India, Polish Ministry of Sci. and Higher Ed., Korea Research Foundation, Ministry of Sci., Ed. and Sports of the Rep. Of Croatia, and RosAtom of Russia.

## References

[1] J. Adams *et al.*, *Phys. Rev. Lett.* **91** (2003) 072304
[2] J. Adams *et al.*, *Phys. Rev. Lett.* **95** (2005) 152301
[3] S. S. Adler *et al.*, *Phys. Rev. Lett.* **91** (2003) 072301
[4] A. Adare *et al.*, *Phys. Rev.* **C 78** (2008) 014901
[5] M. M. Aggarwal *et al.*, *Phys. Rev.* **C 82** (2010) 024912
[6] B. Abelev *et al.*, *Phys. Rev. Lett.* **102** (2009) 052302
[7] H. Stoecker, *Nucl. Phys.* **A 750** (2005) 121
[8] J. Casalderrey-Solana, E. Shuryak and D. Teaney, *Nucl. Phys.* **A 774** (2006) 577
[9] J. Ruppert and B. Muller, *Phys. Lett.* **B 618** (2005) 123
[10] V. Koch, A. Majumder and X. N. Wang, *Phys. Rev. Lett.* **96** (2006) 172302
[11] J. Putschke, *AIP Conf. Proc.* **842** (2006) 119
[12] B. Abelev *et al.*, *Phys. Rev.* **C 80** (2009) 064912
[13] J. Adams *et al.*, *Phys. Rev.* **C 73** (2006) 064907
[14] B. Abelev *et al.*, *Phys. Rev. Lett.* **105** (2010) 022301
[15] J. Adams *et al.*, *Phys. Rev. Lett.* **97** (2006) 162301
[16] A. Adare *et al.*, *Phys. Rev. Lett.* **104** (2010) 252301
[17] K.H. Ackermann *et al.*, *Nucl. Inst. Meth.* **A499** (2003) 624
[18] K.H. Ackermann *et al.*, *Nucl. Phys.* **A661** (1999) 681
[19] J. Adams *et al.*, *Phys. Rev.* **C 72** (2005) 014904
[20] B. B. Back *et al.*, *Phys. Rev. Lett.* **94** (2005) 122303
[21] C. Adler *et al.*, *Phys. Rev. Lett.* **90** (2003) 082302
[22] A. Adare *et al.*, *Phys. Rev. Lett.* **98** (2007) 232302
[23] C. Pruneau, S. Gavin and S. Voloshin, *Nucl. Phys.* **A 802** (2008) 107
[24] J. Adams *et al.*, *Phys. Rev. Lett.* **91** (2003) 172302
[25] H. Agakishiev *et al.*, *Phys. Rev.* **C 83** (2011) 016901
[26] T. Renk *Phys. Rev.* **C 78** (2008) 014903
[27] T. Renk and K. Eskola, *Phys. Rev.* **C 75** (2007) 054910
[28] T. Renk, *Phys. Rev.* **C 74** (2006) 024903
[29] B. Alver *et al.*, *Phys. Rev.* **C 77** (2008) 014906
[30] S. S. Adler *et al.*, *Phys. Rev. Lett.* **91** (2003) 072301
[31] S. S. Adler *et al.*, *Phys. Rev. Lett.* **94** (2005) 232301

# The PHENIX Decadal Plan: Crafting the Future of RHIC

**Christine A. Aidala for the PHENIX Collaboration**
Los Alamos National Laboratory, Los Alamos, NM 87545
E-mail: caidala@bnl.gov

**Abstract.** After ten years of running, the PHENIX Collaboration is starting to lay out the compelling physics that RHIC, as the most versatile hadron collider in the world, will be well positioned to explore beyond the program planned for upgrades already in progress. As the fields that RHIC set out to investigate have advanced and evolved, new questions and directions have arisen at the frontiers of QCD, and we have exciting opportunities before us to continue to confront the challenges and surprises of strong interactions into the next decade. The latest thoughts of the PHENIX Collaboration on this ongoing planning process for the future of RHIC are discussed.

## 1. Studying QCD at RHIC

The Relativistic Heavy Ion Collider (RHIC), as the most versatile hadron collider in the world, has provided an outstanding facility for the study of quantum chromodynamics (QCD) since it began taking data in 2000. Its independent rings allow collisions of a variety of symmetric and asymmetric nuclear species, and past RHIC runs have collided deuterons on gold, copper on copper, and gold on gold. As the world's only polarized proton collider, RHIC has also achieved polarized proton-proton collisions over energies ranging from $\sqrt{s} = 62.4$ GeV to 500 GeV.

These different colliding systems allow the study of a variety of QCD matter. The proton is the simplest stable bound state in QCD; investigating cold nuclear matter reveals how nuclei cannot merely be described as groups of protons and neutrons but rather the roles that collective effects play; the QCD matter formed by central heavy ion collisions, strongly interacting quark-gluon plasma (sQGP), is a deconfined state of QCD matter, too hot for hadronic bound states to exist. Just as many areas of experimental condensed matter research seek to isolate or create a physical system and study its properties, such is the nature of the various programs at RHIC. RHIC was built with a maximum collision energy for heavy ions comfortably above the energy needed to create deconfined QCD matter, and with great flexibility to explore a range of energies below this. Given the complexity of nuclear collisions and QCD systems in general, the RHIC facility was designed from the beginning with $p+p$, $d+A$, and $A+A$ collisions at the same center-of-mass energy per nucleon all as integral parts of the program, enabling a comprehensive program in QCD physics and permitting more complex systems to be understood in terms of simpler ones.

After the tremendous success of the first decade of running RHIC, the community is now looking toward the next ten years. A Decadal Planning document, which represents only a first step in the ongoing process of planning for the future of the facility, was submitted by the

PHENIX Collaboration to BNL Management in September 2010 [1, 2]. As the several fields under study at RHIC have advanced and matured over the past decade, new physics questions have come to the forefront as we start to move into more quantitative investigation of QCD physics.
Questions relevant to the hot, dense QCD matter program at RHIC include:

- Are quarks strongly coupled to the QGP at all distance scales?
- What are the detailed mechanisms for parton-QGP interactions and responses?
- Are there quasiparticles at any scale?
- Is there a relevant screening length in the QGP?
- How is rapid equilibration achieved?

The study of QCD processes in cold nuclear matter (CNM) centers on the following fundamental questions:

- What are the dynamics of partons at very small and very large momentum fraction ($x$) in nuclei, and at high gluon density what are the nonlinear evolution effects (i.e., saturation)?
- What are the pQCD mechanisms that cause energy loss of partons in CNM, and is this intimately related to transverse momentum broadening?

Questions to be addressed mainly within the proton-proton program at RHIC include:

- What is the 3D spatial structure of the nucleon?
- What is the nature of the spin of the nucleon?
- What spin-momentum correlations exist within hadrons and in the process of hadronization?
- What is the role of color interactions in different processes?

Improved and extended machine and detector capabilities will enable us to more fully address these questions over the course of the next decade. An approximate timeline indicating the physics topics, detector upgrades, and accelerator upgrades at RHIC over the next decade is shown in Fig. 1.

## 2. An Even More Flexible Machine for the Future

Quantitative understanding will develop from having a variety of measurements to compare, and upgrades in place or under development for RHIC will enable an even wider range of systems to be studied.

The electron beam ion source (EBIS) upgrade, already available as of 2011, has replaced the Tandem Van de Graaff accelerators as the heavy ion pre-injector for RHIC and will provide even greater flexibility in the choice of nuclear species to accelerate, regardless of their atomic electronegativity. Uranium-uranium collisions, which will create an even higher energy density than gold-gold collisions, are planned for the near future as a first step in exploring these new capabilities.

Research and development are underway to study the prospects for polarized beams of helium-3 at RHIC, which would serve as effective polarized neutron beams, complementing the polarized proton beams already available.

Discussions are ongoing regarding the possibility of replacing the RHIC DX magnets to achieve energies up to $\sqrt{s}$ = 650 GeV for $p+p$ collisions, or $\sqrt{s}$ = 260 GeV per nucleon for Au+Au collisions. While increasing the center-of-mass energy somewhat could be useful in order to increase the production rate of certain probes (e.g. $W$ bosons, heavy flavor), the parton momentum fraction ($x$) range of interest to the nucleon structure program is $x \sim$0.0001–0.5, and many intriguing effects seem to lie in the region $x \gtrapprox 0.01$, so greatly increased center-of-mass energies would be undesirable. In the case of the heavy ion program, moderately higher energies

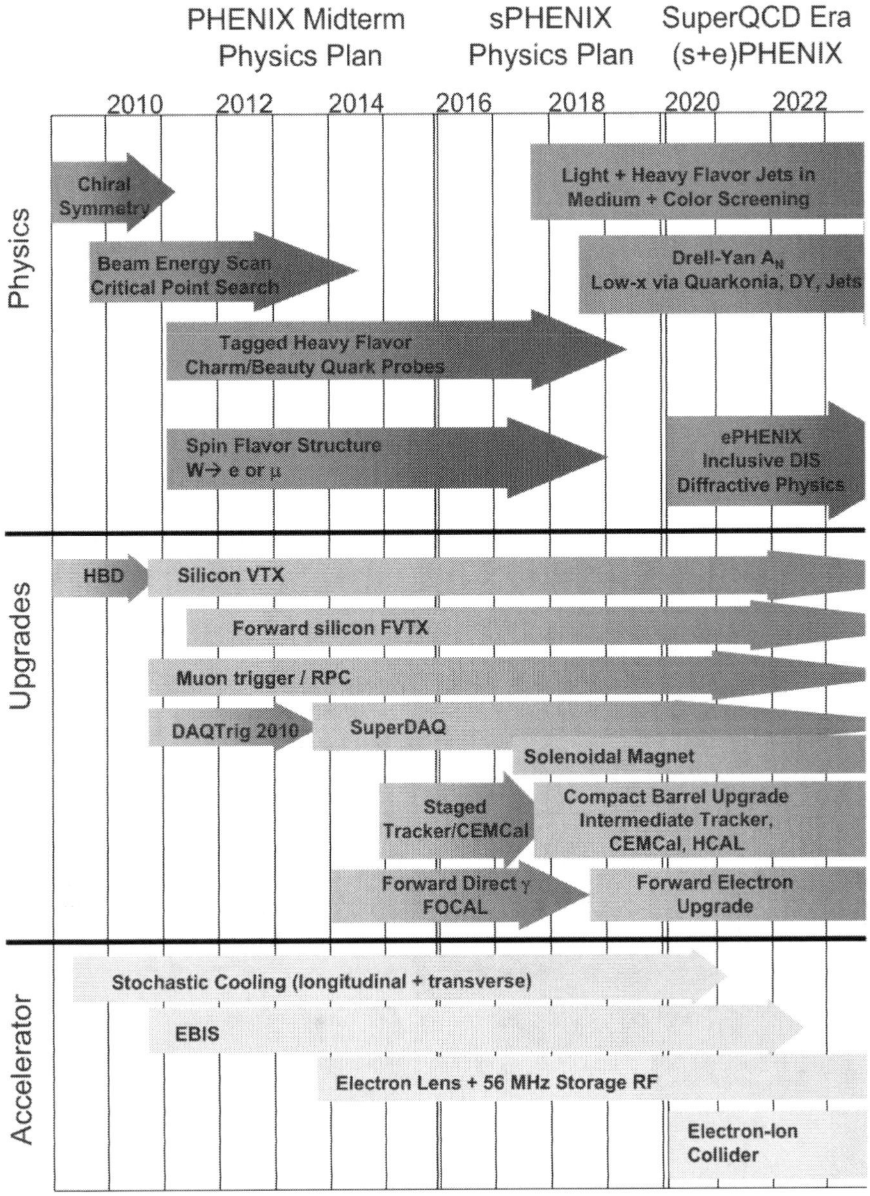


**Figure 1.** Approximate timeline indicating the physics topics, detector upgrades, and accelerator upgrades at RHIC over the next decade, taken from [2].

would again increase the production rate of certain probes, but early data from the Large Hadron Collider (LHC) indicate that even at the significantly higher center-of-mass energies of 2.76 TeV per nucleon achieved for lead-lead collisions there, the deconfined QCD matter created is qualitatively the same as at RHIC.

### 2.1. *Adding an Electron Ring to RHIC*

The field of heavy ion physics already takes advantage of final-state probes that do not interact strongly, such as direct photons and internal conversions of thermal photons, as well as $Z$ bosons at the LHC. By adding an electron ring to RHIC, in the future it will be possible to select electroweak or colored probes not only in the *final* state but also the *initial* state. The availability of an electron and possibly a positron beam at RHIC will open up a wealth of new possibilities. Measuring the energy and angle of the scattered electron allows one to reconstruct the kinematics of the scattered parton. A lepton beam also provides a very clean probe of the initial hadronic or nuclear state, without the complications of strong interactions.

Electron-nucleus collisions at RHIC will further extend the existing battery of simpler collision systems regularly run at RHIC which aid in our understanding of complex heavy ion interactions. As illustrated in the left panel of Fig. 2, the modification of the gluon distribution function in heavy nuclei is extremely uncertain, with depletion factors ranging from $\sim$10% to nearly no gluons at $x \simeq 5 \times 10^{-3}$. This uncertainty is due largely to the limited $Q^2$ range of earlier (fixed-target) nuclear deep-inelastic scattering (DIS) and Drell-Yan experiments, although inclusive pion production data from $d$+Au and $p+p$ collisions at RHIC are also included in recent global analyses of nuclear pdfs such as [3]. High-energy, high-precision $e + A$ data for a range of nuclei would greatly reduce uncertainties on the initial-state gluon distribution, and further $d$ or $p+A$ data with multiple nuclear species would provide some sensitivity to the impact-parameter dependence of the nuclear pdfs as well as confirm and test our understanding of $e+A$ data. The kinematic coverage in $x$ and $Q^2$ for lepton beam energies of 10 and 20 GeV at RHIC compared to previous nuclear DIS experiments is given in Fig. 3.

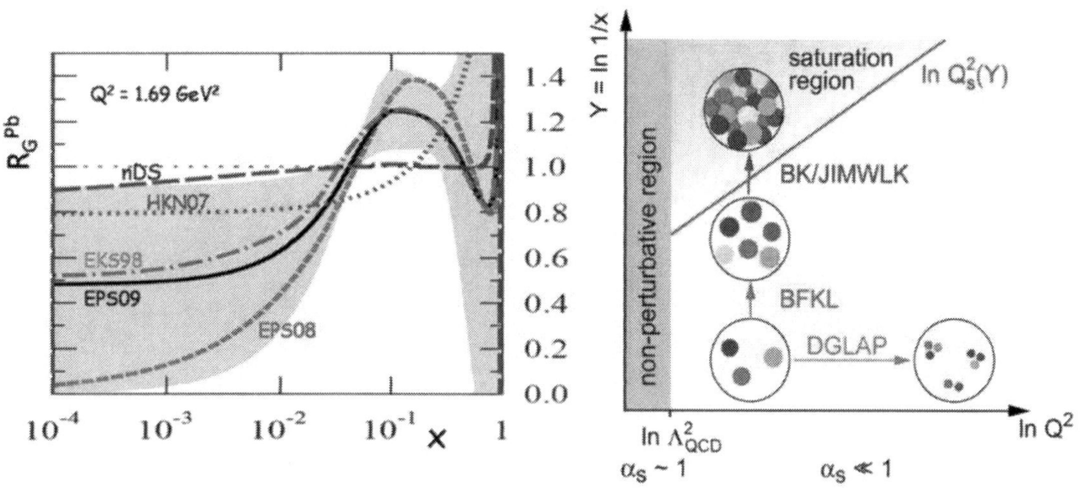


**Figure 2.** (left) Nuclear modification of gluons from a global analysis of deep-inelastic scattering (DIS), Drell-Yan, and PHENIX hadron data by the EPS09 group [3]. (right) Schematic diagram of the different regions in the momentum transfer ($Q^2$) vs momentum fraction ($x$) space, showing the gluon saturation region at large $\ln 1/x$ and small $\ln Q^2$.

The problem of unitarity and the role of nonlinear evolution of parton distributions at low $x$, generally referred to as gluon saturation, are among the key outstanding problems in QCD. Saturation models predict a regime where diagrams in which two low-momentum gluons fuse into one higher-momentum gluon become significant, thus depleting the low-momentum gluons. These effects are present in a nucleon, but become much stronger ($\propto A^{\frac{1}{3}}$) in a nucleus. Recent saturation models assert that a novel semiclassical state, the color glass condensate (CGC), is formed above a critical saturation scale ($Q_s^2$) at low enough momentum [4]. The right panel of Fig. 2 shows a conceptual diagram of the CGC picture. However, the $x$, $Q^2$ region where saturation physics plays a significant role is not yet well constrained. In conjunction with continued $p(d) + A$, and $A + A$ collisions, electron-nucleus collisions could provide detailed information on a gluon saturation regime in nuclei.

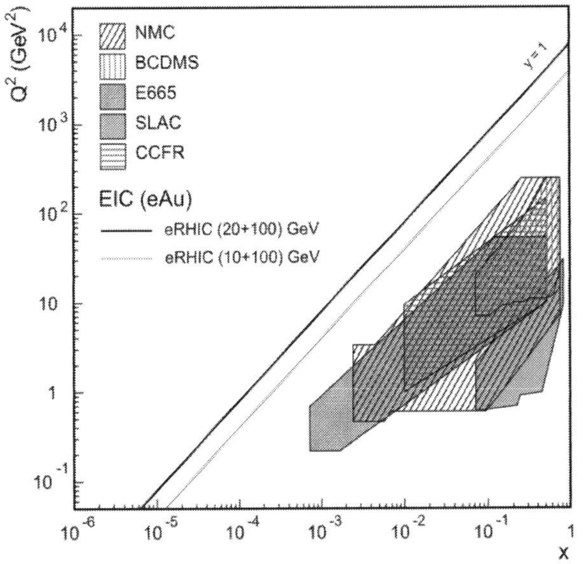


**Figure 3.** The $x - Q^2$ coverage of electron-nucleus collisions at RHIC for two beam-energy combinations compared to different fixed-target experiments.

Through diffractive measurements in electron-nuclear collisions, nuclear density distributions as a function of impact parameter can be studied. Diffraction patterns measured in the processes of coherent vector meson production and deeply-virtual Compton scattering may be inverted to determine the initial nuclear gluon and quark densities. The most sensitive density determination is at the edge of the nucleus, so these measurements will provide key information about the overlap region to which hydrodynamics and energy loss calculations are most sensitive. The Feynman diagram contributing to coherent vector meson production, in which an electron emits a virtual photon that splits into a $q\bar{q}$ pair and subsequently interacts coherently with the nucleus to form the vector meson, is shown in Fig. 4, and Fig. 5 shows vector meson diffraction patterns for gold. The differential cross section for a gluon density assumed to scale with the Woods-Saxon charge distribution with the usual parameters taken for the radius and width is compared to the pattern for a Woods-Saxon distribution with a 10% smaller radius and 20% smaller width. It can be seen that the measured coherent diffraction pattern is very sensitive to the details of the gluon density. Also shown in the figure is the dominant contribution to the cross section coming from incoherent scattering, in which the outgoing nucleus is either broken up or in an excited state.

Proton-proton collisions at RHIC will remain important even after $e + p$ capabilities are in place. The $p + p$ data will serve not only as a simpler system for comparison to heavy ion data, but also for continued study of proton structure, in particular its *color* structure. Study of the color structure of the nucleon and color interactions in hadronic collisions is a nascent field in QCD, for example with predictions of modified universality of certain (T-odd) $k_T$-unintegrated distribution functions due to different color interactions in semi-inclusive deep-inelastic lepton-nucleon scattering versus the Drell-Yan process [5].

Journal of Physics: Conference Series **316** (2011) 012017 doi:10.1088/1742-6596/316/1/012017

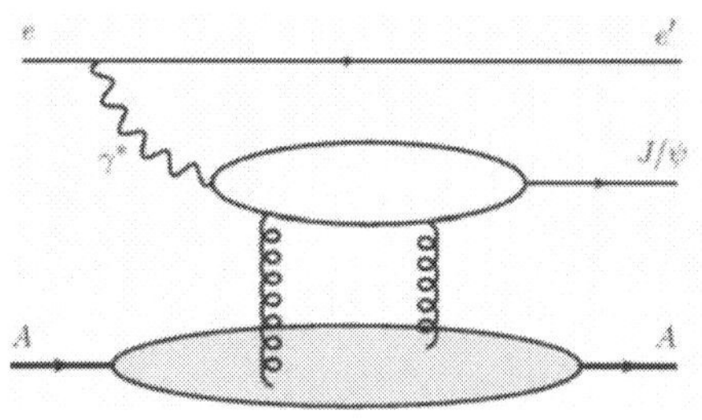


**Figure 4.** Relevant Feynman diagram for diffractive vector meson production (here the $J/\psi$) in $e + A$. The $q\bar{q}$ dipole probes the gluon density in the nucleus $A$ with two gluons: not only can vector meson production probe the mean gluon density in the nucleus but also gluon density correlations.

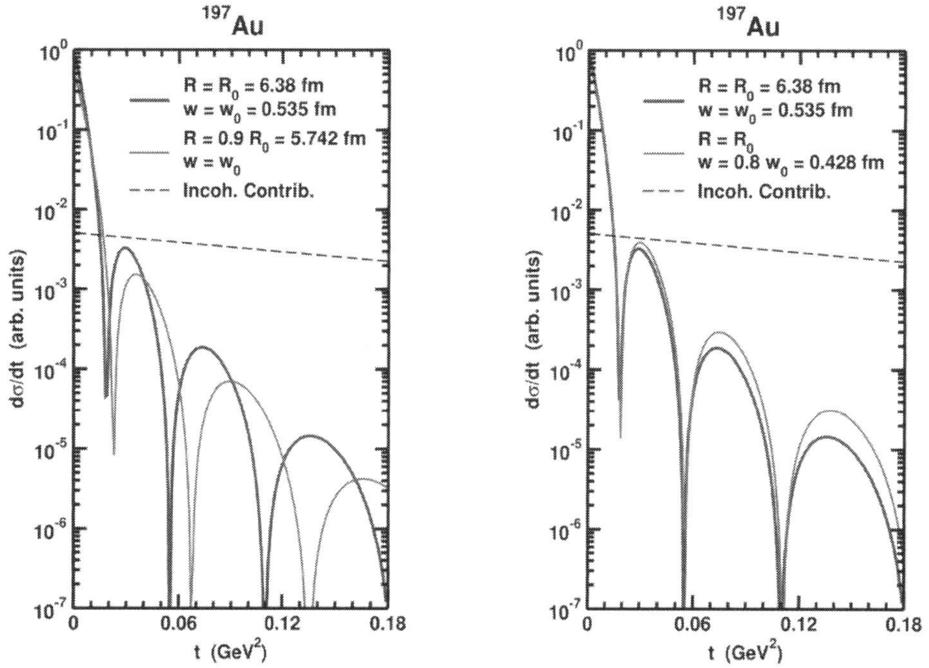


**Figure 5.** Vector meson diffraction patterns for $^{197}$Au. (blue curves) Differential cross section when the gluon density is assumed to scale with the Woods-Saxon charge distribution with the usual parameters taken for the radius and width. Shown are comparisons for a Woods-Saxon distribution with a (left) 10% smaller radius and (right) 20% smaller width. Both plots also show the dominant contribution to the cross section due to incoherent vector meson production.

## 3. The sPHENIX Detector

The different physics questions of interest can be addressed by a variety of experimental observables, and these observables will determine the design of any upgraded detector at RHIC. Figure 6 illustrates heavy ion physics questions expected to remain open once the science program with currently planned or newly commissioned upgrades is completed around 2015. It indicates their correspondence to a variety of observables and the associated detector requirements.

Shown in Fig. 7 is a schematic design for an upgraded detector based on the requirements itemized in Fig. 6. The basic philosophy is to design a compact, uniform, nearly hermetic detector with fast rate capabilities and excellent full jet reconstruction at both mid- and forward rapidities. The recently installed VTX silicon detector at midrapidity as well as the FVTX forward silicon detector, to be installed by the end of 2011, are retained. Hadronic calorimetry at midrapidity provides improved jet energy resolution compared to tracking and

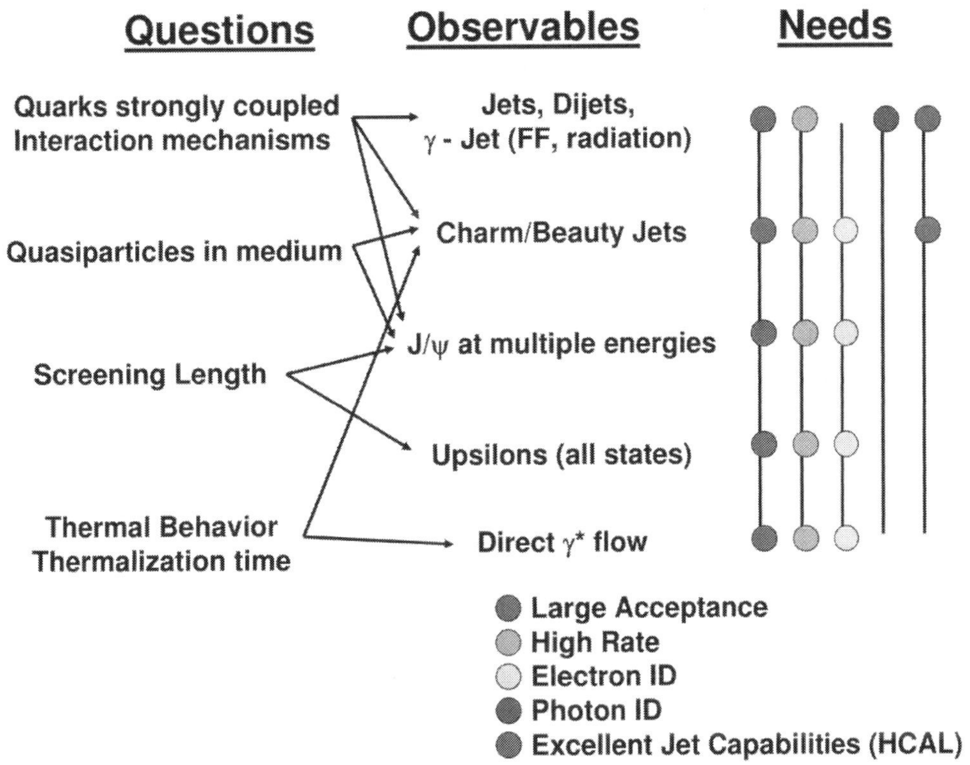


**Figure 6.** A chart illustrating the interesting physics questions after ~2015, their relationship to final-state observables, and the detector requirements needed for these measurements.

electromagnetic calorimetry alone, and it increases the efficiency of jet reconstruction and triggering in order to extend measurements to high transverse momenta, as illustrated in Fig. 8. A forward spectrometer covering up to $\eta \approx 4$ with tracking, particle identification, and both electromagnetic and hadronic calorimetry is envisioned. Electromagnetic calorimetry up to large rapidities in the direction of the electron beam will be crucial to perform measurements in $e+p$ and $e+A$ collisions. A large rapidity spectrometer will also be key for the study of asymmetric nuclear collisions as well as the investigation of many of the striking effects observed at high Feynman-$x$ related to parton dynamics in the nucleon. One of the current PHENIX muon arms, not shown in Fig. 7, would remain.

## 4. Conclusion

As the challenge of QCD continues, RHIC is well positioned to become an even more powerful tool to fulfill advancement to a quantitative era in QCD going into the 2020s. As a dedicated QCD facility with capabilities for both electroweak as well as hadronic and nuclear collisions over a wide range of energies and system sizes, and with control over beam polarization for protons and possibly helium-3, the RHIC complex and its detectors would allow access to an impressive breadth of QCD physics. Work is ongoing within the community to realize such a future.

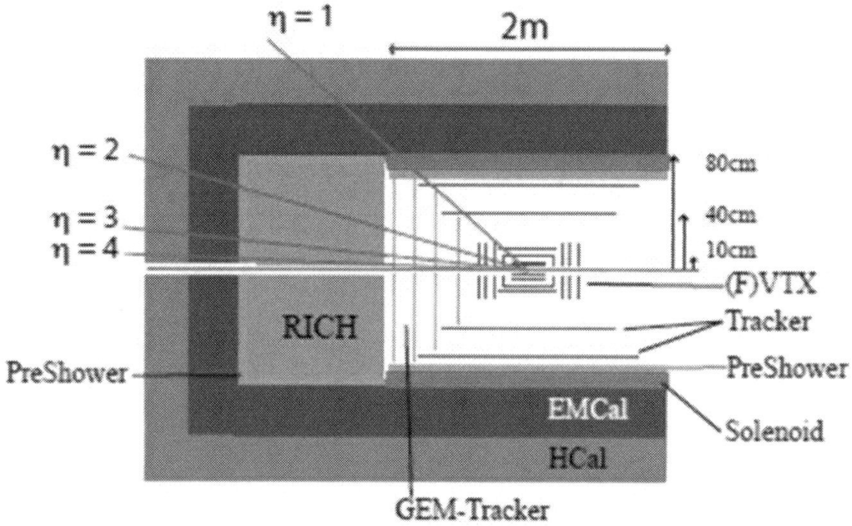


**Figure 7.** A schematic layout of the sPHENIX detector.

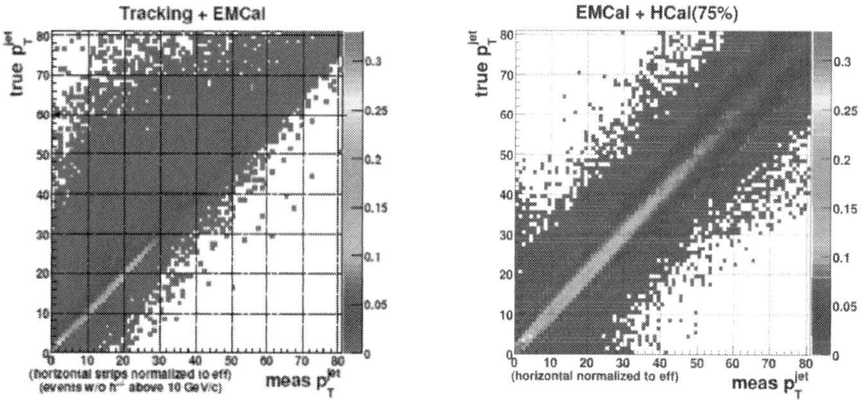


**Figure 8.** Fast simulation response matrices demonstrating the improved correlation between measured and true $p_T$ of jets when tracking (left) is replaced by hadronic calorimetry with $\sigma_E/E$ =75% (right).

## References


[1] Brookhaven National Laboratory nuclear and particle physics website `http://www.bnl.gov/npp/`

[2] PHENIX decadal plan `http://www.phenix.bnl.gov/phenix/WWW/docs/decadal/2010/phenix_decadal10_full_refs.pdf`

[3] Eskola K, Paukkunen H and Salgado C 2009 *JHEP* **0904** 065 (*Preprint* 0902.4154)

[4] Gelis F, Iancu E, Jalilian-Marian J and Venugopalan R 2010 (*Preprint* 1002.0333)

[5] Collins J C 2002 *Phys.Lett.* **B536** 43–48 (*Preprint* hep-ph/0204004)

# Jet Physics in pp and PbPb Collisions with the CMS Experiment

**Edward Wenger, for the CMS Collaboration** [1]
Massachusetts Institute of Technology, Cambridge, Massachusetts, United States
E-mail: edwenger@mit.edu

**Abstract.** The properties of dijets are reviewed for pp and PbPb collisions recorded by the CMS experiment. In particular, a dramatic quenching of jets is observed as a significant imbalance in the transverse momenta of dijets in central PbPb collisions at a center-of-mass energy of 2.76 TeV. The redistribution of the quenched jet energy is studied using the transverse momentum balance of charged tracks projected onto the direction of the leading jet. In contrast to pp collisions, a large fraction of the momentum balance for asymmetric jets is carried by low momentum particles at large angular distance to the jet axis.

## 1. Introduction

Fully-reconstructed dijets are a powerful tool for tomographically probing the hot, dense medium produced in ultra-relativistic collisions of heavy ions. By quantifying the pattern of partonic energy loss and the redistribution of energy in the medium, one can access experimentally the properties of the medium, e.g., the gluon density ($dN_g/dy$) or the transport coefficient ($\hat{q}$).

The energy-loss mechanism has been studied already at RHIC in a more indirect fashion, namely measurements of two-particle angular correlations between charged particles. For charged particles with moderate transverse momentum ($p_T^{trig}$ = 4–6 GeV/$c$, $p_T^{assoc}$ = 2–4 GeV/$c$), the characteristic away-side jet correlation that is prominent in pp collisions is nearly extinguished in central PbPb collisions [1]. While this observation represents strong indirect evidence of jet quenching at RHIC, direct jet reconstruction [2, 3] — especially of the most strongly quenched dijets — is challenging due to fluctuations in the soft underlying event.

At $\sqrt{s_{NN}} = 2.76$ TeV, copious production of high-$E_T$ jets above 100 GeV gives a much larger separation between the hard and soft scales. Even looking at event displays in the first days of the 2010 PbPb run, a significant imbalance was apparent in reconstructed dijets [4, 5]. For the first time, it is possible to reconstruct strongly modified dijets unambiguously above the soft background (see Fig. 1), marking the beginning of a new era of more powerful studies of jet quenching at the LHC.

## 2. Analysis Methods

The dijet analysis presented in these proceedings was performed using the data collected in 2010 from PbPb collisions at a nucleon-nucleon center-of-mass energy of $\sqrt{s_{NN}} = 2.76$ TeV at

[1] The complete CMS collaboration is listed in a recent submission to *Phys. Rev. C* [4], from which the text of these proceedings is largely excerpted.

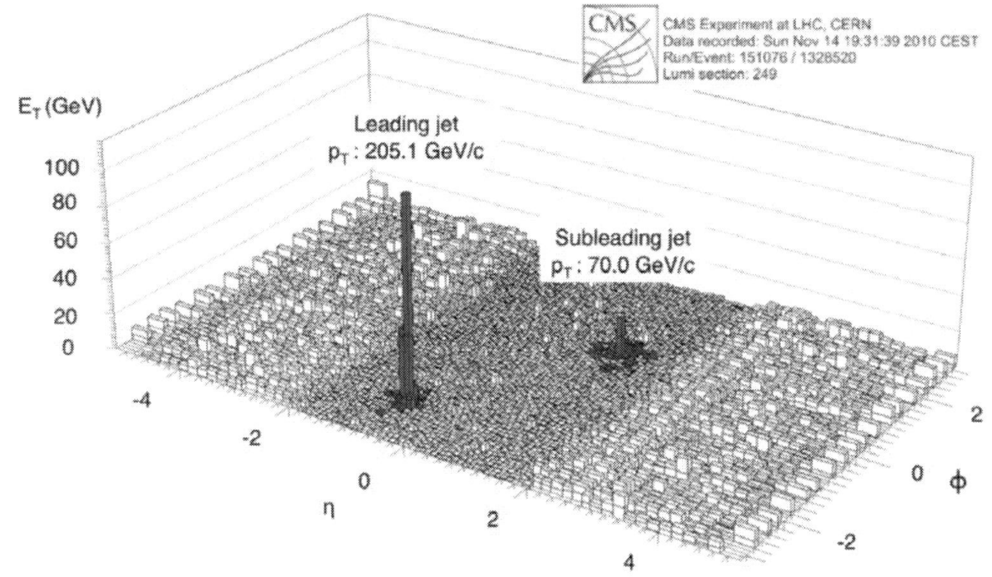


**Figure 1.** Example of an unbalanced dijet in a PbPb collision event at $\sqrt{s_{NN}} = 2.76\,\text{TeV}$. Plotted is the summed transverse energy in the electromagnetic and hadron calorimeters vs. $\eta$ and $\phi$, with the identified jets highlighted in red, and labeled with the corrected jet transverse momentum.

the Compact Muon Solenoid (CMS) detector. The data included in this measurement correspond to a total integrated luminosity of $6.7\,\mu b^{-1}$.

Jets are identified primarily using the energy deposited in the lead-tungstate crystal electromagnetic calorimeter (ECAL) and the brass/scintillator hadron calorimeter (HCAL) covering $|\eta| < 3$. In addition, a steel/quartz-fiber Cherenkov calorimeter, called Hadron Forward (HF), covers the forward rapidities $3 < |\eta| < 5.2$ and is used to determine the centrality of the PbPb collision. The central calorimeters are embedded in a solenoid with 3.8 T central magnetic field. The CMS tracking system, located inside the calorimeter, consists of pixel and silicon-strip layers covering $|\eta| < 2.5$, and provides track reconstruction down to $p_T \approx 100\,\text{MeV}/c$, with a track momentum resolution of about 1% at $p_T = 100\,\text{GeV}/c$. A set of scintillator tiles, the Beam Scintillator Counters (BSC), are mounted on the inner side of the HF calorimeters for triggering and beam-halo rejection. The full details of the CMS detector are described in Ref. [6].

For online event selection, CMS uses a two-level trigger system: Level-1 (L1) and High Level Trigger (HLT). The events for this analysis were selected using an inclusive single-jet trigger that required an L1 jet with $p_T > 30\,\text{GeV}/c$ and an HLT jet with $p_T > 50\,\text{GeV}/c$, where neither $p_T$ value was corrected for the $p_T$-dependent calorimeter energy response. The trigger becomes fully efficient for collisions with a leading jet with corrected $p_T$ greater than $100\,\text{GeV}/c$.

For the analysis of PbPb events, it is important to know the "centrality" of the collision, i.e., whether the overlap of the two colliding nuclei is large or small. In this analysis, the total energy deposited in both HF calorimeters was used to divide the event sample into five centrality classes, corresponding to the most central 10% of the events, the next most central 10% of the events (denoted 10–20%), and further bins corresponding to the 20–30%, 30–50%, and 50–100% selections of the total hadronic cross section.

The baseline jet reconstruction for heavy ion collisions in CMS is performed with an iterative cone algorithm modified to subtract the soft underlying event on an event-by-event basis [7].

Each cone is selected with a radius $\Delta R = \sqrt{\Delta\phi^2 + \Delta\eta^2} = 0.5$ around a seed of minimum transverse energy of 1 GeV. The performance of this algorithm is documented in Ref. [4]. Jet corrections for the calorimeter response have been applied, as determined in studies for pp collisions [8]. When applying the algorithm to PbPb data, the subtracted background energy for $R = 0.5$ jet cones ranges from 6–13 GeV for peripheral events (centrality bins 50–100%) to 90–130 GeV for central collisions (0–10%), before applying jet energy scale corrections.

The basic offline selection of events for the dijet analysis is the presence of a leading calorimeter jet in the pseudorapidity range of $|\eta| < 2$ with a corrected jet $p_T > 120\,\mathrm{GeV}/c$. By selecting these leading jets we avoid possible biases due to inefficiencies close to the trigger threshold. Furthermore, the selection of a rather large leading jet momentum expands the range of jet momentum imbalances that can be observed between the leading and subleading jets, as the subleading jets need a minimum momentum of $p_T > 35$–$50\,\mathrm{GeV}/c$ to be reliably detected above the high-multiplicity underlying event in PbPb collisions. In order to ensure a high quality dijet selection, additional kinematic selection cuts were applied. The azimuthal angle between the leading and subleading jets was required to be at least $2\pi/3$. A minimum $p_T$ of $p_{T,1} > 120\,\mathrm{GeV}/c$ was required for leading jets and $p_{T,2} > 50\,\mathrm{GeV}/c$ for subleading jets. No explicit requirement is made either on the presence or absence of a third jet in the event.

For the analysis of dijet properties in PbPb events, it is crucial to understand how the jet reconstruction is modified in the presence of the high multiplicity of particles produced in the PbPb underlying event. The jet-finding performance was studied using dijets in pp collisions simulated with the PYTHIA event generator (version 6.423, tune D6T) [9], modified for the isospin content of the colliding nuclei [10]. In order to enhance the number of Pythia dijets in the momentum range studied, a minimum $\hat{p}_T$ selection of $80\,\mathrm{GeV}/c$ was used. Lower $\hat{p}_T$ selections, as discussed in [11], were also investigated and found to agree with the $\hat{p}_T = 80\,\mathrm{GeV}/c$ results within uncertainties. The PYTHIA dijet events were processed with the full detector simulation and analysis chain. Additional samples were produced in which the PYTHIA dijet events were embedded into a minimum bias selection of PbPb events at the raw data level. For this embedding procedure, both real PbPb data events (PYTHIA+DATA), and PbPb events simulated with the HYDJET event generator [10] (PYTHIA+HYDJET) were used. The HYDJET parameters were tuned to reproduce the total particle multiplicities at all centralities and to approximate the underlying event fluctuations seen in data. The HYDJET events included the simulation of hard-scattering processes for which radiative parton energy loss was simulated, but collisional energy loss was turned off [10]. Both embedded samples were propagated through the standard reconstruction and analysis chain.

## 3. Calorimeter Jet Imbalance

The goal of this analysis is to characterize possible modifications of dijet properties as a function of centrality in PbPb collisions. One possible medium effect on the dijet properties is a change of the back-to-back alignment of the two partons. This can be studied using the event-normalized differential dijet distribution, $(1/N)(dN/d\Delta\phi_{12})$, versus $\Delta\phi_{12}$. Figure 2 shows distributions of $\Delta\phi_{12}$ between leading and subleading jets which pass the respective $p_T$ selections. In Fig. 2 (a), the dijet $\Delta\phi_{12}$ distributions are plotted for 7 TeV pp data in comparison to the corresponding PYTHIA simulations using the anti-$k_T$ algorithm for jets based on calorimeter information. PYTHIA provides a good description of the experimental data, with slightly larger tails seen in the PYTHIA simulations. It is important to note that the PYTHIA simulations include events with more than two jets, which provide the main contribution to events with large momentum imbalance or $\Delta\phi_{12}$ far from $\pi$.

Figures 2 (b)–(f) show the dijet $\Delta\phi_{12}$ distributions for PbPb data in five centrality bins, compared to PYTHIA+DATA simulations. The distributions for the four more peripheral bins are in good agreement with the PYTHIA+DATA reference, especially for $\Delta\phi_{12} \gtrsim 2$. The three centrality bins spanning 0–30% show an excess of events with azimuthally misaligned dijets ($\Delta\phi_{12} \lesssim 2$), compared with more peripheral events. A similar trend is seen for the PYTHIA+DATA simulations, although the fraction of events with azimuthally misaligned dijets is smaller in the simulation. The centrality dependence of the azimuthal correlation in PYTHIA+DATA can be understood as the result of the increasing fake-jet rate and the drop in jet reconstruction efficiency near the 50 GeV/$c$ threshold from 95% for peripheral events to 88% for the most central events. In PbPb data, this effect is magnified since low-$p_T$ away-side jets can undergo a sufficiently large energy loss to fall below the 50 GeV/$c$ selection criteria.

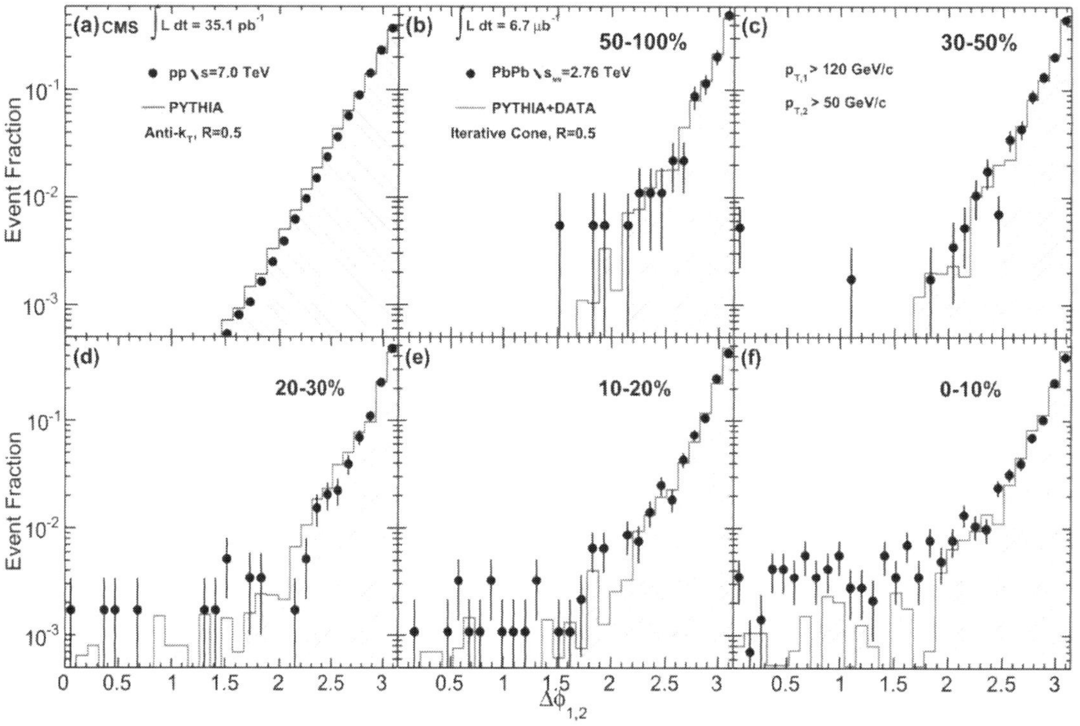


**Figure 2.** $\Delta\phi_{12}$ distributions for leading jets of $p_{T,1} > 120\,\mathrm{GeV}/c$ with subleading jets of $p_{T,2} > 50\,\mathrm{GeV}/c$ for 7 TeV pp collisions (a) and 2.76 TeV PbPb collisions in several centrality bins: (b) 50–100%, (c) 30–50%, (d) 20–30%, (e) 10–20% and (f) 0–10%. Data are shown as black points, while the histograms show (a) PYTHIA events and (b)-(f) PYTHIA events embedded into PbPb data. The error bars show the statistical uncertainties.

To characterize the dijet momentum balance (or imbalance) quantitatively, we use the asymmetry ratio,

$$A_J = \frac{p_{T,1} - p_{T,2}}{p_{T,1} + p_{T,2}} , \tag{1}$$

where the subscript 1 always refers to the leading jet, so that $A_J$ is positive by construction. The use of $A_J$ removes uncertainties due to possible shifts in the overall normalization of the jet energy scale. It is important to note that the subleading jet $p_{T,2} > 50\,\mathrm{GeV}/c$ selection imposes a $p_{T,1}$-dependent limit on the magnitude of $A_J$. For example, for the most frequent leading jets near the 120 GeV/$c$ threshold, this limit is $A_J < 0.41$, while the largest possible $A_J$ for the present dataset is 0.7 for 300 GeV/$c$ leading jets. Dijets in which the subleading jet is lost below the 50 GeV/$c$ threshold are not included in the $A_J$ calculation.

In Fig. 3 (a), the $A_J$ dijet asymmetry observable calculated by PYTHIA is compared to pp data at $\sqrt{s} = 7$ TeV. Again, data and event generator are found to be in excellent agreement. This observation, as well as the good agreement between PYTHIA+DATA and the most peripheral PbPb data shown in Fig. 3 (b), suggests that PYTHIA at $\sqrt{s} = 2.76$ TeV can serve as a good reference for the dijet imbalance analysis in PbPb collisions.

The centrality dependence of $A_J$ for PbPb collisions can be seen in Figs. 3 (b)–(f), in comparison to PYTHIA+DATA simulations. Whereas the dijet angular correlations show only a small dependence on collision centrality, the dijet momentum balance exhibits a dramatic change in shape for the most central collisions. In contrast, the PYTHIA simulations only exhibit a modest broadening, even when embedded in the highest multiplicity PbPb events.

Central PbPb events show a significant deficit of events in which the momenta of leading and subleading jets are balanced and a significant excess of unbalanced pairs. The large excess of unbalanced compared to balanced dijets explains why this effect was apparent even when simply scanning event displays (see Fig. 1). The striking momentum imbalance is also confirmed when studying high-$p_T$ tracks associated with leading and subleading jets, as will be shown in Section 4. It is consistent with a degradation of the parton energy, or jet quenching, in the medium produced in central PbPb collisions.

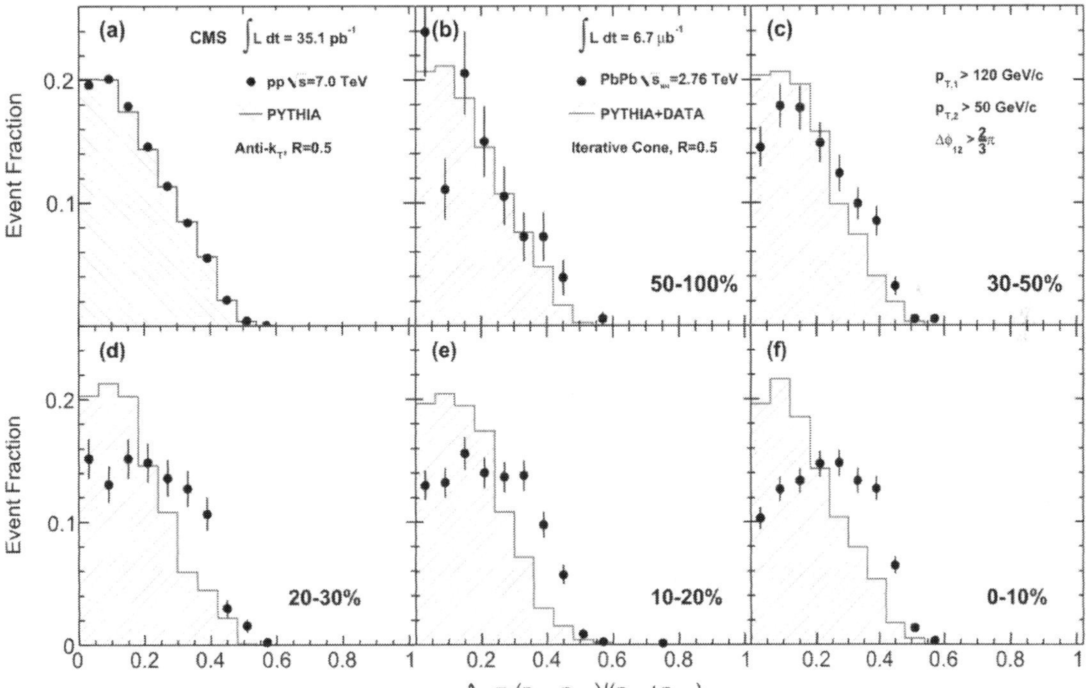


**Figure 3.** Dijet asymmetry ratio, $A_J$, for leading jets of $p_{T,1} > 120$ GeV/$c$, subleading jets of $p_{T,2} > 50$ GeV/$c$, and $\Delta\phi_{12} > 2\pi/3$ for 7 TeV pp collisions (a) and 2.76 TeV PbPb collisions in several centrality bins: (b) 50–100%, (c) 30–50%, (d) 20–30%, (e) 10–20% and (f) 0–10%. Data are shown as black points, while the histograms show (a) PYTHIA events and (b)-(f) PYTHIA events embedded into PbPb data. The error bars show the statistical uncertainties.

Journal of Physics: Conference Series **316** (2011) 012018 doi:10.1088/1742-6596/316/1/012018

## 4. Energy Balance in Charged Tracks

The studies of calorimeter jets show a strong change of the jet momentum balance as a function of collision centrality. This implies a corresponding modification in the distribution of jet fragmentation products, with energy being either transported out of the cone area used to define the jets, or to low-momentum particles that are not measured in the calorimeter jets. By using the information from reconstructed charged particles in the tracker, we have an opportunity to do the first in-depth studies of where the missing energy is redistributed, whether to low-momentum particles or to large angles from the jet axis.

The study of the overall momentum balance in the dijet events is performed using the projection of missing $p_{\mathrm{T}}$ of reconstructed charged tracks onto the leading jet axis. For each event, this projection is calculated as

$$\not{p}_{\mathrm{T}}^{\parallel} = \sum_{\mathrm{i}} -p_{\mathrm{T}}^{\mathrm{i}} \cos\left(\phi_{\mathrm{i}} - \phi_{\mathrm{Leading\ Jet}}\right), \tag{2}$$

where the sum is over all tracks with $p_{\mathrm{T}} > 0.5\,\mathrm{GeV}/c$ and $|\eta| < 2.4$. The results are then averaged over events to obtain $\langle \not{p}_{\mathrm{T}}^{\parallel} \rangle$.

In Fig. 4, $\langle \not{p}_{\mathrm{T}}^{\parallel} \rangle$ is shown as a function of $A_J$ for two centrality bins, 30–100% (left) and 0–30% (right). Results for PYTHIA+HYDJET are presented in the top row, while the bottom row shows the results for PbPb data. Using tracks with $|\eta| < 2.4$ and $p_{\mathrm{T}} > 0.5\,\mathrm{GeV}/c$, one sees that indeed the momentum balance of the events, shown as solid circles, is recovered within uncertainties, for both centrality ranges and even for events with large observed dijet asymmetry, in both data and simulation. This shows that the dijet momentum imbalance is not related to undetected activity in the event due to instrumental (e.g. gaps or inefficiencies in the calorimeter).

The figure also shows the contributions to $\langle \not{p}_{\mathrm{T}}^{\parallel} \rangle$ for five transverse momentum ranges from 0.5–1 GeV/$c$ to $p_{\mathrm{T}} > 8\,\mathrm{GeV}/c$. The vertical bars for each range denote statistical uncertainties. For data and simulation, a large negative contribution to $\langle \not{p}_{\mathrm{T}}^{\parallel} \rangle$ (i.e., in the direction of the leading jet) by the $p_{\mathrm{T}} > 8\,\mathrm{GeV}/c$ range is balanced by the combined contributions from the 0.5–8 GeV/$c$ regions. Looking at the $p_{\mathrm{T}} < 8\,\mathrm{GeV}/c$ region in detail, important differences between data and simulation emerge. For PYTHIA+HYDJET both centrality ranges show a large balancing contribution from the intermediate $p_{\mathrm{T}}$ region of 4–8 GeV/$c$, while the contribution from the two regions spanning 0.5–2 GeV/$c$ is very small. In peripheral PbPb data, the contribution of 0.5–2 GeV/$c$ tracks relative to that from 4–8 GeV/$c$ tracks is somewhat enhanced compared to the simulation. In central PbPb events, the relative contribution of low and intermediate-$p_{\mathrm{T}}$ tracks is actually the opposite of that seen in PYTHIA+HYDJET. In data, the 4–8 GeV/$c$ region makes almost no contribution to the overall momentum balance, while a large fraction of the negative imbalance from high $p_{\mathrm{T}}$ is recovered in low-momentum tracks.

Further insight into the radial dependence of the momentum balance can be gained by studying $\langle \not{p}_{\mathrm{T}}^{\parallel} \rangle$ separately for tracks inside cones of size $\Delta R = 0.8$ around the leading and subleading jet axes, and for tracks outside of these cones. The results of this study for central events are shown in Fig. 5 for the in-cone balance and out-of-cone balance for MC and data. As the underlying PbPb event in both data and MC is not $\phi$-symmetric on an event-by-event basis, the back-to-back requirement was tightened to $\Delta\phi_{12} > 5\pi/6$ for this study.

One observes that for both data and MC an in-cone imbalance of $\langle \not{p}_{\mathrm{T}}^{\parallel} \rangle \approx -20\,\mathrm{GeV}/c$ is found for the $A_J > 0.33$ selection. In both cases this is balanced by a corresponding out-of-cone imbalance. However, in the PbPb data the out-of-cone contribution is carried almost entirely by tracks with $0.5 < p_{\mathrm{T}} < 4\,\mathrm{GeV}/c$, whereas in MC more than 50% of the balance is carried by tracks with $p_{\mathrm{T}} > 4\,\mathrm{GeV}/c$, with a negligible contribution from $p_{\mathrm{T}} < 1\,\mathrm{GeV}/c$. The PYTHIA+HYDJET results are indicative of semi-hard initial or final-state radiation as the underlying cause for large $A_J$ events in the MC study. This has been confirmed by further studies which showed that in

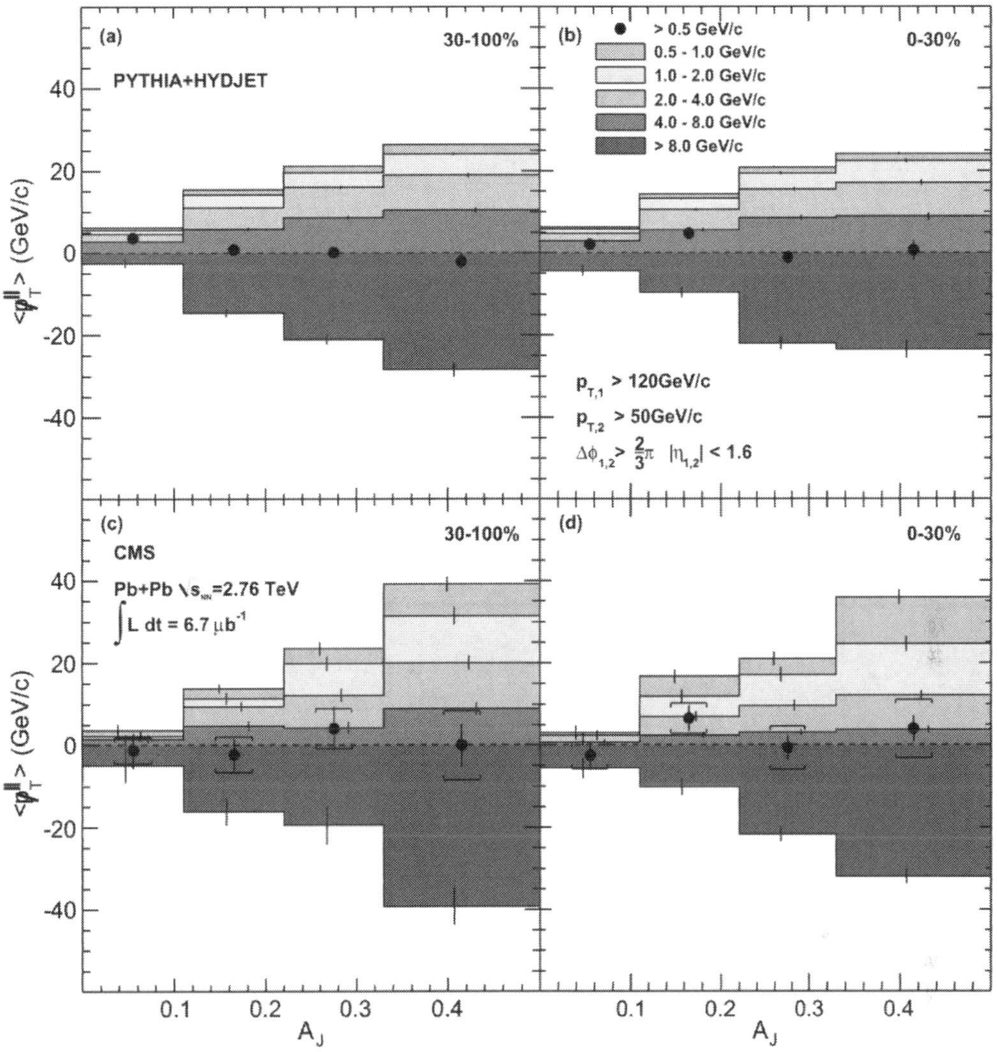


**Figure 4.** Average missing transverse momentum, $\langle \not{p}_{\mathrm{T}}^{\parallel} \rangle$, for tracks with $p_{\mathrm{T}} > 0.5\,\mathrm{GeV}/c$, projected onto the leading jet axis (solid circles). The $\langle \not{p}_{\mathrm{T}}^{\parallel} \rangle$ values are shown as a function of dijet asymmetry $A_J$ for 30–100% centrality (left) and 0–30% centrality (right). For the solid circles, vertical bars and brackets represent the statistical and systematic uncertainties, respectively. Colored bands show the contribution to $\langle \not{p}_{\mathrm{T}}^{\parallel} \rangle$ for five ranges of track $p_{\mathrm{T}}$. The top and bottom rows show results for PYTHIA+HYDJET and PbPb data, respectively. For the individual $p_{\mathrm{T}}$ ranges, the statistical uncertainties are shown as vertical bars.

PYTHIA the momentum balance in the transverse plane for events with large $A_J$ can be restored if a third jet with $p_T > 20\,\text{GeV}/c$, which is present in more than 90% of these events, is included. This is in contrast to the results for large-$A_J$ PbPb data, which show that a large part of the momentum balance is carried by soft particles ($p_T < 2\,\text{GeV}/c$) and radiated at large angles to the jet axes ($\Delta R > 0.8$).

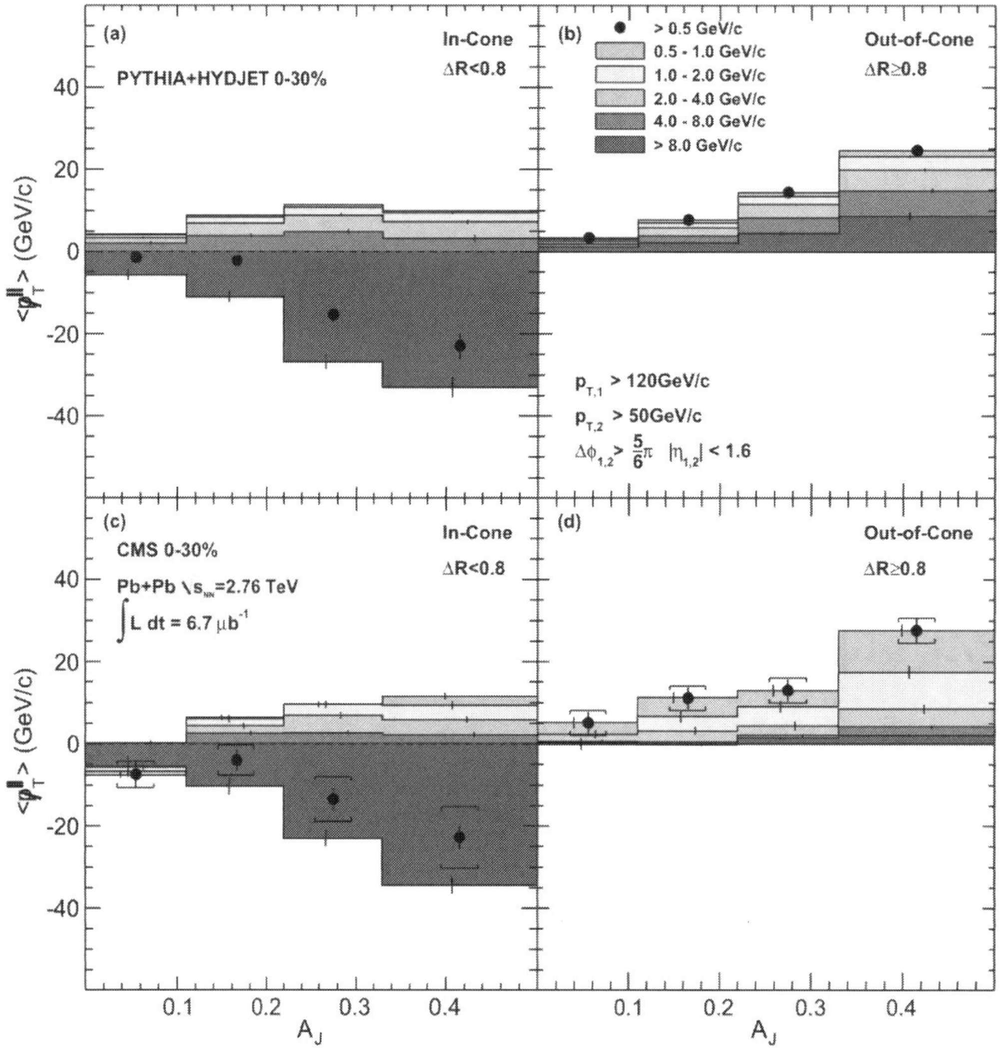


**Figure 5.** Average missing transverse momentum, $\langle \not{p}_T^{\parallel} \rangle$, for tracks with $p_T > 0.5\,\text{GeV}/c$, projected onto the leading jet axis (solid circles). The $\langle \not{p}_T^{\parallel} \rangle$ values are shown as a function of dijet asymmetry $A_J$ for 0–30% centrality, inside ($\Delta R < 0.8$) one of the leading or subleading jet cones (left) and outside ($\Delta R > 0.8$) the leading and subleading jet cones (right). For the solid circles, vertical bars and brackets represent the statistical and systematic uncertainties, respectively. For the individual $p_T$ ranges, the statistical uncertainties are shown as vertical bars.

## 5. Summary

The properties of dijets have been measured in pp and PbPb collisions recorded by the CMS experiment. A significant imbalance in the transverse momenta of dijets is observed in central PbPb collisions. The redistribution of the quenched jet energy is studied using the transverse momentum balance of charged tracks projected onto the direction of the leading jet. The momentum balance is recovered by including charged particles at low $p_T$ and at large angle with respect to the jet axis. In contrast to pp collisions where the rarer asymmetric dijets are balanced by the hard fragmentation products of third jets, a large fraction of the PbPb momentum balance is carried by low momentum particles at large angular distance to the jet axis.

The measurements presented in these proceedings represent the beginning of a new era in jet quenching studies, relying on the unambiguous identification of both partners in copious, asymmetric dijets. Future studies will include more detailed studies of the fragmentation functions and the flavor-dependence of jet quenching (e.g. via gamma-jet correlations, multi-jet events, and heavy-flavor-tagged jets). Together with future studies, these first measurements will put more powerful, quantitative constraints on the transport properties of hot, dense QCD matter through comparison to theoretical models.

## References

[1] **STAR** Collaboration, J. Adams et al., "Evidence from d+Au measurements for final state suppression of high $p_T$ hadrons in Au+Au collisions at RHIC", *Phys. Rev. Lett.* **91** (2003) 072304, arXiv:nucl-ex/0306024. doi:10.1103/PhysRevLett.91.072304.

[2] **STAR** Collaboration, S. Salur, "First Direct Measurement of Jets in $\sqrt{s_{NN}} = 200$ GeV Heavy Ion Collisions by STAR", *Eur. Phys. J.* **C61** (2009) 761–767, arXiv:0809.1609. doi:10.1140/epjc/s10052-009-0880-y.

[3] **STAR** Collaboration, J. Putschke, "First fragmentation function measurements from full jet reconstruction in heavy-ion collisions at $\sqrt{s_{NN}} = 200$ GeV by STAR", *Eur. Phys. J.* **C61** (2009) 629–635, arXiv:0809.1419. doi:10.1140/epjc/s10052-009-0904-7.

[4] **CMS** Collaboration, S. Chatrchyan et al., "Observation and studies of jet quenching in PbPb collisions at $\sqrt{s_{NN}} = 2.76$ TeV", arXiv:1102.1957.

[5] **ATLAS** Collaboration, G. Aad et al., "Observation of a Centrality-Dependent Dijet Asymmetry in Lead-Lead Collisions at $\sqrt{s_{NN}} = 2.76$ TeV with the ATLAS Detector at the LHC", *Phys. Rev. Lett.* **105** (2010) 252303, arXiv:1011.6182.

[6] **CMS** Collaboration, "The CMS experiment at the CERN LHC", *JINST* **3** (2008) S08004. doi:10.1088/1748-0221/3/08/S08004.

[7] O. Kodolova, I. Vardanian, A. Nikitenko et al., "The performance of the jet identification and reconstruction in heavy ions collisions with CMS detector", *Eur. Phys. J.* **C50** (2007) 117. doi:10.1140/epjc/s10052-007-0223-9.

[8] **CMS** Collaboration, CMS Collaboration, "Determination of the Jet Energy Scale in CMS with pp Collisions at $\sqrt{s} = 7$ TeV", *CMS Physics Analysis Summary* **CMS-PAS-JME-10-010** (2010).

[9] T. Sjöstrand, S. Mrenna, and P. Skands, "PYTHIA 6.4 Physics and Manual", *JHEP* **05** (2006) 026 (tune D6T with PDFs CTEQ6L1 used for 2.76 TeV, tune Z2 for pp 7 TeV), arXiv:hep-ph/0603175.

[10] I. P. Lokhtin and A. M. Snigirev, "A model of jet quenching in ultrarelativistic heavy ion collisions and high-$p_T$ hadron spectra at RHIC", *Eur. Phys. J.* **C45** (2006) 211, arXiv:hep-ph/0506189. doi:10.1140/epjc/s2005-02426-3.

[11] M. Cacciari, G. P. Salam, and G. Soyez, "Fluctuations and asymmetric jet events in PbPb collisions at the LHC", arXiv:1101.2878.

# Measurement of identified charged hadron spectra with the ALICE experiment at the LHC

**Leonardo Milano**
Dipartimento di Fisica Sperimentale dell'Universitá and Sezione INFN, Turin, Italy
E-mail: milano@to.infn.it

**Abstract.** The ALICE experiment features multiple particle identification systems. The measurement of the identified charged hadron $p_t$ spectra in proton-proton collisions at $\sqrt{s} = 900$ GeV will be discussed. In the central rapidity region ($|\eta| < 0.9$) particle identification and tracking are performed using the Inner Tracking System (ITS), which is the closest detector to the beam axis, the Time Projection Chamber (TPC) and a dedicated time-of-flight system (TOF). Particles are mainly identified using the energy loss signal in the ITS and TPC. In addition, the information from TOF is used to identify hadrons at higher momenta. Finally, the kink topology of the weak decay of charged kaons provides an alternative method to extract the transverse momentum spectra of charged kaons. This combination allows to track and identify charged hadrons in the transverse momentum ($p_t$) range from 100 MeV/c up to 2.5 GeV/$c$. Mesons containing strange quarks ($K^0_S$, $\phi$) and both singly and doubly strange baryons ($\Lambda$, $\overline{\Lambda}$, and $\Xi^+ + \Xi^-$) are identified by their decay topology inside the TPC detector. Results obtained with the various identification tools above described and a comparison with theoretical models and previously published data will be presented.

## 1. Introduction

The transverse momentum spectra and yields of identified particles at mid-rapidity from the first pp collisions are presented. Data are collected in the autumn of 2009, during the commissioning of the LHC, at $\sqrt{s} = 900$ GeV. The evolution of particle production in pp collisions with collision energy is studied by comparing data from previous experiments.

The $\pi^+$, $\pi^-$, $K^+$, $K^-$, p, and $\overline{p}$ distributions, identified via several independent techniques utilizing specific energy loss, $dE/dx$, information from the Inner Tracking System (ITS) and the Time Projection Chamber (TPC), velocity measurements in the Time-Of-Flight array (TOF) are reported. The combination of these methods provides particle identification over the transverse momentum range 0.1 GeV/c $< p_t <$ 2.5 GeV/c. Charged kaons, identified via kink topology of their weak decays in the TPC, provide a complementary measurement over a similar $p_t$ range. All reported particle yields are for primary particles, namely those directly produced in the collision including the products of strong and electromagnetic decays but excluding weak decays of strange particles.

## 2. Experimental setup and data analysis

### 2.1. The ALICE detector

The ALICE detector and its performance are described in detail in [1]. For the analyses described the following detectors are used: the ITS, the TPC and the TOF detector. These detectors are positioned in a solenoidal magnetic field of $B = 0.5$ T and have a common pseudo-rapidity coverage of $-0.9 < \eta < 0.9$. Two forward scintillator hodoscopes (VZERO) are used for triggering purposes. They are placed on either side of the interaction region, covering regions $2.8 < \eta < 5.1$ and $-3.7 < \eta < -1.7$.

- The Inner Tracking System - The ITS is the closest of the central barrel detectors to the beam axis. It is composed of six cylindrical layers of silicon detectors. The two innermost layers are equipped with pixel detectors (SPD), followed by two layers of drift detectors (SDD) and two layers of double-sided silicon strip detectors (SSD). The innermost layer is at 3.9 cm from the beam axis, while the outer layer is at 43.0 cm. The four layers equipped with SDD and SSD also provide a measurement of the specific energy loss $\mathrm{d}E/\mathrm{d}x$. The ITS is also used as a standalone tracker to reconstruct charged particles with momenta below 200 MeV/$c$ that are deflected or decay before reaching the TPC, and to recover tracks crossing dead regions of the TPC.
- The Time Projection Chamber - The TPC is the main tracking device. It is a large volume, high granularity, cylindrical detector with an outer radius of 2.78 m and a length of 5.1 m. The active volume extends from 0.85 m to 2.47 m in radius. It covers $2\pi$ in azimuth and $|\eta| < 0.9$ in polar angle for the full radial track length. Accepting one third of the full radial track length extends the range to $|\eta| < 1.5$. The 90 $\mathrm{m}^3$ drift volume is filled with a Ne (85.7%), $CO_2$ (9.5%), and $N_2$ (4.8%) gas mixture. A high voltage central membrane splits the drift region in two halves, resulting in a maximal drift time of 94 $\mu$s. The TPC has a good particle identification capability thanks to the measurement of energy loss of particles in the gas mixture.
- The Time-Of-Flight Detector - The TOF detector consists of 18 azimuthal sectors, each containing 91 Multi-gap Resistive Plate Chambers (MRPCs) distributed in five gas-tight modules. It is positioned at 370-399 cm from the beam axis. The region $260° < \phi < 320°$ at $\eta \sim 0$ is not covered in order to minimize the material in front of the Photon Spectrometer, which is not used in this analysis. The MRPC detectors are installed with a projective geometry along the beam direction, minimizing the variation of the flight path of particles across the sensitive area of the detector. Each MRPC is segmented into 96 read-out pads ($2.5 \times 3.5$ $\mathrm{cm}^2$ size).

### *2.2. Event selection and normalization*

The data presented were collected during the commissioning of the LHC at CERN in the autumn of 2009, with pp collisions at $\sqrt{s} = 900$ GeV. The collider was run with four bunches per beam, resulting in two bunch crossings per beam circulation period (89 $\mu$s) at the ALICE interaction point. The remaining two bunches per beam were not collided at ALICE, and served to estimate the contribution of beam-gas interactions. The average event rate was a few Hz, so the fraction of pile-up events was negligible.

## 3. $\pi^+$, $\pi^-$, $K^+$, $K^-$, p, and $\bar{p}$ identification in ALICE

The $\mathrm{d}E/\mathrm{d}x$ and TOF signals are used for particle identification as a function of the momentum $p$, whereas the final spectra are given as a function of the transverse momentum $p_t$.

In the case of the TPC and ITS analyses, particles were identified via the specific energy loss $\mathrm{d}E/\mathrm{d}x$. Unique identification on a track-by-track basis is possible in regions of momentum where the bands are clearly separated from each other. In overlapping areas, particle identification is still possible on a statistical basis using fits to the energy loss distribution in each $p_t$-bin. The fits are performed on the distribution of the difference between the measured and the expected energy deposition for tracks within the selected rapidity range $|y| < 0.5$. The calculated expected energy loss depends on the measured track momentum $p$ and the assumed mass for the particle.

Journal of Physics: Conference Series **316** (2011) 012019 doi:10.1088/1742-6596/316/1/012019

The procedure is therefore repeated three times for the entire set of tracks, assuming the pion, kaon, and proton mass.

In the TPC analysis, the difference $[\mathrm{d}E/\mathrm{d}x]_{\mathrm{meas}} - [\mathrm{d}E/\mathrm{d}x(pid, p_{tot})]_{\mathrm{calc}}$ is used. For the ITS the difference of the logarithm of the measured and calculated energy deposit $\ln[\mathrm{d}E/\mathrm{d}x_{\mathrm{meas}}] - \ln[\mathrm{d}E/\mathrm{d}x(pid,p_{tot})_{\mathrm{calc}}]$ is taken to suppress the non-Gaussian tails originating from the smaller number of $\mathrm{d}E/\mathrm{d}x$ measurements.

In the case of the TOF, the identification is based on the time-of-flight information. The procedure for the extraction of the raw yields differs slightly from the one used for TPC and ITS.

### *3.1. Particle identification in the ITS*

In both the ITS standalone (track reconstruction only using the ITS) and in the ITS-TPC (reconstruction is performed using both the ITS and the TPC) analyses, the $\mathrm{d}E/\mathrm{d}x$ measurement from the SDD and the SSD is used to identify particles. The standalone tracking result extends the momentum range to lower $p_{\mathrm{t}}$ than can be measured in the TPC, while the combined tracking provides a better momentum resolution.

For each track, $\mathrm{d}E/\mathrm{d}x$ is calculated using a truncated mean: the average of the lowest two points in case four points are measured, or a weighted sum of the lowest (weight 1) and the second lowest point (weight 1/2), in case only three points are measured.

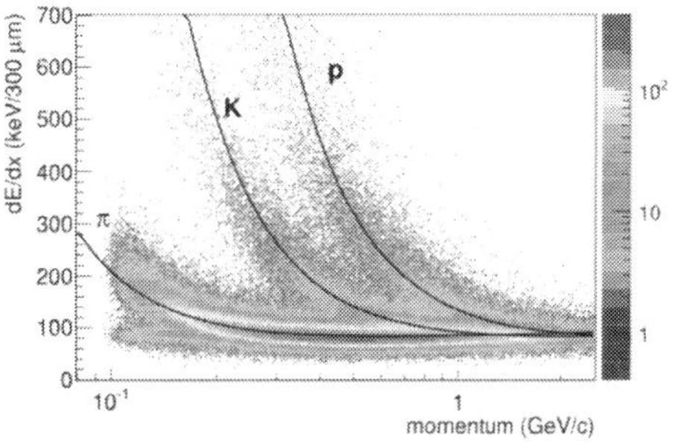


**Figure 1.** Specific energy loss $\mathrm{d}E/\mathrm{d}x$ vs. momentum in pp collisions at $\sqrt{s} = 0.9$ TeV for ITS standalone tracks measured with the ITS. The solid lines are a parametrization (from [2]) of the detector response based on the Bethe-Bloch formula.

Figure 1 shows the truncated mean $\mathrm{d}E/\mathrm{d}x$ for the sample of ITS standalone tracks along with the PHOBOS parametrization of the most probable value [2].

For the ITS standalone track sample, the histograms are fitted with three Gaussians and the integral of the Gaussian centered at zero is used as the raw yield of the corresponding hadron species.

For the ITS-TPC combined track sample, a non-Gaussian tail is visible. This tail is a remnant of the tail of the Landau distribution for energy loss. It was verified using simulations that the shape and size of the tail are compatible with the expectations for a truncated mean using two out of four samples.

Examples of $\mathrm{d}E/\mathrm{d}x$ distributions are shown in Fig. 2 for negative tracks using the kaon mass hypothesis in a given $p_{\mathrm{t}}$ interval ( 400-450 MeV/$c$) for both ITS standalone tracks (upper panel) and ITS-TPC combined tracks (lower panel).

*Efficiency correction* The raw hadron yields extracted from the fits to the $\mathrm{d}E/\mathrm{d}x$ distributions are corrected for the reconstruction efficiency determined from Monte Carlo simulations, applying the same analysis criteria to the simulated events as to the data. Secondary particles from interactions in the detector material and strange particle decays have been subtracted from the yield of both simulated and real data. The fraction of secondaries after applying the track impact-parameter cut depends on the hadron species and amounts to 1-3% for pions and 5-10%

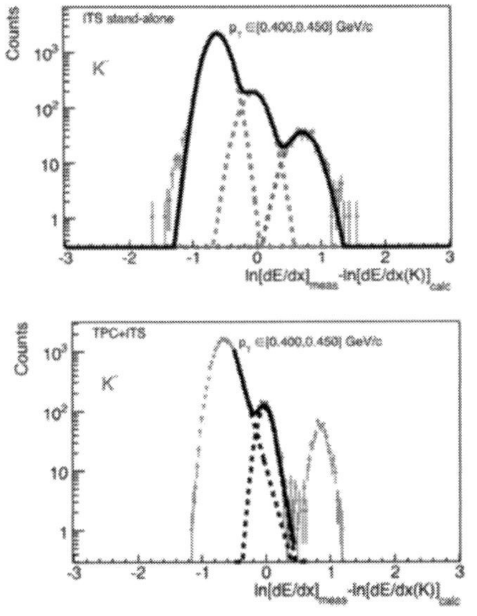


**Figure 2.** Distribution of $\ln[\mathrm{d}E/\mathrm{d}x]_{\mathrm{meas}}-\ln[\mathrm{d}E/\mathrm{d}x(\mathrm{K})]_{\mathrm{calc}}$ measured with the ITS in the $p_{\mathrm{t}}$-range 400-450 MeV/$c$, using the kaon mass hypothesis. The upper panel shows the result for ITS standalone tracks, while the lower panel shows the ITS-TPC combined result. The lines indicate fits as described in the text.

for protons depending on $p_{\mathrm{t}}$. The secondary-to-primary ratio has been estimated by fitting the measured track impact-parameter distributions with three components, prompt particles, secondaries from strange particle decays and secondaries produced in the detector material for each hadron species. Alternatively, the contamination from secondaries have been determined using Monte Carlo samples, after rescaling the $\Lambda$ yield to the measured values [3]. The difference between these two procedures is about 3% for protons and is negligible for other particles.

### *3.2. Particle identification in the TPC*

Particle identification is based on the specific energy deposit of each particle in the drift gas of the TPC (up to 159 independent measurements for tracks crossing the whole TPC volume), shown in Fig. 3 as a function of momentum for positive and negative charges. The solid curves show the calibration curves obtained by fitting the ALEPH parametrization of the Bethe-Bloch curve [4] to the data points in regions of clear separation.

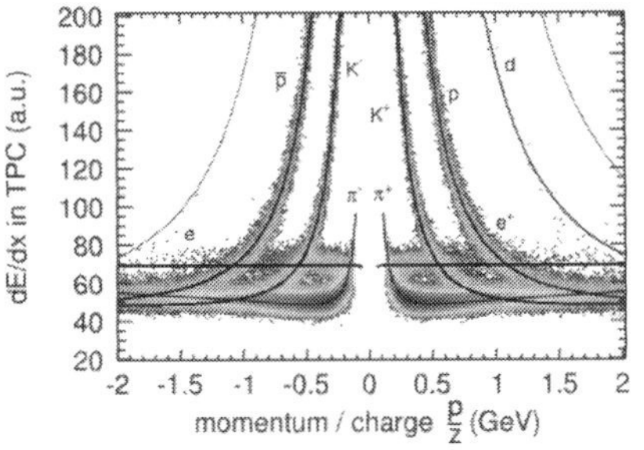


**Figure 3.** Specific energy loss $\mathrm{d}E/\mathrm{d}x$ vs. momentum in pp collisions at $\sqrt{s}=0.9$ TeV for tracks measured with the ALICE TPC. The solid lines are a parametrization of the Bethe-Bloch curve [4].

As in the case of the ITS, a truncated-mean procedure is used to determine $\mathrm{d}E/\mathrm{d}x$ (60% of the points are kept). This reduces the Landau tail of the $\mathrm{d}E/\mathrm{d}x$ distribution to the extent that it is very close to a Gaussian distribution.

One example of the $\mathrm{d}E/\mathrm{d}x$ distribution in a specific $p_{\mathrm{t}}$ bin in the kaon mass hypothesis is shown in Fig. 4. The peak centered at zero refers to kaons and the other peaks are from other particle species. As the background in all momentum bins is negligible, the integrals of the Gaussian give the raw yields.

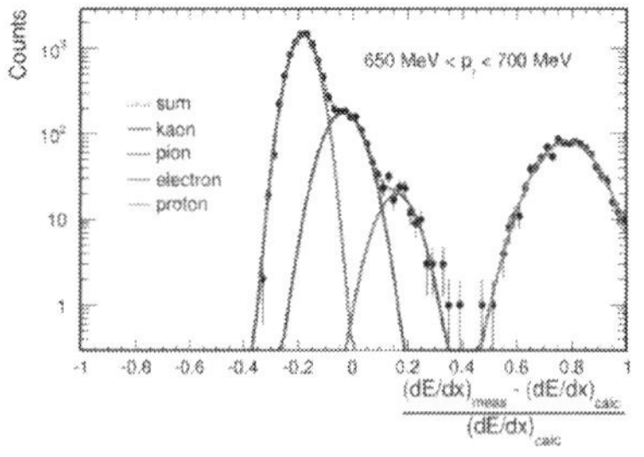


**Figure 4.** Distribution of $([dE/dx]_{\text{meas}} - [dE/dx(\text{kaon})]_{\text{calc}})$ / $[dE/dx(\text{kaon})]_{\text{calc}}$ measured with the TPC in a given $p_t$ -bin showing the separation power. The solid lines are Gaussian fits to the distributions.

*Efficiency correction* The raw hadron spectra are corrected for the reconstruction efficiency, determined by doing the same analysis on MonteCarlo events. The efficiency is calculated by comparing the number of reconstructed particles to the number of charged primary particles from PYTHIA in the chosen rapidity range. The range with a reconstruction efficiency lower than 60% (for pions and protons) is omitted for the analysis corresponding to a low-$p_t$ cut-off of 200 MeV/$c$ for pions, 250 MeV/$c$ for kaons, and 400 MeV/$c$ for protons.

Protons are corrected for the contamination of secondaries from material and of feed down from weak decays. The feed down was determined by two independent methods as described in 3.1.

### *3.3. Particle Identification with the Time Of Flight detector*

Particles reaching the TOF system are identified by measuring simultaneously their momentum and velocity. The flight path of the track $L$ is defined as the distance along the track trajectory between the point of its closest approach to the event vertex and the TOF sensitive surface. The time of flight is calculated for different mass hypotheses by summing up, at each tracking step, the time-of-flight increments $\Delta t_k = \Delta l_k \sqrt{p_k^2 + m_i^2}/p_k$, with $p_k$ being the local value of the track momentum, and $\Delta l_k$ the track-length increment along its trajectory.

From the reconstructed flight path $L$ and the measured time of flight $t_{\text{TOF}}$, the velocity $\beta = L/t_{\text{TOF}}$ is obtained (as displayed in Fig. 5) as a function of the momentum $p$ at the vertex. Clearly visible are the bands corresponding to charged pions, kaons and protons. The width of the bands reflects the observed overall time-of-flight resolution of about 180 ps, which has contributions from the TOF time resolution, the accuracy of the reconstructed flight path and the uncertainty on the event start time, $t_0^{ev}$.

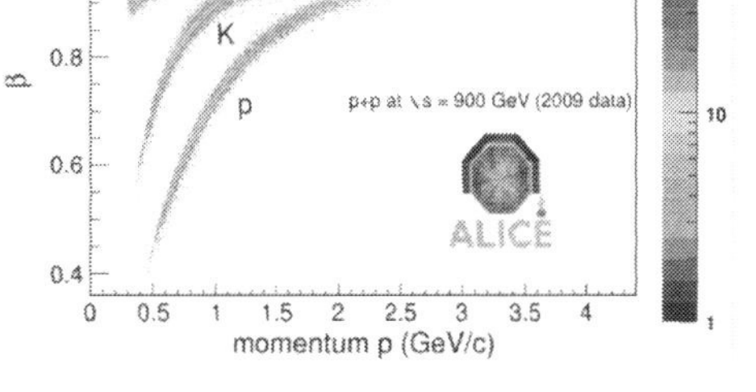


**Figure 5.** $\beta$ vs. momentum of the tracks in the TOF.

The yields of $\pi$, K and p are obtained from the simultaneous fit of the distribution of the time difference $S$ between measured $t_{\mathrm{TOF}}$ and the average between the calculated time averaging for pions and kaons

$$S = t_{\mathrm{TOF}} - (t^{\pi}_{\mathrm{calc}} + t^{\mathrm{K}}_{\mathrm{calc}})/2. \tag{1}$$

The choice of this variable is motivated by the observation that the identification procedure is optimized when pions and kaons have a symmetric treatment. Moreover, extracting the yield for different species in a simultaneous fit guarantees that the resulting number of pions, kaons and protons matches the total number of tracks in the given momentum bin.

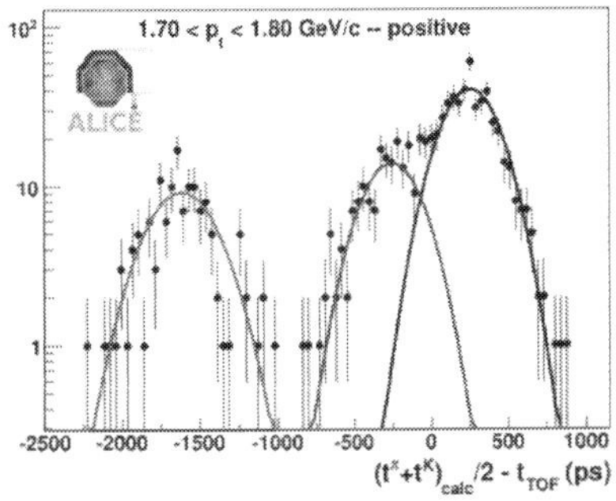


**Figure 6.** Distribution of the time difference between the TOF signal and the average of the expected times for pions and kaons in a given momentum bin for positively charged particles. The fits are performed using Gaussian shapes.

The distribution of the variable $S$ is reported in Fig. 6 for positive particles in a given transverse momentum bin. The distribution shows a Gaussian behavior for each particle species and the lines indicate the result of a three-Gaussian fit used to extract the raw yields.

*Efficiency Correction* As the track selection used in the TOF analysis is the same as the one described in the TPC analysis (subsection 3.2), the same tracking corrections are applied. In the case of TOF analysis, an additional correction is needed in order to take into account that only a fraction of the particles reconstructed by the TPC are associated to a signal in TOF. This matching efficiency includes all sources of track losses in the propagation from the TPC to the TOF (geometry, decays and interactions with the material) and its matching with a TOF signal (the TOF intrinsic detector efficiency, the effect of dead channels and the efficiency of the track-TOF signal matching procedure).

## 4. Particle identification in ALICE via topological criteria

### *4.1. Kaon Identification using their decay in the TPC*

Charged kaons can also be identified by their weak decay (kink topology) inside the TPC detector. These tracks are rejected in the previously described TPC analysis. The decay channels with the highest branching ratio (B.R.) for kaons are the two-body decays

(1) $\mathrm{K}^{\pm} \rightarrow \mu^{\pm} + \nu_{\mu}$ , (B.R. 63.55%)
(2) $\mathrm{K}^{\pm} \rightarrow \pi^{\pm} + \pi^{0}$ , (B.R. 20.66%).

Three-body decays with one charged daughter track (B.R. 9.87%) as well as three-body decays into three charged pions (B.R. 5.6%) are also detected.

The identification of kaons from kink topology and its separation from pion decay is based on the decay kinematics. The transverse momentum of the daughter with respect to the mother's direction, $q_{\mathrm{t}}$, has an upper limit of 236 MeV/$c$ for kaons and 30 MeV/$c$ for pions for the two-body decay to $\mu + \nu_{\mu}$. The corresponding upper limit for the two-body decay (2) $\mathrm{K}^{\pm} \rightarrow \pi^{\pm} + \pi^{0}$ is 205 MeV/$c$. All three limits can be seen as the edges of the peaks in Fig. 7, which shows the

Journal of Physics: Conference Series **316** (2011) 012019 doi:10.1088/1742-6596/316/1/012019

$q_t$ distribution of all measured kinks inside the selected volume and rapidity range $|y| < 0.7$. Selecting kinks with $q_t > 40$ MeV/$c$ removes the majority of $\pi$-decays as shown by the dashed (before) and solid (after) histograms.

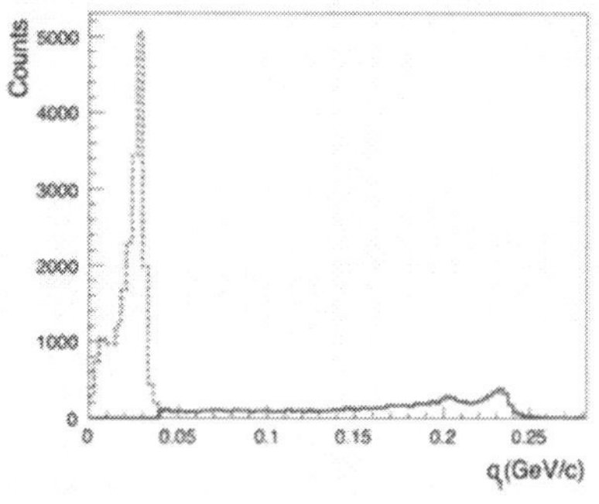


**Figure 7.** $q_t$ distribution in pp collisions at $\sqrt{s} = 0.9$ TeV of the daughter tracks with respect to mother momentum for reconstructed kinks inside the analyzed sample. The dashed (solid) histograms show the distribution before (after) applying the $q_t > 40$ MeV/$c$ cut.

The invariant mass for the decay into $\mu^{\pm} + \nu_{\mu}$ is calculated from the measured difference between the mother and daughter momentum, their decay angle, assuming zero mass for the neutrino.

### *4.2. Topological reconstruction of* $K^0_S$, $\Lambda$, $\overline{\Lambda}$

The measurement of $K^0_S$, $\Lambda$ and $\overline{\Lambda}$ is based on the reconstruction of the secondary vertex (V0) associated to their weak decay. The V0 finding procedure starts with the selection of secondary tracks, i.e. tracks having a sufficiently large impact parameter with respect to the primary vertex. All possible combinations between two secondary tracks of opposite charge are then examined. They are accepted as V0 candidates only if the DCA between them is smaller than 0.5 cm. The minimization of the distance between the tracks is performed numerically using helix parameterizations in 3D. The V0 vertex position is a point on the line connecting the points of closest approach between the two tracks. Its distance from each daughter track is taken to be proportional to the precision of the track parameter estimations. Once their position is determined, only the V0 candidates located inside a given fiducial volume are accepted.

### *4.3. Topological reconstruction of* $\Xi^+$ *and* $\Xi^-$

The $\Xi^+$ and $\Xi^-$ particles are identified via their "cascade" decay topology. The cascade finding procedure starts from the V0 finding procedure for the $\Lambda(\overline{\Lambda})$ daughter but with less stringent selection criteria. This is done to increase the efficiency and to allow for the fact that the daughter $\Lambda$'s do not have to point back to the primary vertex.

The V0 candidates found within the $\Lambda$ mass window ($1116 \pm 6$ MeV/$c^2$) are combined with all possible secondary tracks (bachelor candidates) with the exception of both V0 daughter tracks. A cut on the impact parameter of the bachelor track is applied to reject the primary particles which increase the combinatorial background.

A V0-bachelor association is performed if the distance of closest approach between the bachelor track and the V0 trajectory (DCA between V0 and bachelor track) is small (less than $3cm$). Finally, this cascade candidate is selected if its reconstructed momentum points back to the primary vertex (cosine of cascade pointing angle). The cascade finding is limited to the fiducial region used for V0 reconstruction).

### *4.4. Reconstruction of* $\phi$

The $\phi$ resonance is reconstructed through its principal decay channel $\phi \to K^+K^-$. With a $c\tau$ of 45 fm, its decay vertex is indistinguishable from the primary collision vertex. Therefore the selection criteria adopted for the candidate daughter tracks are the ones used for primaries.

A crucial issue for the $\phi$ reconstruction, as for any strongly decaying resonance, is the combinatorial background determination. In the present analysis PID is used to select kaons, rejecting most of the background while leading to a very small loss in efficiency. For this purpose, tracks are selected if the PID information from the TPC is compatible with a kaon signal and using the TOF signal when available.

### *4.5. Efficiency corrections*

The efficiency corrections are obtained by analyzing Monte Carlo (MC) events in exactly the same way as for the real events. Little dependence is found on the several MC generators which are used. Therefore the corrections presented here are obtained using the event generator PYTHIA 6.4 (tune D6T) [5] and GEANT3 [6] for particle transport through the ALICE detectors.

## 5. Results

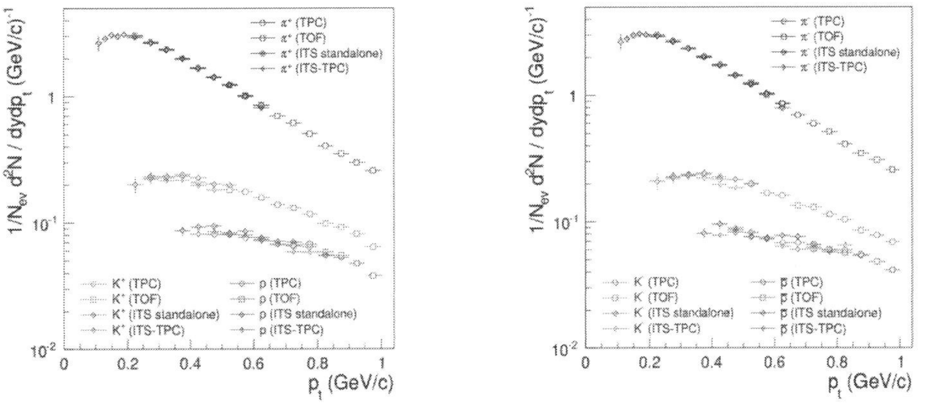


**Figure 8.** Transverse momentum spectra $\mathrm{d}^2N/(\mathrm{d}p_\mathrm{t}\mathrm{d}y)$ for $|y| < 0.5$ of positive (left) negative (right) hadrons from the various analyses. Only systematic errors are plotted.

Spectra from ITS standalone, ITS-TPC, TPC and TOF analyses are reported in Figure 8. The analyses from the different detectors use a slightly different sample of tracks and have largely independent systematics (mainly coming from the PID method and the contamination from secondaries). The spectra have been averaged, using the systematic errors as weights. From this weighted average, the combined, $p_\mathrm{t}$-dependent, systematic error is derived. The combined spectra have an additional overall normalization error, coming primarily from the uncertainty on the material budget (3%) and from the normalization procedure (2%). Combined spectra are reported in Figure 9.

The combined spectra shown in Fig. 9 are fitted with the Lévy (or Tsallis) function (see e.g. [7])

$$\frac{\mathrm{d}^2N}{\mathrm{d}p_\mathrm{t}\mathrm{d}y} = p_\mathrm{t} \times \frac{\mathrm{d}N}{\mathrm{d}y}\frac{(n-1)(n-2)}{nC(nC+m_0(n-2))}\left(1+\frac{m_\mathrm{t}-m_0}{nC}\right)^{-n} \quad (2)$$

with the fit parameters $C$, $n$ and the yield $\mathrm{d}N/\mathrm{d}y$. This function gives a good description of the spectra and has been used to extract the total yields and the $\langle p_\mathrm{t}\rangle$.

The $p_\mathrm{t}$ spectra for $\mathrm{K}^0_\mathrm{S}$, $\Lambda$, $\overline{\Lambda}$, $\phi$ and $\Xi^+ + \Xi^-$ are shown in Fig. 10 along with the Lévy fits. When comparing the different spectra, it is found that the inverse slope parameter $T$ increases

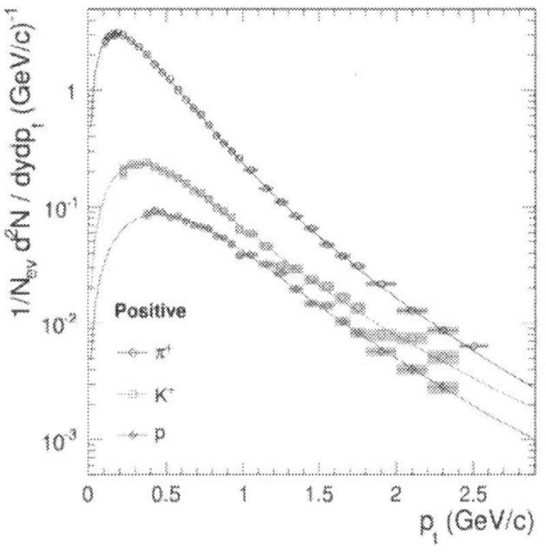


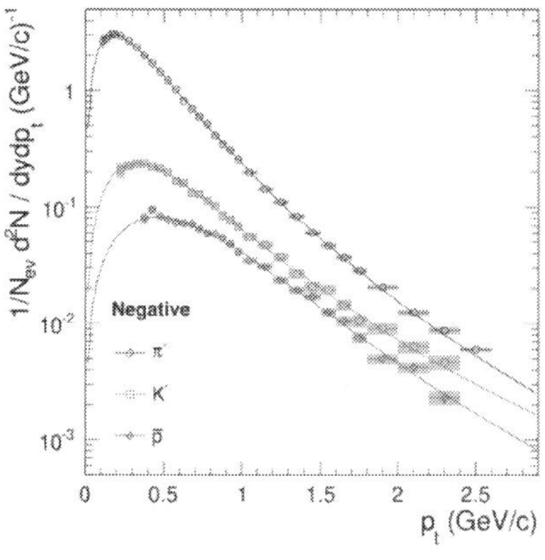


**Figure 9.** Transverse momentum spectra of positive (left) and negative (right) hadrons from pp collisions at $\sqrt{s}$= 900 GeV. Grey bands: total $p_t$-dependent error (systematic plus statistical); normalization systematic error (3.6%) not shown. The curves represent fits using a Lévy function.

with the mass of the particle. For example, it changes from 168±5 MeV for $K^0_S$ to 229±15 MeV for $\Lambda$ when the Lévy fit is used. The $\Xi^+$ and $\Xi^-$ apparently do not follow this trend. However, this is most likely because the very limited statistics do not allow for a well-constrained fit.

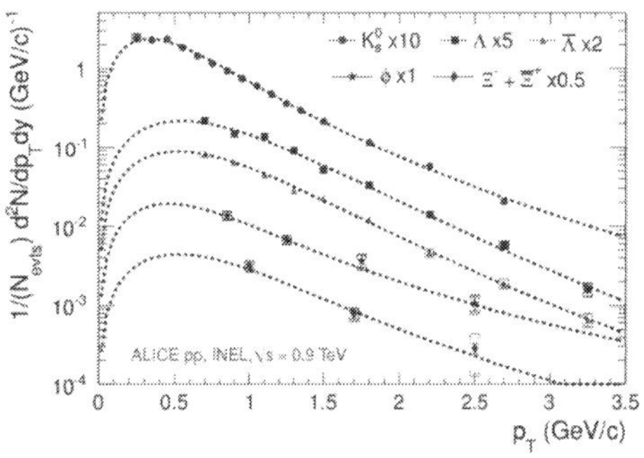


**Figure 10.** Particle spectra (corrected yields) as a function of $p_t$ for $K^0_S$(circles), $\Lambda$(squares), $\overline{\Lambda}$(triangles), $\phi$ (stars) and $\Xi^+ + \Xi^-$(diamonds). The data points are scaled for visibility and plotted at the centre of the bins. The dotted curves show Lévy fits.

Figure 11 shows the $p_t$-dependence of the $(K^+ + K^-)/(\pi^+ + \pi^-)$ ratio and also the measurements by the E735 [11] and STAR Collaborations [9]. It can be seen that the observed increase of K/$\pi$ with $p_t$ does not depend strongly on collision energy.

A comparison with event generators shows that at $p_t$ > 1.2 GeV/c, the measured K/$\pi$ ratio is larger than any of the model predictions.

In Figure 12, the measured $(p + \overline{p})$ / $(\pi^+ + \pi^-)$ ratio is compared to results at $\sqrt{s}$= 200 GeV from the PHENIX Collaboration [12]. Both measurements are feed-down corrected. At low $p_t$, there is no energy-dependence of the p/$\pi$ ratio visible, while at higher $p_t$> 1 GeV/c, the p/$\pi$ ratio is larger at $\sqrt{s}$= 900 GeV than at $\sqrt{s}$= 200 GeV energy.

Event generators seem to separate into two groups, one with high p/$\pi$ ratio (PYTHIA CSC and D6T), which agree better with the data and one group with a lower p/$\pi$ ratio (PHOJET and PYTHIA Perugia0), which are clearly below the measured values. These comparisons can be used for future tunes of the event generators.

Using the particle integrated yields, a comparison with STAR feed-down corrected particle ratios at $\sqrt{s}$ = 0.2 TeV [13] is shown in Fig. 13. With the centre of mass energy increasing

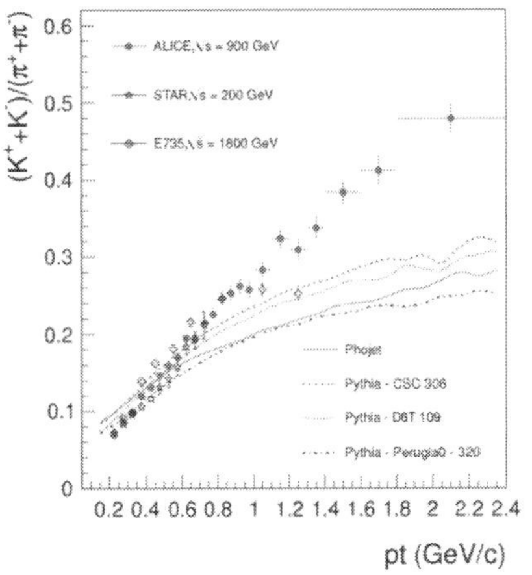


**Figure 11.** Ratio of $(K^+ + K^-)/(\pi^+ + \pi^-)$ as a function of $p_t$ from pp collisions at $\sqrt{s}$= 900 GeV (statistical errors only). Values from the E735 Collaboration [11] and the STAR Collaboration also are given. The dashed and dotted curves refer to calculations using PYTHIA and PHOJET at $\sqrt{s}$= 900 GeV.

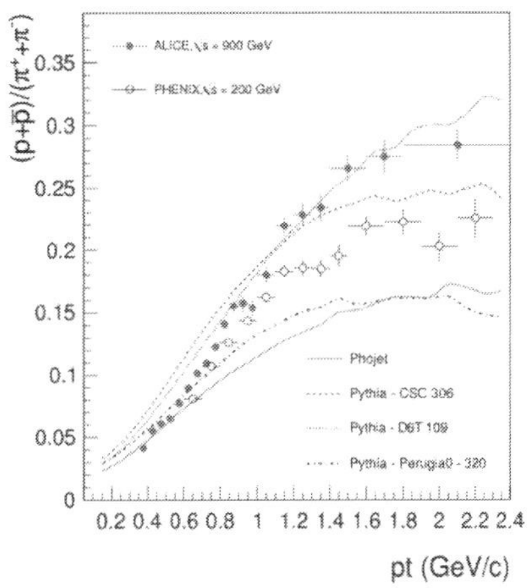


**Figure 12.** Ratio $(p + \bar{p})$ / $(\pi^+ + \pi^-)$ as a function of $p_t$ from pp collisions at $\sqrt{s}$= 900 GeV (statistical errors only). Values from the PHENIX Collaboration [12] also are given. The dashed and dotted curves refer to calculations using PYTHIA and PHOJET at $\sqrt{s}$= 900 GeV.

from $\sqrt{s}$ = 0.2 TeV to 0.9 TeV the measured ratios are similar except the $\bar{p}/\pi^-$ ratio which decreases slightly from $0.068 \pm 0.011$ to $0.051 \pm 0.005$. The strange to non-strange particle ratios seem to increase but stay compatible within uncertainties: the $K^-/\pi^-$ from $0.101 \pm 0.012$ to $0.121 \pm 0.013$ and the $\Lambda/\pi^+$ from $0.027 \pm 0.004$ to $0.032 \pm 0.003$.

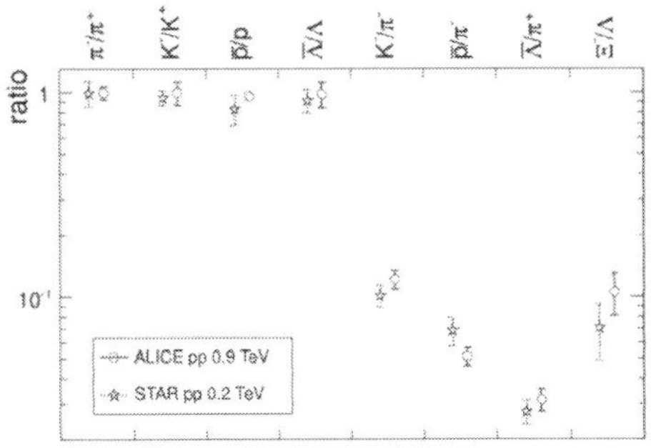


**Figure 13.** Ratios of integrated yields including $\pi^{(\pm)}$, $K^{(\pm)}$, p and $\bar{p}$ performed with the ALICE experiment [14, 15] and compared with STAR values for pp collisions at $\sqrt{s}$ = 0.2 TeV [13]. All ratios are feed-down corrected. Statistical and systematic uncertainties are added in quadrature.

The yields and $\langle p_t \rangle$ obtained with the ALICE experiment are compared for each particle with existing data at the same energy and also with results at lower and higher energies. The various experiments differ in acceptance and event selection (i.e. NSD or INEL) but the dependence of $\langle p_t \rangle$ with respect to these variables is found to be negligible. Consequently the $\langle p_t \rangle$ values are directly comparable, whereas the comparison of the yields can require further scaling because of different (pseudo)rapidity coverages. Figure 14 reports $\langle p_t \rangle$ measurements along with those of the STAR experiment [13, 16]. It is remarkable that the $\langle p_t \rangle$ remains close to the ISR parameterization [17] although the collision energy increased by a factor 36.

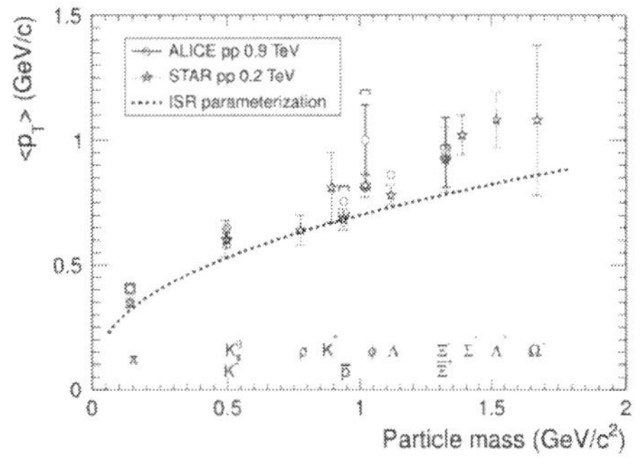


**Figure 14.** $\langle p_t \rangle$ vs. particle mass for the measurements performed with the ALICE experiment and compared with STAR values for pp collisions at $\sqrt{s}$ = 0.2 TeV [13, 16] and the ISR parameterization [17]. Both statistical (vertical error bars) and systematic (brackets) uncertainties are shown for ALICE data.

## 6. Conclusions

The first analysis of transverse momentum spectra of identified hadrons, $\pi^+$, $\pi^-$, $K^+$, $K^-$, p, $\overline{p}$, $K^0_S$, $\Lambda$, $\overline{\Lambda}$, $\Xi^+$, $\Xi^-$and $\phi$ in pp collisions at $\sqrt{s}$ = 900 GeV with the ALICE detector is presented. Various particle identification techniques have been used and this allows us to cover a broad momentum range. Agreement in the $K/\pi$ ratio is seen when comparing to $\overline{p}$p collisions at the Tevatron and Sp$\overline{p}$S. Comparing our results with similar measurement from the STAR Collaboration using pp collisions at $\sqrt{s}$= 200 GeV the shape of the spectra shows an increase of the hard component, but we observe only a slight increase of the mean-$p_t$ values.

The integrated yields and average transverse momenta of mesons containing strange quarks ($K^0_S$ and $\phi$) and hyperons ($\Lambda$, $\overline{\Lambda}$ and $\Xi$) have been compared with earlier data collected in pp and p$\overline{p}$ interactions at various energies. These results provide a useful baseline for comparisons with recent tunes of the PYTHIA model and a reference for future measurements in heavy-ion collisions at the LHC.

These studies demonstrate the precision with which ALICE can measure charged hadrons, resonances and topologically reconstructed weakly decaying particles. Measurements of these particles is a substantial part of the ALICE program in both pp and PbPb collisions.

## References

[1] ALICE Collaboration, K. Aamodt *et al.*, JINST 3 **64**, (2008) S08002.
[2] PHOBOS collaboration, Back, B. B. *et al.*, Phys. Rev. C **75**, (2007) 024910.
[3] ALICE Collaboration, K. Aamodt *et al.*, Eur. Phys. J. C **71**, (2011) 1594.
[4] W. Blum and L. Rolandi, Particle Detection with Drift Chambers, Springer, Berlin, 1998.
[5] T. Sjöstrand, S. Mrenna and P. Z. Skands, J. High Energy Phys. 0605 (2006) 026.
[6] R. Brun *et al.*, 1985 "GEANT3 User Guide", CERN Data Handling Division DD/EE/841; R. Brun *et al.*, 1994 "CERN Program Library Long Write- up, W5013, GEANT Detector Description and Simulation Tool".
[7] C. Tsallis and Constantino, J. Stat. Phys. **52** (1988) 479-487.
[8] PHENIX Collaboration, Adare A. *et al.*, Phys.Rev.D **83** (2011) 052004.
[9] STAR Collaboration, B. I. Abelev *et al.*, Phys. Rev. C **79** (2009) 034909.
[10] UA5 Collaboration, Ansorge, R. E. *et al.*, Phys. Lett. B **199** (1987) 311-316.
[11] E735 Collaboration, T. Alexopoulos *et al.*, Phys. Rev. D **48** (1993) 984-997.
[12] PHENIX Collaboration, Adler, Stephen Scott *et al.*, Phys. Rev. C **74** (2006) 024904.
[13] STAR Collaboration, B. I. Abelev *et al.*, Phys. Rev. C **75**, (2007) 064901.
[14] ALICE Collaboration, K. Aamodt *et al.*, Phys. Rev. Lett. **105**, (2010) 072002.
[15] ALICE Collaboration, K. Aamodt *et al.*, "Production of pions, kaons and protons in pp collision at $\sqrt{s} = 900\ GeV$ with ALICE at the LHC", accepted by Eur. Phys. J. C.
[16] STAR Collaboration, J. Adams *et al.*, Phys. Rev. Lett. **97**, (2006) 132301.
[17] M. Bourquin and J. M. Gaillard, Nucl. Phys. B **114**, (1976) 334.

Journal of Physics: Conference Series **316** (2011) 012020 doi:10.1088/1742-6596/316/1/012020

# Transition temperature and the equation of state from lattice QCD, Wuppertal-Budapest results

**Szabolcs Borsanyi[1], Gergely Endrodi[2], Zoltan Fodor[1,3], Christian Hoelbling[1], Sandor Katz[3], Stefan Krieg[1], Claudia Ratti[4,5] and Kalman Szabo[1]**

[1] Dep. of Physics, Wuppertal University, Gaussstr. 20, D-42119 Wuppertal, Germany
[2] Inst. für Theoretische Physik, Regensburg University, D-93040 Regensburg, Germany.
[3] Inst. for Theoretical Physics, Eötvös University, Pázmány 1, H-1117 Budapest, Hungary
[4] Dip. di Fisica Teorica, Università di Torino, via Giuria 1, I-10125 Torino, Italy
[5] INFN, Sezione di Torino

E-mail: `fodor@phys.uni-wuppertal.de`

**Abstract.** The QCD transition is studied on lattices up to $N_t = 16$. The chiral condensate is presented as a function of the temperature, and the corresponding transition temperature is extracted. The equation of state is determined on lattices with $N_t = 6, 8, 10$ and at some temperature values with $N_t = 12$. The pressure and the trace anomaly are presented as functions of the temperature in the range 100 ...1000 MeV . Using the same configurations we determine the continuum extrapolated phase diagram of QCD on the $\mu - T$ plane for small to moderate chemical potentials. Two transition lines are defined with two quantities, the chiral condensate and the strange quark number susceptibility.

## 1. Introduction

The study of QCD thermodynamics is receiving increasing interest in recent years. A systematic approach to determine the properties of the deconfinement phase transition is through lattice QCD. Lattice simulations indicate that the transition at vanishing chemical potential is merely an analytic crossover [1]. (This transition is similar to the water-vapor transition at high pressures, see Fig. 1. Note, that even in this case different definitions for the transition temperature give different numerical values.) Some interesting quantities that can be extracted from lattice simulations are the transition temperature $T_c$, the QCD equation of state and, for small chemical potentials, the phase diagram in the $\mu - T$ plane: we review the results on these observables that have been obtained by our collaboration using the staggered stout action with physical light and strange quark masses, thus $m_s/m_{ud} \simeq 28$ [2, 3]. For all details we refer the reader to Refs. [4, 5, 6]. For a pedagogical review see Ref [7]

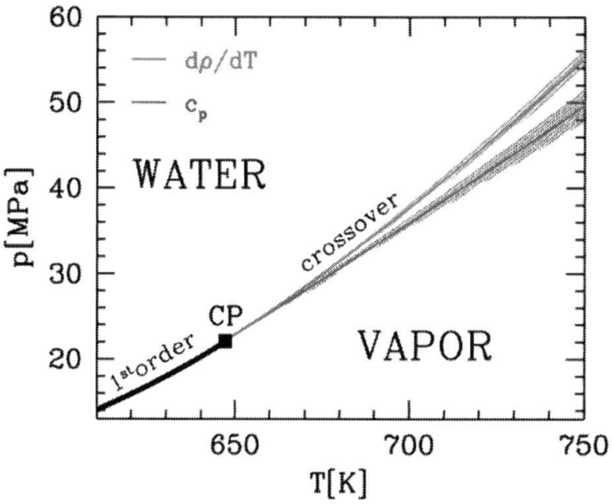


**Figure 1.** The phase diagram of water around its critical point (CP). For pressures below t he critical value ($p_c$) the transition is first order, for $p > p_c$ values there is a rapid cross-over. In the cross-over region the critical temperatures defined from different quanti ties are not necessarily equal. This can be seen for the temperature derivative of the densit y ($d\rho/dT$) and the specific heat ($c_p$). The bands show the non-negligible experimental uncertainties.

## 2. QCD transition and its charactiristic scale (transition temperature)

We present here the results for several quantities. We study strange susceptibility, the Polyakov-loop and the chiral condensate, and extract the value of $T_c$ associated to these observables. The $T_c$ values are different which reflects the nature of the crossover transition. For details we refer the reader to Ref. [4].

The strange susceptibility does not need any additional renormalization. The renormalization procedure of the Polyakov loop was given in [3]. The temperature dependences of the strange susceptibility and the Polyakov loop are shown on figure 2. The chiral condensate is defined as

$$\langle\bar{\psi}\psi\rangle_q = T\partial \ln Z/(\partial m_q V) \qquad \text{for } q = u, d, s \tag{1}$$

It is an indicator for the remnant of the chiral transition, since it rapidly changes around $T_c$. Its renormalization is given in [1]. A similar quantity can be defined as

$$\Delta_{l,s} = \frac{\langle\bar{\psi}\psi\rangle_{l,T} - m_l/m_s\langle\bar{\psi}\psi\rangle_{s,T}}{\langle\bar{\psi}\psi\rangle_{l,0} - m_l/m_s\langle\bar{\psi}\psi\rangle_{s,0}} \tag{2}$$

for $l = u, d$. Since the results at different lattice spacings are essentially on top of each other, we connect them to lead the eye (see figure 3). The value of $T_c$ that we obtain from the inflection point of the latter observable is $T_c = 157(3)(3)$. The lattice results can be compared with the predictions of the hadron resonce gas model. As it is shown on figure 4 the two results agree upto quite high temperatures.

It is also instructive to compare the peresent results obtained on $N_t$=6,8,10,12 and 16 with the results of the HotQCD Collaboration (c.f. [4]). Figure 5 shows this comparison. As it can be seen the results of the HotQCD Collaboration are getting closer and closer to our predictions. The long standing discrepancy is disappearing.

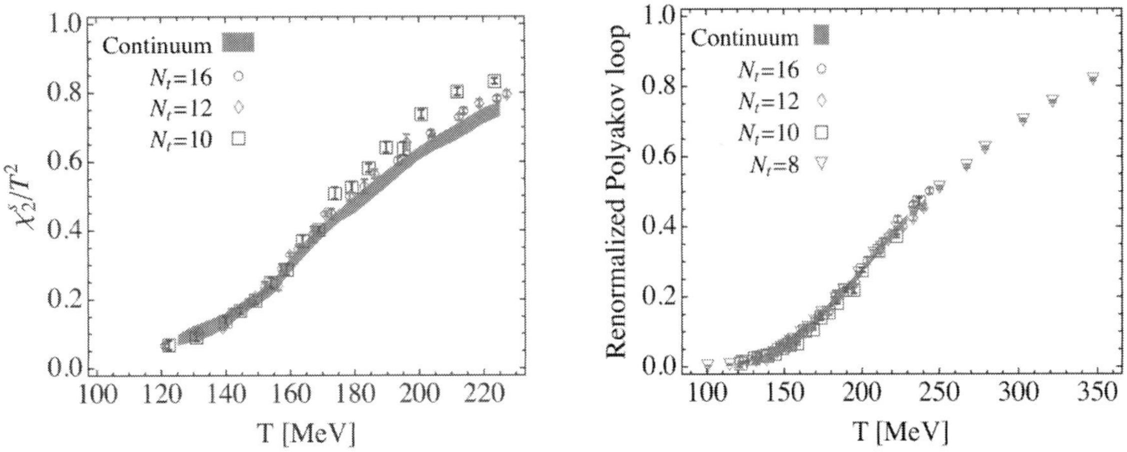


**Figure 2.** Left: strange quark number susceptibility as a function of the temperature. Right: renormalized Polyakov loop as a function of the temperature. In both figures, the d ifferent symbols correspond to different $N_t$. The gray band is the continuum extrapolated result.

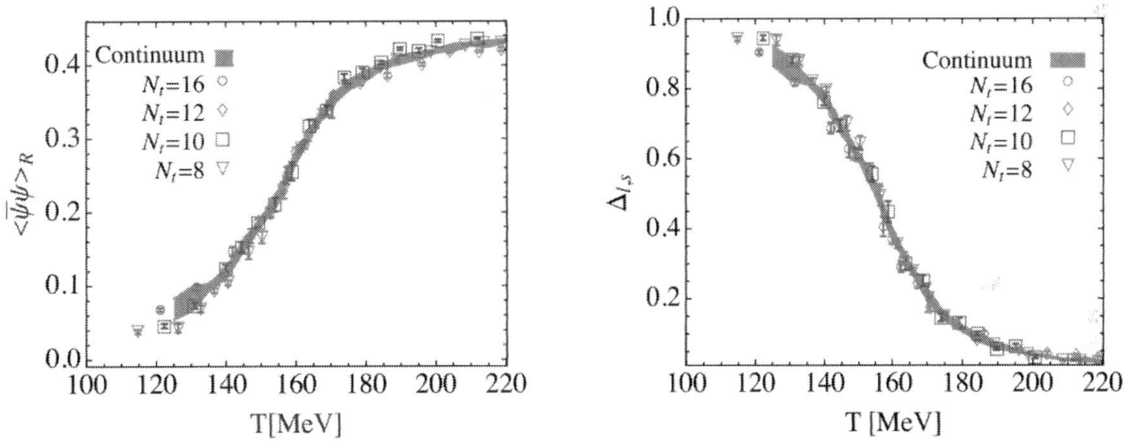


**Figure 3.** Left: renormalized chiral condensate $\langle\bar{\psi}\psi\rangle_R$ defined in Ref. [4]. Right: subtracted chiral condensate $\Delta_{l,s}$ defined in Eq. (2). In both figures, the different symbols correspond to different $N_t$. The gray band is our continuum estimate.

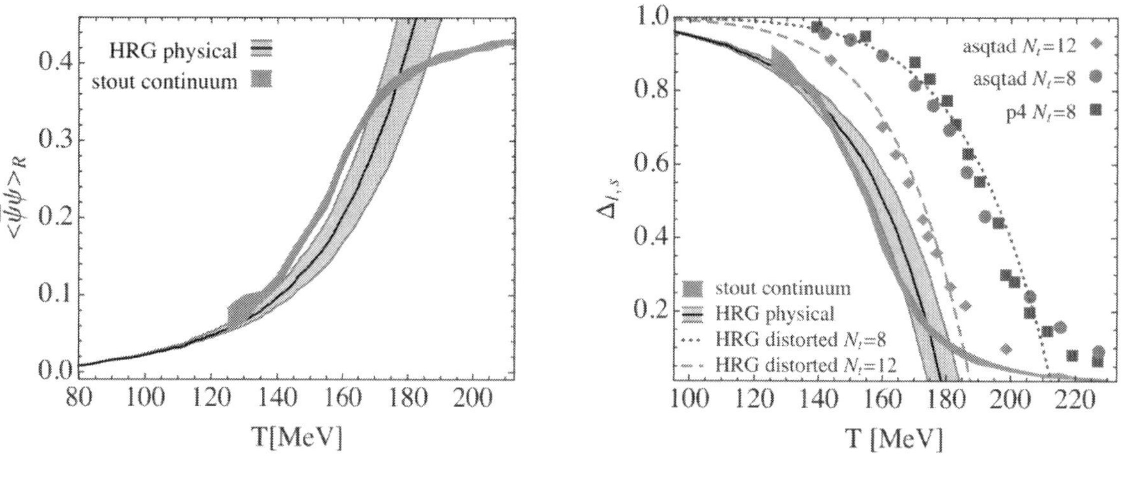


**Figure 4.** Left: Renormalized chiral condensate as defined in Ref. [4]. Right: Subtracted chiral condensate $\Delta_{l,s}$ as defined in eq. (2), as a function of the temperature. Gray bands are the continuum results of our collaboration, obtained with the stout action. Full symbols are obtained with the asqtad and p4 actions [8, 9]. In both panels, the solid line is the HRG model result with physical masses. The error band corresponds to the uncertainty in the quark mass-dependence of hadron masses. The dashed lines are the HRG+$\chi$PT model result with distorted masses, which take into account the discretization effects and heavier quark masses used in [8, 9] for $N_t = 8$ and $N_t = 12$.

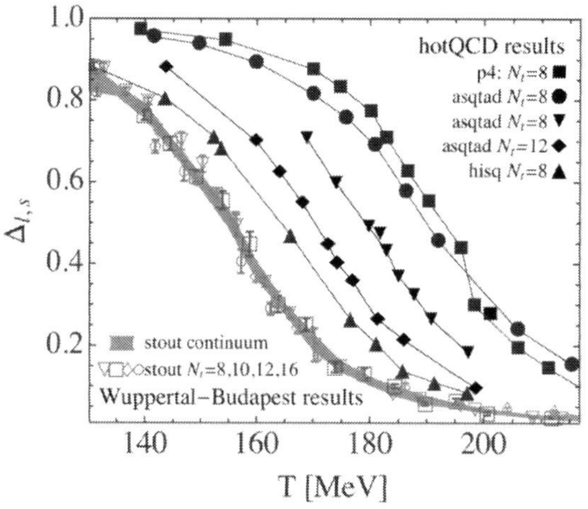


**Figure 5.** The subtracted chiral condensate $\Delta_{l,s}$ as a function of the temperature. We show a comparison between stout, asqtad, p4 and HISQ [8, 9] results. Our results are shown by colored open symbols, whereas the hotQCD results are shown by full black symbols. The gray band is our continuum result, the thin lines for the hotQCD data are intended to lead the eye. Our stout results were all obtained by the physical pion mass of 135 MeV. The full dots and squares correspond to $m_\pi = 220$ MeV, the full triangles and diamonds correspond to $m_\pi = 160$ MeV of the hotQCD collaboration.

Journal of Physics: Conference Series **316** (2011) 012020 doi:10.1088/1742-6596/316/1/012020

## 3. QCD equation of state

Next we present our results regarding the equation of state. The details of this calculation can be found in [5]. It is important to emphasize that quark masses were set to their physical values and we used quite fine lattices upto $N_t$=12. We have explicitely showed that four our systems the finite volume corrections are under control. Figure 6 shows two systems. One of them with a volume V the other one with a volume of 8·V. As it can be seen for the whole temperature range, there are practically no finite volume corrections, the two curves are lying on top of each other. On figure 7 (left) the $T$ dependence of the trace anomaly is shown for the $2+1$ flavor system. We have results at four different lattice spacings. Results show essentially no dependence on "$a$", they all lie on top of each other. Only the coarsest $N_t = 6$ lattice shows some deviation around $\sim 300$ MeV. On the same figure, we zoom in to the transition region. Here we also show the results from the Hadron Resonance Gas model: a good agreement with the lattice results is found up to hight $T$.

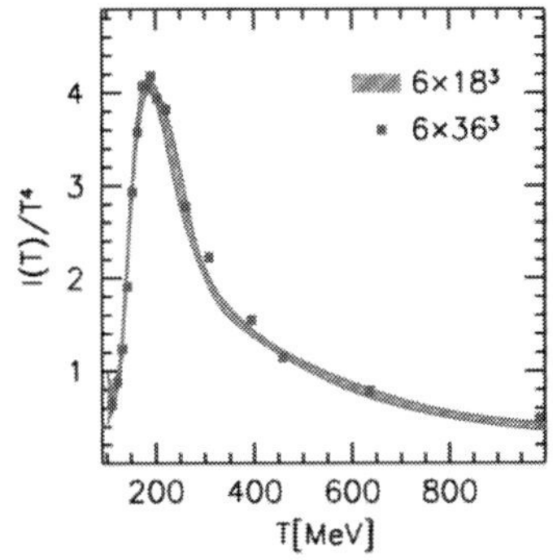


**Figure 6.** The trace anomaly on lattices with different spatial volumes: $N_s/N_t = 3$ (red band) and $N_s/N_t = 6$ (blue points).

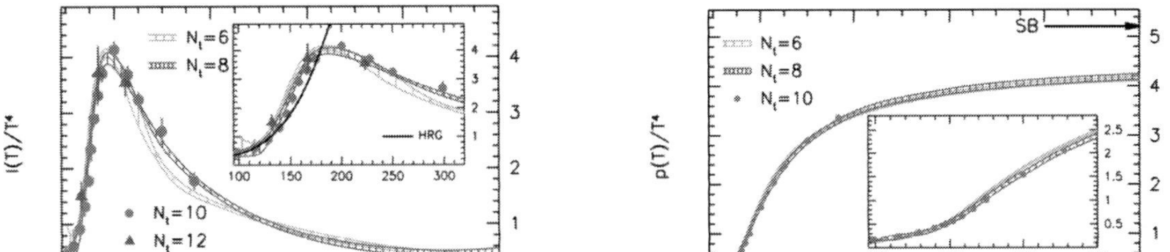


**Figure 7.** Left: The trace anomaly $I = \epsilon - 3p$ normalized by $T^4$ as a function of the temperature on $N_t = 6, 8, 10$ and 12 lattices. Right: The pressure normalized by $T^4$ as a function of the temperature on $N_t = 6, 8$ and 10 lattices. The Stefan-Boltzmann limit $(p/T^4)_{SB} \approx 5.209$ is indicated by an arrow. For our highest temperature $T = 1000$ MeV the pressure is almost 20% below this limit.

In order to obtain the pressure, we determine its partial derivatives with respect to the bare lattice parameters. $p$ is then rewritten as a multidimensional integral along a path in the space

Journal of Physics: Conference Series **316** (2011) 012020 doi:10.1088/1742-6596/316/1/012020

of bare parameters. The result is shown on figure 7 (right). To obtain the EoS for various $m_\pi$, we simulate for a wide range of bare parameters on the plane of $m_{u,d}$ and $\beta$ ($m_s$ is fixed to its physical value). Having obtained this large set of data we generalize the integral method and include all possible integration paths into the analysis [5, 10]. We remove the additive divergence of $p$ by subtracting the same observables measured on a lattice, with the same bare parameters but at a different $T$ value. Here we use lattices with a large enough temporal extent, so it can be regarded as $T = 0$. Figures 8 and 9 show the energy density, the entropy density and the speed of sound, respectively.

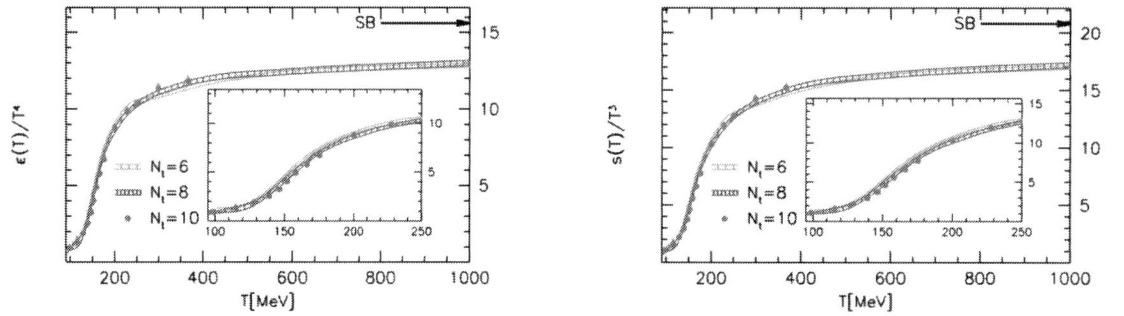


**Figure 8.** The energy density and the entropy normalized by $T^4$ and $T^3$, respectively, on $N_t = 6, 8$ and 10 lattices. The Stefan-Boltzmann limits $(\epsilon/T^4)_{SB} = 3(p/T^4)_{SB}$ and $(s/T^3)_{SB} = 4(p/T^4)_{SB}$ are indicated by an arrow.

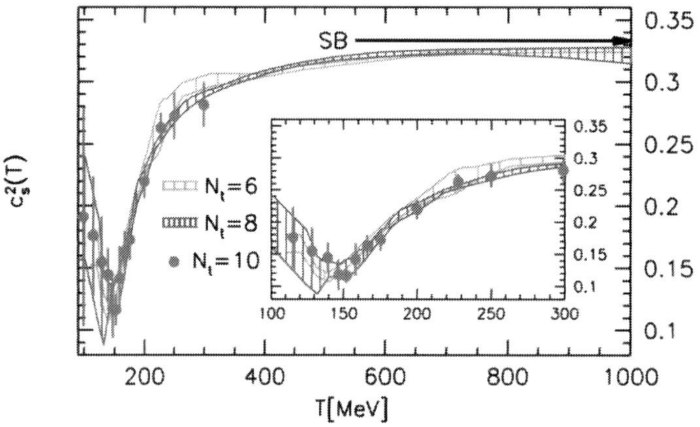


**Figure 9.** The squared of the speed of sound as a function of the temperature on $N_t = 6, 8$ and 10 lattices. The Stefan-Boltzmann limit is $c^2_{s,SB} = 1/3$ indicated by an arrow.

It is also instructive to compare the peresent results obtained on $N_t$=6,8,10 and 12 with the results of the HotQCD Collaboration (c.f. [4]). Figure 10 shows this comparison for the trace anomaly. As it can be seen the results of the HotQCD Collaboration are still quite far away from our results. Their peak position is about at a 20 MeV higher temperature, whereas their peaks heights are about 50% larger than ours. The clarification of this discrepancy remains for the future.

Journal of Physics: Conference Series **316** (2011) 012020 doi:10.1088/1742-6596/316/1/012020

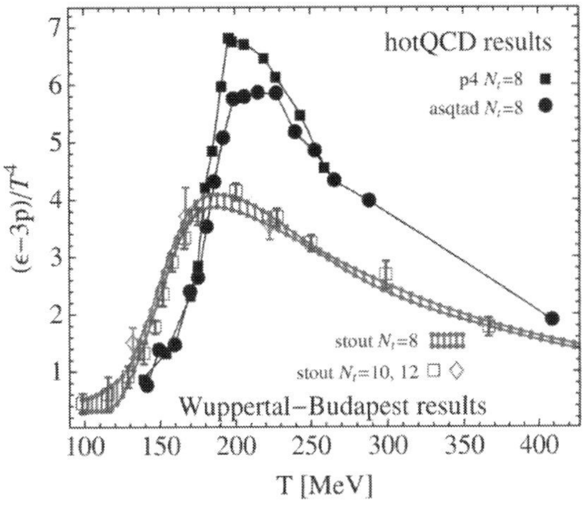


**Figure 10.** The normalized trace anomaly obtained in our study is compared to recent results from the "hotQCD" collaboration [8, 11].

## 4. The QCD phase diagram at nonzero quark density

We provided results for the transition temperatures $T_c$ at vanishing chemical potential ($\mu = 0$). Now we move out to the $\mu \neq 0$ plane (for the details see [6]). There are several possibilities. Figure 11 shows two scenarios. In principle it is possible that the transition temperatures (e.g. the one related to the chiral condensate and the one related to the strange susceptibility) are getting closer and closer to each other as the chemical potential increases. One might even end up with a critical endpoint. It is also possible that the transition is getting weaker and weaker and the difference between two transition temperatures increases as we increase the chemical potential. In this section we determine the leading order behaviour of the transition lines. Note, that this order will be not able to tell the difference between these two scenarios, thus it can not prove or exclude the existence of the critical endpoint. To that end one has to determine higher order terms.

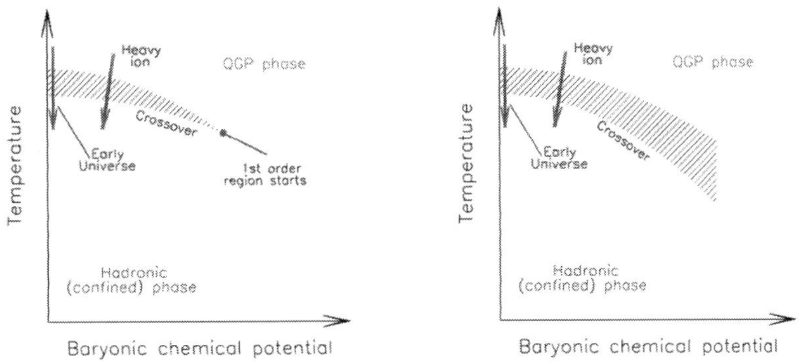


**Figure 11.** Two possible scenarios for the QCD phase diagram on the $\mu - T$ plane, defined using a given observable. The left panel shows a phase diagram with a transition growing stronger and possibly even turning into a real, first-order phase transition at a critical endpoint. The right panel on the other hand corresponds to a scenario with a weakening transition and no critical endpoint. The paths corresponding to systems describing the early Universe and a heavy ion collision are also shown by the arrows. Note that different observables may lead to different scenarios.

As we will se as one increases $\mu$ the transition temperature $T_c(\mu^2)$ decreases. Let us parameterize the transition line in the vicinity of the vertical $\mu$=0 axis as $T_c(\mu^2) = T_c(1 - \kappa \cdot \mu^2/T_c^2)$.

In order to determine the transition temperature as a function of $\mu$ we use two quantities which are monotonic in the transition region and do not depend on $\mu$ for zero or infinite temperatures. The transition temperature is defined as the temperature value at which these observables take their value as given by the inflection points of the curves at $\mu$=0.

The two observables we use are the renormalized chiral condensate and the normalized strange quark number susceptibility. In order to measure the $\mu$ dependence of these quantities we apply reweighting. Since our lattices are quite large the full reweighting method is quite expensive. Therefore, we truncate the $\mu$ dependence of the weights at $\mu^2$ order (this truncation of the original [12, 13] method is usually called the Taylor method; for a recent application see [14]).

The strange quark number susceptibility is the second derivative of the partition function with respect to the strange chemical potential. It needs no renormalization, since it is related to a conserved current. It is useful to normalize it by $T^2$, which provides a dimensionless combination. It is easy to see that at $T = 0$ its value is 0, whereas for infinitely large temperatures it approaches the Stefan-Boltzmann limit of 1.

For the chiral condensate we apply here a slightly different renormalization prescription than what was used for the determination of $T_c$. We cancel the additive divergences by subtracting the $T = 0$ contribution, while the multiplicative divergence due to the derivative with respect to the mass can be eliminated with a multiplication by the bare quark mass. Then, in order to have a dimensionless combination the whole expression can be divided by the fourth power of some dimensionful mass scale. In this work we use the T=0 pion mass for the normalization. This observable has also $\mu$ independent limiting values at zero and at infinitely high temperatures (at T=0 this is true for chemical potetials smaller than the baryon mass, well within our applicability region).

Our final result for the phase diagram on the $\mu$-T plane is shown in figure 12. The crossover region's extent changes little as the chemical potential increases. The two definitions give slightly different curves for $T_c(\mu)$. In this order of the $T_c(\mu)$ curve we do not see the critical enpoint. One should emphasize, that this is not a signal for non-existence of the critical point. This order neither can exclude nor prove the existence of the critical endpoint. It is useful to compare the whole picture to the freeze-out curve which summarizes experimental results on the T–$\mu$ points where hadronization of the quark-gluon plasma was observed. This curve is expected to lie in the interior of the crossover region, as is indicated by our results as well.

## Acknowledgements
Work supported in part by the EU grant (FP7/2007-2013)/ERC no. 208740. The work of C. R. is supported by funds provided by the Italian Ministry of Education, Universities and Reserach under the Firb Research Grant RBFR0814TT. The calculations have been carried out on the computer clusters at the Universities of Wuppertal and Budapest using GPU programming [15] and on the BG/P system at Forschungszentrum Julich.

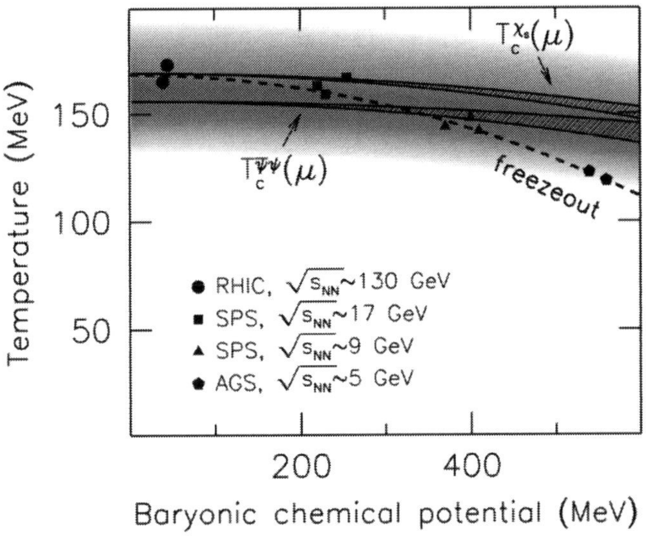


**Figure 12.** The crossover transition between the 'cold' and 'hot' phases is represented by the coloured area (blue and red correspond to the transition regions obtained from the chiral condensate and the strange susceptiblity, respectively). The lower solid band shows the result for $T_c(\mu)$ defined through the chiral condensate and the upper one through the strange susceptibility. The width of the bands represent the statistical uncertainty of $T_c(\mu)$ for the given $\mu$ coming from the error of the curvature $\kappa$ for both observables. The dashed line is the freeze-out curve from heavy ion experiments [16]. Also indicated are with different symbols the individual measurements of the chemical freeze-out from RHIC, SPS (Super Proton Synchrotron) and AGS (Alternating Gradient Synchrotron), respectively. The center of mass energies $\sqrt{s_{NN}}$ for each are shown in the legend.

## References

[1] Y. Aoki, G. Endrodi, Z. Fodor, S. D. Katz and K. K. Szabo, Nature **443**, 675 (2006)
[2] Y. Aoki, Z. Fodor, S. D. Katz and K. K. Szabo, Phys. Lett. B **643**, 46 (2006)
[3] Y. Aoki, S. Borsanyi, S. Durr, Z. Fodor, S. D. Katz, S. Krieg and K. K. Szabo, JHEP **0906**, 088 (2009)
[4] S. Borsanyi et al., [Wuppertal-Budapest Collaboration], JHEP **1009** (2010) 073
[5] S. Borsanyi *et al.*, JHEP **1011**, 077 (2010) [arXiv:1007.2580 [hep-lat]].
[6] G. Endrodi, Z. Fodor, S. D. Katz, K. K. Szabo, JHEP **1104** (2011) 001. [arXiv:1102.1356 [hep-lat]].
[7] Z. Fodor, S. D. Katz, [arXiv:0908.3341 [hep-ph]].
[8] A. Bazavov *et. al.*, *Phys. Rev.* **D80** (2009) 014504, [arXiv:0903.4379]
[9] A. Bazavov and P. Petreczky, [arXiv:1005.1131]
[10] G. Endrodi, Comput. Phys. Commun. **182**, 1307 (2011) [arXiv:1010.2952 [physics.comp-ph]].
[11] M. Cheng *et. al.*, *Phys. Rev.* **D81** (2010) 054504, [arXiv:0911.2215]
[12] Z. Fodor, S. D. Katz, Phys. Lett. **B534** (2002) 87-92. [hep-lat/0104001].
[13] Z. Fodor, S. D. Katz, JHEP **0404** (2004) 050. [hep-lat/0402006].
[14] O. Kaczmarek, F. Karsch, E. Laermann, C. Miao, S. Mukherjee, P. Petreczky, C. Schmidt, W. Soeldner *et al.*, Phys. Rev. **D83** (2011) 014504. [arXiv:1011.3130 [hep-lat]].
[15] G. I. Egri, Z. Fodor, C. Hoelbling, S. D. Katz, D. Nogradi, K. K. Szabo, Comput. Phys. Commun. **177** (2007) 631-639. [hep-lat/0611022].
[16] J. Cleymans and K. Redlich, Phys. Rev. Lett. **81**, 5284 (1998) [arXiv:nucl-th/9808030].

Journal of Physics: Conference Series **316** (2011) 012021 doi:10.1088/1742-6596/316/1/012021

# Reaction Plane and Beam Energy Dependence Of The Balance Function at RHIC

**Hui Wang for the STAR collaboration**

W216, National Superconducting Cyclotron Laboratory, Michigan State University, East Lansing, MI, 48824-1321, USA

E-mail: wang@nscl.msu.edu

**Abstract.** The balance function, which measures the correlation between opposite sign charge pairs, is sensitive to the mechanisms of charge formation and the subsequent relative diffusion of the balancing charges. The study of the balance function can provide information about charge creation time as well as the subsequent collective behavior of particles. In this paper, we present a reaction-plane-dependent balance function study for Au+Au collisions at $\sqrt{s_{\rm NN}} = 200$ GeV and compare with results from recent three particle correlation measurements. We also report balance functions for relative pseudorapidity ($\Delta\eta$), relative rapidity ($\Delta y$), and relative azimuthal angle ($\Delta\phi$) from the recent RHIC beam energy scan data.

## 1. Introduction

The study of correlations and fluctuations can provide information about the properties of the hot and dense matter created in relativistic heavy-ion collisions. one such observable, the balance functions, which measures the correlation between the opposite sign charge pairs, are sensitive to the mechanisms of charge formation and the subsequent relative diffusion of the balancing charges [1]. Due to conservation laws like electric charge conservation, particles and their anti-particles are pair produced and correlated initially in coordinate space, if a delayed hadronization occurs, the lower temperature and less expansion and diffusion will result in a narrower charge balance function.

The reaction-plane-dependent balance function in azimuthal angle can be written as

$$B(\phi, \Delta\phi) = \frac{1}{2}\{\frac{\Delta_{+-}(\phi, \Delta\phi) - \Delta_{++}(\phi, \Delta\phi)}{N_{+}(\phi)} + \frac{\Delta_{-+}(\phi, \Delta\phi) - \Delta_{--}(\phi, \Delta\phi)}{N_{-}(\phi)}\}. \quad (1)$$

Here $N_{+(-)}(\phi)$ is the total number of positive(negative) particles that have an azimuthal angle $\phi$ with respect to the event plane, $\Delta_{+-}(\phi, \Delta\phi)$ represents the total number of pairs summed over all events where the first (positive) particle has an azimuthal angle $\phi$ with respect to event plane and the second (negative) particle has a relative azimuthal angle $\Delta\phi$ with respect to the first particle. Similarly we can express $\Delta_{++}(\phi, \Delta\phi)$, $\Delta_{-+}(\phi, \Delta\phi)$ and $\Delta_{--}(\phi, \Delta\phi)$.

On the other hand, it has been discussed recently that the hot and dense matter created in heavy ion collision may form metastable domains where the parity is locally violated, this possible local parity violation [2] coupled with strong magnetic field produced by passing nuclei in such a collision could cause a charge separation across the reaction plane in non-central

Journal of Physics: Conference Series **316** (2011) 012021 doi:10.1088/1742-6596/316/1/012021

collisions called the chiral magnetic effect (CME) [3, 4, 5]. One observable proposed to measure the Chiral Magnetic Effect is the three point correlator [6] ,where $\phi$ is the azimuthal angle of a particle and $\psi_{RP}$ is the reaction plane angle.

$$\gamma_{\alpha,\beta} =< \cos(\phi_\alpha + \phi_\beta - 2\psi_{RP}) > . \tag{2}$$

## 2. Experimental Method

The data used in this analysis is from Au+Au collisions at $\sqrt{s_{NN}} = 200$ , 62.4, 39, 11.5, and 7.7 GeV taken by the STAR experiment. The Time Projection Chamber (TPC) was used as the main detector for charged particle tracking. All tracks were required to have a distance of closest approach (DCA) to the measured event vertex of less than 3 cm. For all charged particles, a transverse momentum cut of $0.2 < p_t < 2.0$ GeV/$c$ was applied while for identified particles, we used a $p_t$ cut of $0.2 < p_t < 0.6$ GeV/$c$. The pseudorapidity cut used is $|\eta| < 1.0$. Particle identification was done by selecting particles whose specific energy losses($dE/dx$) were within two standard deviations of the energy-loss predictions for a given particle type and momentum. In addition, electrons were excluded from the analysis for all cases by specific energy loss inside the TPC. To correct for differences between the acceptances for positive and negative particles, a mixed event subtraction is used for all balance function results throughout this paper.

To determine the event plane angle, the second Fourier harmonic in the azimuthal angle distribution is used. We used all tracks from the TPC while a $p_t$-weight method was used to maximize the event plane resolution. Also the $\phi$ weight method is applied to flatten the event plane distribution.

## 3. Results

### *3.1. Event-Plane-Dependent Balance Function at $\sqrt{s_{NN}}$ = 200 GeV*

Figure 1 shows $\phi = 0°$ (in-plane), $\phi = 45°$, and $\phi = 90°$ (out-of-plane) balance function for 40-50% centrality only. The in-plane balance function is narrower than the out-of-plane balance function, which is caused by the stronger collective flow in-plane: charge pairs are created closely in space and time due to conservation laws, this correlation would remain to the final stage if they are not diffused, which gives a narrower balance function. Figure 1 also shows the $\phi = 45°$ balance function, which is asymmetric and peaked at negative $\Delta\phi$. Because of the strong elliptic flow established at non-central collision, if charge pairs are emitted with an angle respect to the reaction plane, the correlation would be stronger on the in-plane side compare to out-of-plane side. Also shown are the blast-wave model calculations [7], we can see that for all three cases here, the blast-wave model agree well with experimental data.

To quantify the collective flow effect on balance function, we also study the weighted average cosine, $c_b(\phi)$, and sine, $s_b(\phi)$, extracted from the balance functions.

$$\begin{aligned} c_b(\phi) &= \frac{1}{z_b(\phi)} \int d\Delta\phi B(\phi, \Delta\phi) \cos(\Delta\phi), \\ s_b(\phi) &= \frac{1}{z_b(\phi)} \int d\Delta\phi B(\phi, \Delta\phi) \sin(\Delta\phi), \\ z_b(\phi) &= \int d\Delta\phi B(\phi, \Delta\phi). \end{aligned} \tag{3}$$

$c_b(\phi)$ represents the width of balance function. If charges are created at the same point and did not diffuse due to strong collective flow, $c_b(\phi)$ would be close to unity. $s_b(\phi)$ is an odd function of $\Delta\phi$, so it can quantify the asymmetry of balance function. Figure 2 shows $c_b(\phi)$

and $s_b(\phi)$ for Au+Au collisions at $\sqrt{s_{NN}}$ = 200 GeV. In the figure, $c_b(\phi)$ is closer to unity in the 0-5% centrality bin, which is due to a stronger collective flow in central collisions, while in mid-peripheral and peripheral collisions, $c_b(\phi)$ shows a difference between the in-plane and out-of-plane balance functions, which is caused by stronger elliptic flow in-plane. For $s_b(\phi)$, it reaches maximum at $\phi$=135°, 315° and minimum at $\phi$=45°, 225°, which demonstrates that charged pairs are more correlated on the in-plane side when emitted with an angle $\phi$ with respect to the reaction plane.

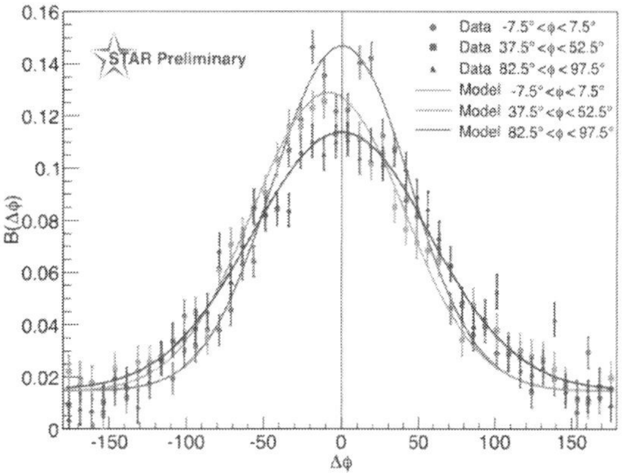


**Figure 1.** (Color online) The balance function for $\phi = 0°$ (in-plane), $\phi = 45°$, and $\phi = 90°$ (out-of-plane) particles pairs. The 40-50% centrality bin is shown. The points are from the experimental data (not corrected for event plane resolution), while solid lines represent the blast-wave model calculations [7].

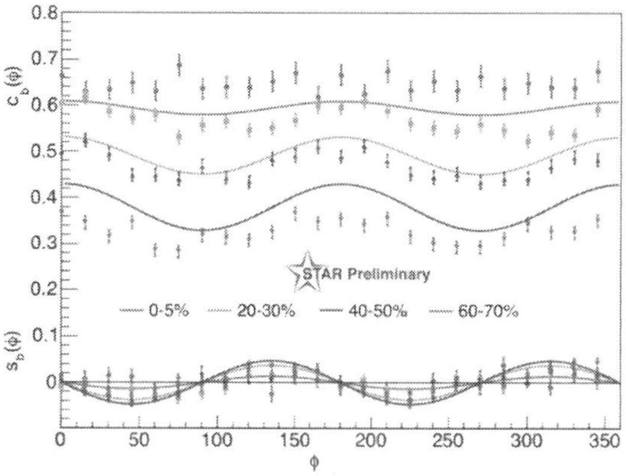


**Figure 2.** (Color online) The weighted average cosine and sine of balance function, four centralities are shown here. The points are from the experimental data (not corrected for event plane resolution), while solid lines are from the blast-wave model [7].

Figure 2 also shows a comparison with the blast-wave model [7]. The blast-wave model includes a breakup temperature $T_{kin}$, the maximum collective velocities in the in-plane and out-of-plane directions, the spatial anisotropy of the elliptic shape by fitting STAR published $v_2$ and spectra data [8], this model also assume local charge conservation and initial separation of balancing charges at freeze-out by fitting experimental results [9]. The difference between data and the blast-wave model could due to the finite event plane resolution for the data.

The difference between the same-sign and opposite-sign three point correlator $\gamma_{\alpha\beta}$ can be expressed as [7]

$$\gamma_p = \frac{1}{2}(2\gamma_{+-} - \gamma_{++} - \gamma_{--}) = \frac{2}{M}[v_2 < c_b(\phi) > +v_{2c} - v_{2s}], \quad (4)$$

where

$$v_{2c} =< c_b(\phi)\cos(2\phi) > -v_2 < c_b(\phi) >,$$
$$v_{2s} =< s_b(\phi)\sin(2\phi) >,$$

and the bracket represents

$$< f(\phi) >= \frac{1}{M}\int d\phi \frac{dM}{d\phi} z_b(\phi) f(\phi).$$

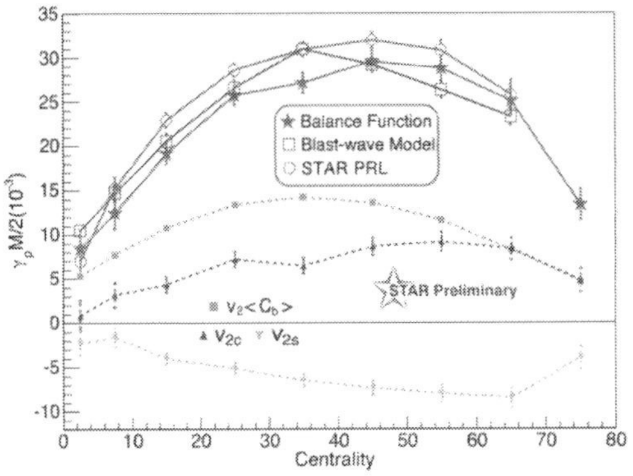


**Figure 3.** (Color online) parity observable $\gamma_P$ scaled by experimental multiplicity

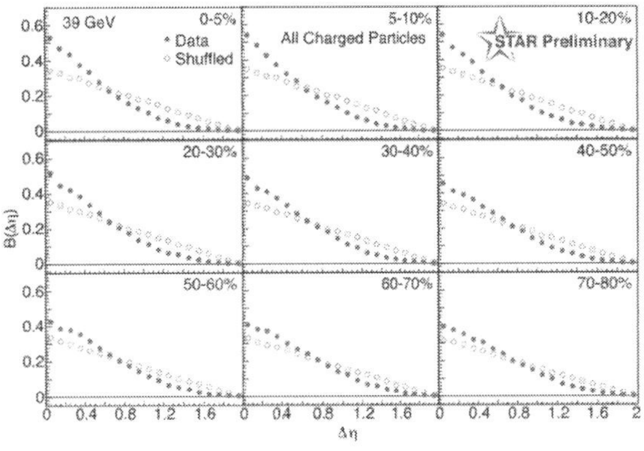


**Figure 4.** (Color online) The balance function in terms of $\Delta\eta$ for all charged particle pairs from Au + Au collisions at $\sqrt{s_{\rm NN}}$ = 39 GeV for nine centrality bins.

In this equation, $v_2 \langle c_b(\phi) \rangle$ will be positive if there are more charge pairs in-plane than out-of-plane, $v_{2c}$ will be positive if the charge pairs are more correlated in-plane than out-of-plane, while $v_{2s}$ will be negative if the charge pairs are more correlated on the in-plane side.

Figure 3 shows the parity observable calculate from balance function as well as its three components. All data points are corrected for the event-plane resolution here. To compare with previous results, we also plot the $\gamma_P$ from STAR published data [10] scaled by the measured uncorrected multiplicity in the same plot. Mathematically, the balance function result should equal the one from $\gamma_P$ and they do agree well. We can also see that, A thermal blast-wave model [7] incorporating local charge conservation and flow reproduces most of the signal.

*3.2. Balance Function at $\sqrt{s_{\rm NN}}$ = 39 GeV*

Previously, STAR has published measurements of the balance function from Au + Au, d + Au, and p + p collisions at $\sqrt{s_{\rm NN}}$ = 200 GeV [9]. In that paper, a simplified(Eq.(5), no reaction-plane-dependence) balance function was used to measure the charge pairs correlation in terms of relative pseudorapidity, $\Delta\eta$, relative rapidity, $\Delta y$, relative azimuthal angle, $\Delta\phi$, and invariant relative momentum, $q_{inv}$

$$B = \frac{1}{2}\{\frac{\Delta_{+-} - \Delta_{++}}{N_+} + \frac{\Delta_{-+} - \Delta_{--}}{N_-}\}. \quad (5)$$

It has been shown that the balance function in terms of $\Delta\eta$ and $\Delta\phi$ for all charged particles and in terms of $\Delta y$ for charged pions narrow in central Au + Au collisions. Here we perform a similar measurement from Au+Au collisions at $\sqrt{s_{\rm NN}}$ = 39 GeV.

Figure 4 shows the balance function in terms of $\Delta\eta$ for all charged particles for nine centrality bins. The balance function for central collisions is much more narrower compare to the one from peripheral collisions. The balance function for shuffled events is much wider than the balance functions from the measured data. The shuffled events are created by randomly shuffling the charges of the particles in each event. By doing this, all charge-momentum correlations are removed in shuffled events, which could produce the widest balance function within the same experimental acceptance.

The balance function in terms of $\Delta\phi$ would be more sensitive to transverse flow effect. Figure 5 shows the balance function in terms of $\Delta\phi$ for all charged particles for nine centrality bins. The data shows a peak at around $\Delta\phi = 0$ in central collision while it is almost flat in peripheral collisions. This is due to strong radial flow in central collision, when charge pairs are created, they are strongly correlated in space and time due to conservation laws, if a flow of perfect liquid established in the system, charges would remain correlated throughout the whole process which would provide this observed peak at $\Delta\phi = 0$. Also due to STAR's uniform $\eta$ acceptance, the balance function for shuffled events are always flat for all centralities.

Study of identified particle's balance function could reveal the charge production mechanisms of different particle species. Figure 6 show the balance functions for identified charged-pion pairs for nine centrality bins in terms of the relative rapidity. The balance function for experimental data gets narrower in central collisions, while balance functions for shuffled events are much more wider. The dip of balance function for $\Delta y < 0.2$ should due to interpair correlations (HBT, etc) according to model predictions [11].

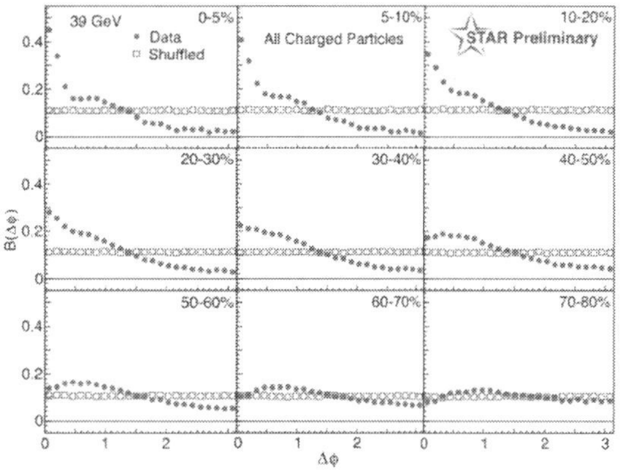


**Figure 5.** The balance function in terms of $\Delta\phi$ for all charged particle pairs from Au + Au collisions at $\sqrt{s_{\rm NN}} = 39$ GeV for nine centrality bins.

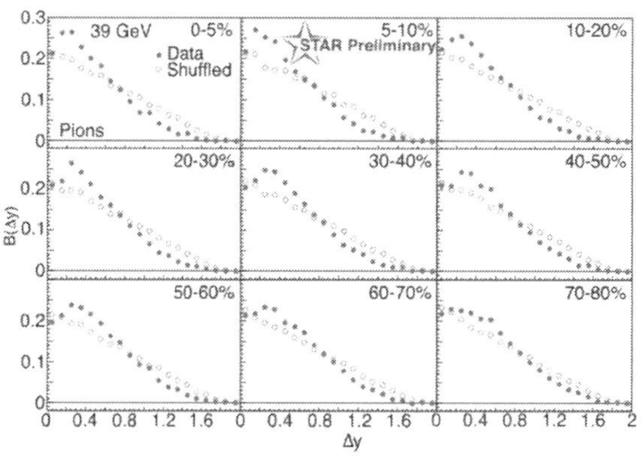


**Figure 6.** (Color online) The balance function in terms of $\Delta y$ for identfied pion pairs from Au + Au collisions at $\sqrt{s_{\rm NN}} = 39$ GeV for nine centrality bins.

### *3.3. Beam Energy Dependence of the Balance Function Widths*

To quantify the correlation measured from balance function, we calculate the weighted averages of $<\Delta\eta>$, $<\Delta\phi>$, and $<\Delta y>$ using Eq.(6). To remove the interpair correlation, the weighted average is calculated within $0.1 \leq \Delta\eta \leq 2.0$ for $<\Delta\eta>$ and $0.2 \leq \Delta y \leq 2.0$ for $<\Delta y>$

$$<\Delta\eta>= \frac{\Sigma_i B(\Delta\eta_i)\Delta\eta_i}{\Sigma_i B(\Delta\eta_i)}. \tag{6}$$

The left panel of Figure 7 shows the balance function widths for all charged particle in terms of the number of participating nucleons at $\sqrt{s_{\rm NN}} = 200$ , 62.4, 39, 11.5, and 7.7 GeV. For both $\Delta\eta$ and $\Delta\phi$, the balance function widths narrow in central collisions for all energies presented here. The upper right panel of Figure 7 shows the energy dependence of the balance function widths of $\Delta\eta$ for most central events(0-5%). The observed $<\Delta\eta>$ from data shows a smooth decrease with increasing beam energy, which is consistent with the effect that radial flow is stronger at higher collision energies. A UrQMD [12] model calculation incorporating STAR acceptance filter show the similar trend but over predict the value, which is due to early

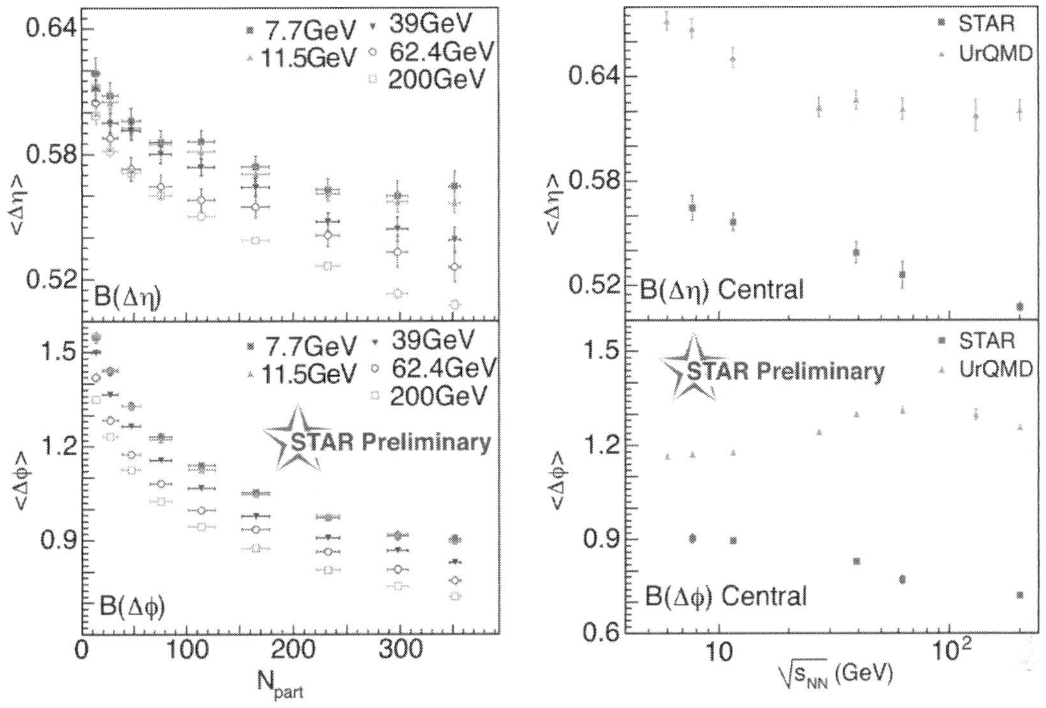


**Figure 7.** The balance function widths in terms of $\Delta\eta$, $\Delta\phi$ for all charged particle pairs from Au + Au collisions. Left panel shows centrality dependence, while the right panel shows beam energy dependence of most central evens (0-5%).

hadronization time and less flow in UrQMD. An early hadronization time would cause a wider balance function, while relative diffusion and interaction after freeze-out could further broaden it. The lower right panel of Figure 7 shows the same thing for $< \Delta\phi >$. The experimental data shows a decreasing balance function widths with increasing beam energy, while UrQMD's width shows a little increase with increasing beam energy. This could due to final state interaction of produced particles in the UrQMD model.

Figure 8 shows widths of the balance function for identified charged pions and identified charged kaons for both centrality dependence and beam energy dependence. For identified pions, the measured balance function widths get narrower in both central collisions and higher energies, the UrQMD model shows a similar energy dependence but over predicts the signal. In contrast, the widths of the measured balance function for identified kaons show little centrality or beam energy dependence, while it is consistent with the predictions of UrQMD model. This lack of centrality and beam energy dependence could indicate the different production mechanism of kaons: they are produced mainly at the beginning of the collision rather than during a later hadronization stage. However, the balance function for kaons is also affected by the decay $\phi \rightarrow K^+ + K^-$, a further study of balance function for $q_{inv}$ could help to distinguish the contribution from $\phi$ decay.

## 4. Summary

In this paper, we have presented new results from the reaction-plane-dependent balance function, the reaction-plane-dependent balance function analysis gives the same difference between the like-sign and unlike-sign charge dependent azimuthal correlations as the three point correlator

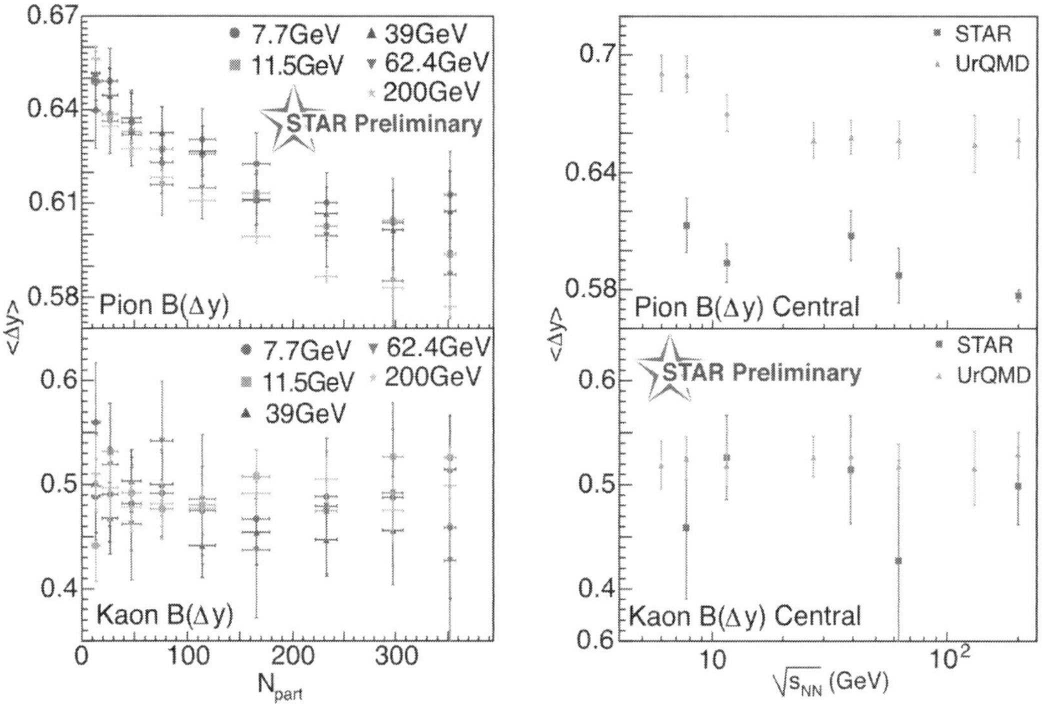


**Figure 8.** (Color online) The balance function widths in terms of $\Delta y$ for identified pions and kaons from Au + Au collisions. Left panel shows centrality dependence, while the right panel shows beam energy dependence of most central evens (0-5%).

results published by STAR. A thermal blast-wave model incorporating local charge conservation and flow can reproduce most of the difference between like- and unlike-sign charge-dependent azimuthal correlation.

We also measured the none event-plane-dependent balance function for Au+Au collisions at $\sqrt{s_{\rm NN}}$ = 200 , 62.4, 39, 11.5, and 7.7 GeV for all charged particles, identified charged pions, and identified charged kaons. We find that the balance functions in terms of $\Delta\eta$ and $\Delta\phi$ for all charged particles narrow both in central collisions and higher collision energies, which is consistent with the delayed hadronization picture. The UrQMD model reproduced the trend for energy dependence, but over predicted the value of balance function widths. For identified kaons, the balance function widths show little centrality or beam energy dependence, which indicate that kaons might be created early in the collision. However, a detail study using balance function for $q_{inv}$ is necessary to remove the effect from $\phi$ decay.

## 5. Acknowledgements

We thank S. Schlichting, S Pratt for enlightening discussions. We thank the RHIC Operations Group and RCF at BNL, the NERSC Center at LBNL, and the Open Science Grid consortium for providing resources and support. This work was supported in part by the Offices of NP and HEP within the US Department of Energy Office of Science, the US NSF, the Sloan Foundation; the DFG cluster of excellence Origin and Structure of the Universe of Germany; CNRS/IN2P3, STFC, and EPSRC of the United Kingdom; FAPESP CNPq of Brazil; Ministry of Education and Science of the Russian Federation; NNSFC, CAS, MoST, and MoE of China; GA and MSMT of the Czech Republic; FOM and NWO of the Netherlands; DAE, DST, and CSIR of

India; Polish Ministry of Science and Higher Education; Korea Research Foundation; Ministry of Science, Education, and Sports of the Republic of Croatia; and the Russian Ministry of Science and Technology and RosAtom of Russia.

## References

[1] S.A. Bass, P. Danielewicz, and S. Pratt, Phys. Rev. Lett. **85**, 2689 (2000).
[2] D. Kharzeev, R. D. Pisarski and M. H. G. Tytgat, Phys. Rev. Lett. **81**, 512 (1998).
[3] D. Kharzeev, Phys. Lett. B **633** 260 (2006).
[4] D. Kharzeev, L. McLerran, and H. Warringa, Nucl. Phys. A **803** 227 (2008).
[5] K. Fukushima, D. Kharzeev, and H. Warringa, Phys. Rev. D **78** 074033 (2008).
[6] S. A. Voloshin, Phys. Rev. C **70** 057901 (2004).
[7] S. Schlichting and S. Pratt, Phys. Rev. C **83**, 014913 (2011).
[8] B. I. Abelev *et al.* [STAR Collaboration], Phys. Rev. C **72**,14904 (2005).
[9] B. I. Abelev *et al.* [STAR Collaboration], Phys. Rev. C **82**, 024905 (2010).
[10] B. I. Abelev *et al.* [STAR Collaboration], Phys. Rev. Lett. **103**, 251601 (2009).
[11] S. Pratt and S. Cheng, Phys. Rev. C **68**, 014907 (2003).
[12] S. A. Bass, M. Belkacem, *et al.* Prog. Part. Nucl. Phys. **41** (1998) 225-370

# A Study of the QCD Critical Point Using Particle Ratio Fluctuations

**Terence J Tarnowsky (for the STAR Collaboration)**
National Superconducting Cyclotron Laboratory, Michigan State University, East Lansing, MI 48824, USA

E-mail: tarnowsk@nscl.msu.edu

**Abstract.** Dynamical fluctuations in global conserved quantities such as baryon number, strangeness, or charge may be observed near a QCD critical point. Results from new measurements of dynamical $K/\pi$ and $p/\pi$ ratio fluctuations are presented. The commencing of a QCD critical point search at RHIC has extended the reach of possible measurements of dynamical $K/\pi$ and $p/\pi$ ratio fluctuations from Au+Au collisions to lower energies. The STAR experiment has performed a comprehensive study of the energy dependence of these dynamical fluctuations in Au+Au collisions at the energies $\sqrt{s_{NN}}$ = 7.7, 11.5, and 39 GeV. New results are compared to previous measurements and to theoretical predictions from several models.

## 1. Introduction

Fluctuations and correlations are well known signatures of phase transitions. In particular, the quark/gluon to hadronic phase transition may lead to significant fluctuations [1]. In 2010, the Relativistic Heavy Ion Collider (RHIC) began a program to search for the QCD critical point. This involves an "energy scan" of Au+Au collisions from top collision energy ($\sqrt{s_{NN}}$ = 200 GeV) down to energies as low as $\sqrt{s_{NN}}$ = 7.7 GeV [2]. This critical point search will make use of the study of correlations and fluctuations, particularly those that could be enhanced during a phase transition that passes close to the critical point [3]. Particle ratio fluctuations are an observable that has already been studied as a function of energy and system-size. Combining these previous measurements with new results from the energy scan will provide additional information to assist in locating the QCD critical point.

Dynamical particle ratio fluctuations, specifically fluctuations in the $K/\pi$ and $p/\pi$ ratio, can provide information on the quark-gluon to hadron phase transition [4, 5, 6]. The variable used to measure these fluctuations is referred to as $\nu_{dyn}$. $\nu_{dyn}$ was originally introduced to study net charge fluctuations [7, 8]. $\nu_{dyn}$ quantifies deviations in the particle ratios from those expected for an ideal statistical Poissonian distribution. The definition of $\nu_{dyn,K/\pi}$ (describing fluctuations in the $K/\pi$ ratio) is,

$$\nu_{dyn,K/\pi} = \frac{< N_K(N_K-1) >}{< N_K >^2} + \frac{< N_\pi(N_\pi-1) >}{< N_\pi >^2} - 2\frac{< N_K N_\pi >}{< N_K >< N_\pi >}, \quad (1)$$

where $N_K$ and $N_\pi$ are the number of kaons and pions in a particular event, respectively. In this proceeding, $N_K$ and $N_\pi$ are the total charged multiplicity for each particle species. A formula similar to (1) can be constructed for $p/\pi$ and other particle ratio fluctuations. By definition, $\nu_{dyn}$

= 0 for the case of a Poisson distribution of kaons and pions. It is also largely independent of detector acceptance and efficiency in the region of phase space being considered [8]. An in-depth study of $K/\pi$ fluctuations in Au+Au collisions at $\sqrt{s_{NN}}$ = 200 and 62.4 GeV was previously carried out by the STAR experiment [9].

Earlier measurements of particle ratio fluctuations utilized the variable $\sigma_{dyn}$ [10],

$$\sigma_{dyn} = sgn(\sigma^2_{data} - \sigma^2_{mixed})\sqrt{|\sigma^2_{data} - \sigma^2_{mixed}|}\ , \qquad (2)$$

where $\sigma$ is the relative width of the $K/\pi$ or $p/\pi$ distribution in either real data or mixed events. It has been shown that $\nu_{dyn}$ is a first order expansion of $\sigma_{dyn}$ about the mean of its denominator [11, 12, 13]. The two variables are related as $\sigma^2_{dyn} \approx \nu_{dyn}$.

## 2. Experimental Analysis

The data presented here for $K/\pi$ and $p/\pi$ fluctuations was acquired by the STAR experiment at RHIC from minimum bias (MB) Au+Au collisions at center-of-mass collision energies ($\sqrt{s_{NN}}$) of 200, 130, 62.4, 39, 19.6, 11.5, and 7.7 GeV [14]. The main particle tracking detector at STAR is the Time Projection Chamber (TPC) [15]. All detected charged particles in the pseudorapidity interval $|\eta| < 1.0$ were measured. The transverse momentum ($p_T$) range for pions and kaons was $0.2 < p_T < 0.6$ GeV/$c$, and for protons was $0.4 < p_T < 1.0$ GeV/$c$. Charged particle identification involved measured ionization energy loss ($dE/dx$) in the TPC gas and total momentum ($p$) of the track. The energy loss of the identified particle was required to be less than two standard deviations ($2\sigma$) from the predicted energy loss of that particle at a particular momentum. A second requirement was that the measured energy loss of a pion/kaon was more than $2\sigma$ from the energy loss prediction of a kaon/pion, and similarly for $p/\pi$ measurements. The centralities used in this analysis account for 0-10, 10-20, 20-30, 30-40, 40-50, 50-60, 60-70, and 70-80% of the total hadronic cross section.

Data was also analyzed using the recently completed Time of Flight (TOF) detector [16]. The TOF is a multi-gap resistive plate chamber (MRPC) detector. Particle identification was carried out using the time-of-flight information for a track along with its momentum, determined by the TPC. From these two quantities the particle's mass can be calculated. Utilizing the TOF allows the identified particle momentum reach to be extended. TOF identification extends the $p_T$ range of pions and kaons from, $0.6 < p_T < 1.4$ GeV/$c$ and protons from $1.0 < p_T < 1.8$ GeV/$c$.

## 3. Results and Discussion

The STAR experiment has recently acquired data at three new energies. The first phase of the QCD critical point search involved Au+Au collisions at $\sqrt{s_{NN}}$ = 39, 11.5, and 7.7 GeV. STAR has measured dynamical $K/\pi$ and $p/\pi$ fluctuations at all these energies.

Figure 1(a) shows the measurement of $\nu_{dyn}$ for $K/\pi$ fluctuations as a function of centrality, scaled by the uncorrected charged particle multiplicity, $dN/d\eta$, from Au+Au collisions at $\sqrt{s_{NN}}$ = 39 GeV. The measured $dN/d\eta$ is a raw value, not corrected for efficiency and acceptance. Scaling by $dN/d\eta$ removes the $1/N_{ch}$ dependence of $\nu_{dyn}$ [7], where $N_{ch}$ is the charged particle multiplicity. The data points include charged particles identified using TPC $dE/dx$ for low momenta and mass-squared ($m^2$) by the STAR TOF for higher momenta. The errors shown are statistical. Dynamical $K/\pi$ fluctuations are always positive. In peripheral collisions at higher energies ($\sqrt{s_{NN}}$ = 62.4 and 200 GeV), $\nu_{dyn,K/\pi}$ was found to scale linearly with $dN/d\eta$. A linear fit to the data points and its parameters is shown for $dN/d\eta > 50$. As seen in Figure 1(a), this linear scaling is broken in very peripheral collisions for $dN/d\eta$ less than 50 charged particles.

Figure 1(b) shows the measurement of $\nu_{dyn,p/\pi}$ as a function of centrality, scaled by the uncorrected charged particle multiplicity, $dN/d\eta$, from Au+Au collisions at $\sqrt{s_{NN}}$ = 39 GeV.

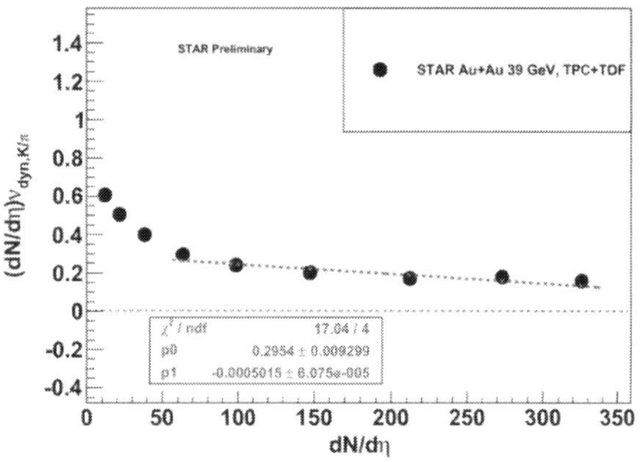


(a)

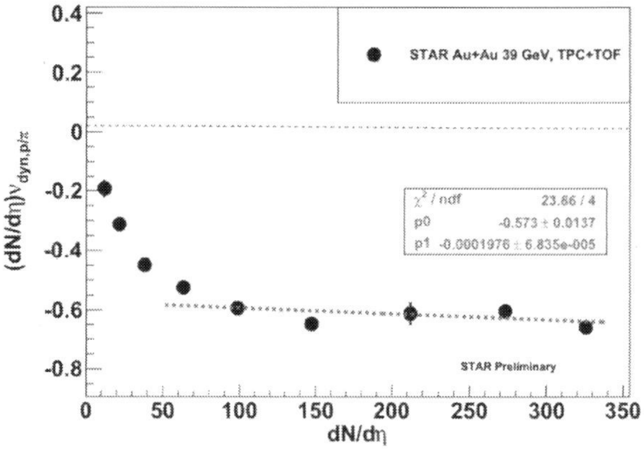


(b)

**Figure 1.** Results for the measurement of $\nu_{dyn,K/\pi}$ (a) and $\nu_{dyn,p/\pi}$ (b) scaled by uncorrected charged particle multiplicity, $dN/d\eta$, as measured by the STAR TPC+TOF (black stars) from central 0-5% Au+Au collisions at $\sqrt{s_{NN}}$ = 39 GeV. Errors shown are statistical.

A linear fit to the data for $dN/d\eta > 50$ is also included. Dynamical $p/\pi$ fluctuations are always negative due to correlated production of protons and pions from resonances such as the $\Delta$. Similar to $\nu_{dyn,K/\pi}$ at $\sqrt{s_{NN}}$ = 39 GeV, dynamical $p/\pi$ fluctuations are approximately centrality independent until $dN/d\eta < 50$, where they become less negative in the peripheral bins.

Accounting for track matching efficiency and detector acceptance, the TOF detection efficiency is approximately $\frac{2}{3}$ that of the TPC. Differences between the two data samples are still under investigation.

Figure 2(a) shows the measurement of $\nu_{dyn,K/\pi}$ as a function of centrality, scaled by the uncorrected charged particle multiplicity, $dN/d\eta$, from Au+Au collisions at $\sqrt{s_{NN}}$ = 11.5 GeV. The data points include charged particles identified using TPC $dE/dx$ for low momenta and mass$^2$ by the STAR TOF for higher momenta. The errors shown are statistical. A linear fit to the data points and its parameters is shown for $dN/d\eta > 40$. In these early results, there is no apparent scaling with $dN/d\eta$ for $\nu_{dyn,K/\pi}$ at $\sqrt{s_{NN}}$ = 11.5 GeV. The magnitude of the dynamical $K/\pi$ fluctuations at $\sqrt{s_{NN}}$ = 11.5 GeV have a similar range from peripheral to central collisions as those shown in Figure 1(a) at $\sqrt{s_{NN}}$ = 39 GeV.

Figure 2(b) shows the measurement of $\nu_{dyn,p/\pi}$ as a function of centrality, scaled by the uncorrected charged particle multiplicity, $dN/d\eta$, from Au+Au collisions at $\sqrt{s_{NN}}$ = 11.5 GeV. A linear fit to the data for $dN/d\eta > 40$ is also included. The multiplicity scaled $\nu_{dyn,p/\pi}$ at $\sqrt{s_{NN}}$ = 11.5 GeV does not show a rapid change as a function of centrality in peripheral collisions, as the same observable does at $\sqrt{s_{NN}}$ = 39 GeV in Figure 1(b). In central collisions, the multiplicity scaled $\nu_{dyn,p/\pi}$ at $\sqrt{s_{NN}}$ = 11.5 GeV measures $\approx$ -0.8, whereas in central collisions at $\sqrt{s_{NN}}$ = 39 GeV it reaches a value of $\approx$ -0.6.

Figure 3(b) includes the measurement of dynamical $p/\pi$ fluctuations in Au+Au collisions at $\sqrt{s_{NN}}$ = 7.7 GeV. This is currently the lowest energy data available at RHIC. As with dynamical fluctuations at $\sqrt{s_{NN}}$ = 11.5 and 39 GeV, $\nu_{dyn,p/\pi}$ is plotted as a function of centrality and scaled by the uncorrected charged particle multiplicity, $dN/d\eta$. At this low energy, the antibaryon/baryon ratio is $\sim 0.01$ [17]. Thus, the measurement of $p/\pi$ fluctuations at this energy is dominated by protons. Measurements using both TPC and TOF particle identification are

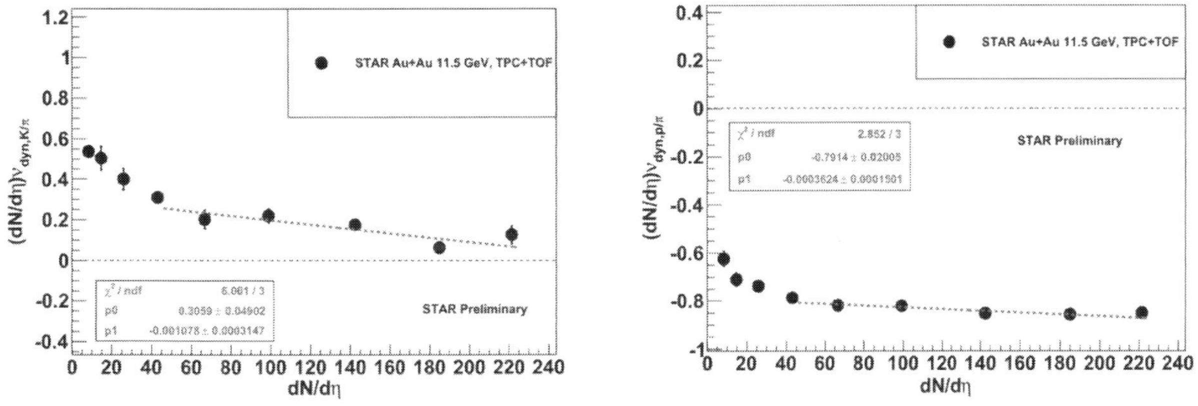


(a) (b)

**Figure 2.** Results for the measurement of $\nu_{dyn,K/\pi}$ (a) and $\nu_{dyn,p/\pi}$ (b) scaled by uncorrected charged particle multiplicity, $dN/d\eta$, as measured by the STAR TPC+TOF (black circles) from central 0-5% Au+Au collisions at $\sqrt{s_{NN}}$ = 11.5 GeV. Errors shown are statistical.

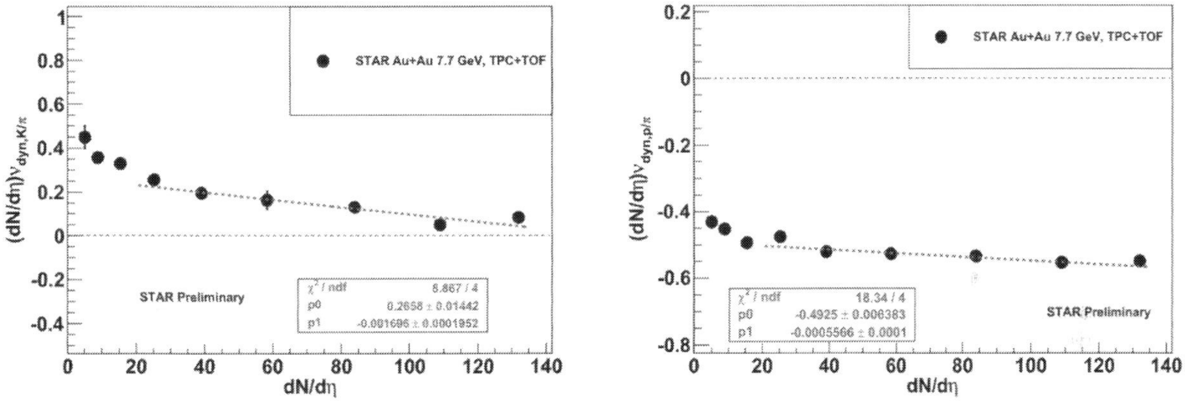


(a) (b)

**Figure 3.** Results for the measurement of $\nu_{dyn,K/\pi}$ (a) and $\nu_{dyn,p/\pi}$ (b) scaled by uncorrected charged particle multiplicity, $dN/d\eta$, as measured by the STAR TPC+TOF (black circles) from central 0-5% Au+Au collisions at $\sqrt{s_{NN}}$ = 7.7 GeV. A linear fit to the TPC data and its parameters are also shown. Errors shown are statistical.

strongly negative and approximately independent of centrality. This can be compared to Figures 1(b) and 2(b) where the scaled $\nu_{dyn,p/\pi}$ exhibits a centrality dependence in peripheral collisions. In central collisions, it reaches a magnitude of ≈ -0.6, which is the same value seen at $\sqrt{s_{NN}}$ = 39 GeV. The dynamics of particle production change with energy and additional studies of charged species dependence may provide information on the behavior of these scalings.

Figure 4(a) shows the measured dynamical $K/\pi$ fluctuations as a function of incident energy, expressed as $\nu_{dyn,K/\pi}$. The measured fluctuations from the STAR experiment are plotted as the black stars, measurements from the NA49 experiment [18] as blue squares, a transport model prediction from UrQMD (red open stars), using the STAR experimental acceptance. The

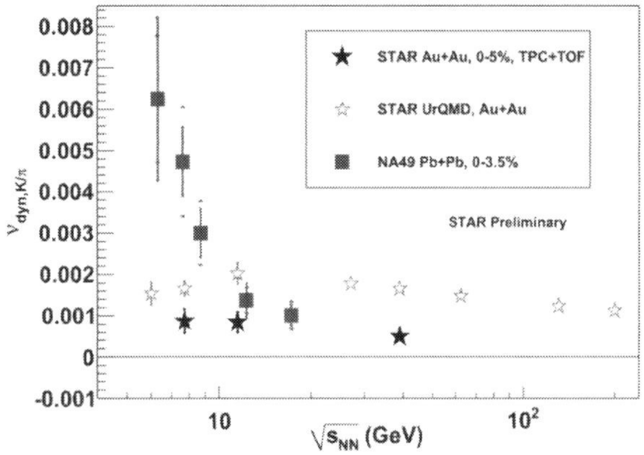


(a)

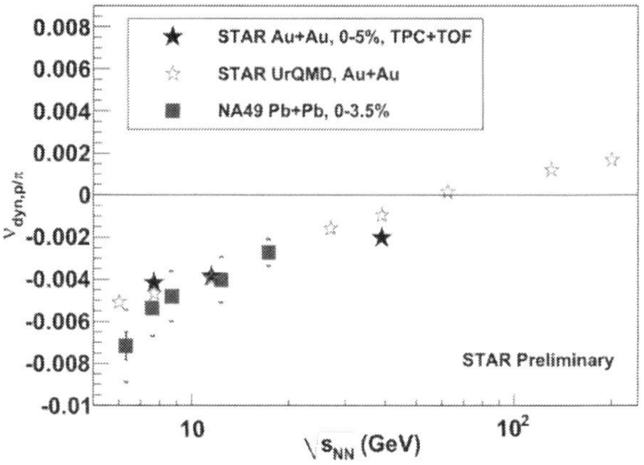


(b)

**Figure 4.** Results for the measurement of $\nu_{dyn,K/\pi}$ (a) and $\nu_{dyn,p/\pi}$ (b) as measured by the STAR TPC+TOF (black stars) from central 0-5% Au+Au collisions at $\sqrt{s_{NN}}$ = 7.7-39 GeV. Also shown are the same measurements from the NA49 experiment (solid blue squares) from central 0-3.5% Pb+Pb collisions. Model predictions from UrQMD using the STAR experimental acceptance (red open stars) are also included.

published NA49 results have been converted to $\nu_{dyn}$ using the relation $\sigma^2_{dyn} \approx \nu_{dyn}$. STAR measures dynamical $K/\pi$ fluct'uations that are approximately independent of collision energy from $\sqrt{s_{NN}}$ = 7.7-39 GeV. The NA49 experiment observes a strong decrease with increasing incident energy for central 0-3.5% Pb+Pb collisions (solid blue squares). Particle production mechanisms change with energy, so an energy independent fluctuation requires further study to determine systematically which components of $\nu_{dyn}$ are changing. UrQMD does not demonstrate a strong energy dependence, while another transport model (HSD, not shown) replicates the large increase in fluctuations at low energy observed by NA49. Both models treat resonance decay differently, which is one possibility to explain the different predictions.

Figure 4(b) shows the measured dynamical $p/\pi$ fluctuations as a function of incident energy, expressed as $\nu_{dyn,p/\pi}$. The measured fluctuations from the STAR experiment are plotted as the black stars, measurements from the NA49 experiment [18] at the SPS as blue squares, and a transport model prediction from UrQMD (open red stars), using the STAR experimental acceptance. The published NA49 results have been converted to $\nu_{dyn}$ using the relation $\sigma^2_{dyn} \approx \nu_{dyn}$. STAR measures a general trend from larger, negative values of $p/\pi$ fluctuations at lower energies, which then increases towards zero at $\sqrt{s_{NN}}$ = 39 GeV. There is good agreement between the two different experiments at the lowest energies measured by STAR ($\sqrt{s_{NN}}$ = 7.7 and 11.5 GeV). The model prediction reproduce the qualitative trend shown in the data, but over predicts the magnitude at higher energies and crosses from negative to positive values. One contribution to this occurs as the models transition from associated to pair particle production at higher energies.

## 4. Summary

Results on dynamical particle ratio ($K/\pi$ and $p/\pi$) fluctuations from Au+Au collisions at $\sqrt{s_{NN}}$ = 7.7, 11.5, and 39 GeV have been presented. The dynamical fluctuations scaled by the measured charged particle multiplicity have been shown as a function of collision centrality. There is no apparent scaling with charged particle multiplicity for $K/\pi$ fluctuations in peripheral collisions at any energy, nor for $p/\pi$ fluctuations at $\sqrt{s_{NN}} = 11.5$ and 39 GeV. However, multiplicity scaled $\nu_{dyn,p/\pi}$ at $\sqrt{s_{NN}} = 7.7$ GeV is almost constant as a function of centrality. The STAR experiment has measured dynamical $K/\pi$ fluctuations that are approximately energy independent at these three energies and $p/\pi$ fluctuations that increase from larger negative values toward zero with increasing energy, starting from $\sqrt{s_{NN}} = 7.7$ GeV.

These are the first measurements of dynamical particle ratio fluctuations in Au+Au collisions below injection energy at RHIC. Additional energies will be analyzed as part of the QCD critical point search at RHIC, where STAR will be able to precisely measure dynamical $K/\pi$ and $p/\pi$ fluctuations, providing a comprehensive picture of these fluctuations from $\sqrt{s_{NN}} = 7.7$-200 GeV. Dynamical fluctuations and their scaling properties will be studied for all separate charged particle ratio combinations.

## References

[1] V. Koch, arXiv:0810.2520 [nucl-th] (2008).
[2] M. M. Aggarwal *et al.* (STAR Collaboration), arXiv:1007.2613v1 [nucl-ex]
[3] M. M. Aggarwal *et al.* (STAR Collaboration), Phys. Rev. Lett. **105**, 022302 (2010).
[4] M. I. Gorenstein, M. Gazdzicki and O. S. Zozulya, Phys. Lett. B **585**, 237 (2004).
[5] V. Koch, A. Majumder and J. Randrup, Phys. Rev. C **72**, 064903 (2005).
[6] V. Koch, A. Majumder and J. Randrup, Phys. Rev. Lett. **95**, 182301 (2005).
[7] C. Pruneau, S. Gavin and S. Voloshin, Phys. Rev. C **66**, 044904 (2002).
[8] J. Adams *et al.* (STAR Collaboration), Phys. Rev. C **68**, 044905 (2003).
[9] B. I. Abelev *et al.* (STAR Collaboration), Phys. Rev. Lett. **103**, 092301 (2009).
[10] C. Alt *et al.* (NA49 Collaboration), Phys. Rev. C **79**, 044910 (2009).
[11] S. Jeon and V. Koch, arXiv:hep-ph/0304012v1 (2003).
[12] G. Baym, and H. Heiselberg, Phys. Lett. B **469**, 7 (1999).
[13] S. Das, Ph.D. thesis, 2005.
[14] K. H. Ackermann *et al.*, Nucl. Instr. and Meth. A **499**, 624 (2003).
[15] M. Anderson *et al.*, Nucl. Instr. and Meth. A **499**, 659 (2003).
[16] W. J. Llope (STAR TOF Group), Nucl. Instr. and Meth. B **241**, 306 (2005).
[17] C. Alt *et al.* (NA49 Collaboration), arXiv:nucl-ex/0512033v2
[18] C. Alt *et al.* (NA49 Collaboration), Phys. Rev. C **79**, 044910 (2009).

Journal of Physics: Conference Series **316** (2011) 012023 doi:10.1088/1742-6596/316/1/012023

# Upsilon Production and Upsilon + Hadron Correlations at STAR

**Matthew C. Cervantes for the STAR Collaboration**
Texas A&M University, Cyclotron Institute, College Station, Texas USA
E-mail: mcervantes@comp.tamu.edu

**Abstract.** The cross-section for the $\Upsilon$(1S+2S+3S) di-electron decay channel is measured at mid-rapidity in the STAR detector, and the preliminary results for azimuthal correlations between $\Upsilon$(1S+2S+3S) and hadrons are presented, using p+p and d+Au collisions at the RHIC center-of-mass energy $\sqrt{s}$ = 200 GeV from Runs 6, 8 and 9. The STAR measurements establish a baseline for future $\Upsilon$ measurements in Au+Au collision systems and provide additional opportunities to study the $\Upsilon$ production mechanism.

## 1. Introduction

The use of heavy quarkonia as a tool for probing the medium formed at the Relativistic Heavy Ion Collider (RHIC). In particular, searches for the 'suppression' of the J/$\psi$ meson, induced by Debye screening of the QCD potential between the charm anti-charm bound states, was initially suggested as being an unambiguous signature of QGP formation [1]. Theoretical scrutiny found that competing effects in the QGP can suppress [2-3] or enhance [4-6] the yield of J/$\psi$ particles; it is also possible that these competing effects counterbalance their respective contributions to the final measured J/$\psi$ yield.

Given the inescapable complexity that arises with the use of J/$\psi$ as a QGP probe, increasing interests in the $\Upsilon$ meson and its excited states (2S,3S) for the studying of the QGP has come into focus. The combination of RHIC's low $b\bar{b}$ production cross-section (1.9 $\mu$b [7]) and the smallness of the interaction cross-section of bottomonium with hadrons in RHIC collisions [8] predicts that suppression due to absorption by hadronic co-movers is relatively negligible in these models. Further, theoretical calculations suggest that the $\Upsilon$ appears to be less susceptible to the competing effects inherent with J/$\psi$ measurements - of the two competing in-medium mechanisms, $\Upsilon$ suppression is expected to be the dominant effect observed, not enhancement (recombination) [9]. As noted, a primary objective of the heavy-ion collision experiment at RHIC is to produce and study the QGP. The use of primordial J/$\psi$ and $\Upsilon$ mesons created during these heavy-ion collisions, as probes of the QGP, is one of the many analysis-techniques used to study the QGP. The use of heavy quarkonia leads to the question - what is the mechanism(s) by which heavy quarkonia production proceeds? The significance of the production mechanism of heavy quarkonium, before being affected by the QGP cannot be overstated and remains an open issue, apart from the QGP-driven use of J/$\psi$ and $\Upsilon$ in the STAR heavy flavor program.

The systematics of the prompt production of heavy quarkonium is not fully described by current models, e.g. the Color Singlet Model (CSM) and the Color Octet Model (COM). Historically, CSM calculations have grossly under-predicted production cross-sections, but recent development with higher-order corrections can describe data better [10]. The COM has had success in explaining the $p_T$-spectra of heavy quarkonia, but the spin-alignment (polarization) prediction disagrees with experimental data [11]. Quarkonium production in hadronic collisions at RHIC energy is dominated by direct production via gluon fusion followed by CSM and/or COM transitions [12]. In Cu+Cu collisions, STAR observes a lack of expected suppression in high-$p_T$ $J/\psi$ $R_{AA}$ - the only hadron exibiting this behavior in RHIC heavy-ion collisions. Figure 1 shows the high-$p_T$ $J/\psi$ $R_{AA}$ compared with various model predictions. The lack of $J/\psi$ suppression at high-$p_T$ suggests that $J/\psi$ cannot exist in a colored state on a long enough timescale to be affected by the medium, leading to an interpretation that the production is possibly not dominated via a color channel [13].

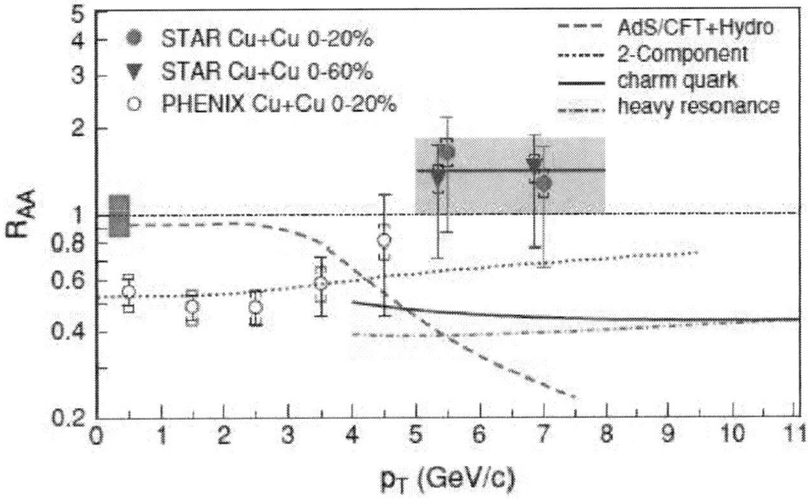


**Figure 1.** $J/\psi$ suppression in Cu+Cu data at $\sqrt{s} = 200$ GeV [13].

The STAR collaboration has contributed to the world data on bottomonium production and toward the understanding of the production mechanism. Here we review $\Upsilon$ results from Run-6 and Run-8 measurements. We also preview the Run-9 $\Upsilon$ invariant mass spectrum and present an additional technique for looking into the production mechanism of heavy quarkonia. Finally, we highlight future STAR capabilities for $\Upsilon$ reconstruction via the di-muon channel.

## 2. STAR at RHIC

RHIC is a two-ring, quasi-circular particle accelerator measuring 3.8 km in circumference, that accelerates and guides particle beams at collision-energies up to $\sqrt{s_{NN}} = 200$ GeV for Au+Au collisions and $\sqrt{s_{NN}} = 500$ GeV for p+p collisions.

The STAR experiment is home to the STAR (Solenoid Tracker At RHIC) detector. STAR is a multi-component detector, with each component designed to provide the specialized capabilities necessary for data collection in the high-energy, high luminosity-driven collision environment at RHIC. Two of the largest sub-detector components in STAR are the Time Projection Chamber (TPC), used for charged-particle tracking, and the Barrel Electro-Magnetic Calorimeter (BEMC), used for energy deposition related measurements. These two core-detector components enclose a significant geometrical volume, increasing the ability to study heavier (larger opening-angle) vector mesons such as $\Upsilon \to e^+e^-$: each detector has full-azimuthal ($0 \leq \phi \leq 2\pi$) coverage, and each detector has substantial rapidity acceptance ($|\eta_{TPC}| < 1.4$ and $|\eta_{BEMC}| < 1.0$).

Journal of Physics: Conference Series **316** (2011) 012023 doi:10.1088/1742-6596/316/1/012023

## 3. Data and Analysis

The data recorded was selected using both Level-0 (L0) and Level-2 (L2) triggers. The L0 and L2 triggers at STAR for $\Upsilon$ reconstruction are detailed below.

The L0 trigger is a fast hardware-level decision, designed to make an acceptance decision for every RHIC bunch crossing. The L0 is a high-tower trigger, checking the condition for at least one BEMC tower with energy above a pre-defined threshold.

The L2 trigger is a software-level decision, acting as a topological trigger which checks for the condition of another BEMC energy deposit, indicative of a two-body decay. It is designed to find towers with an energy similar to the L0 threshold, and then use these energy-similar towers as the 'seeds' to form L2 clusters. The L2 algorithm takes these pairs of clusters and calculates the invariant mass, under the approximation $M \simeq \sqrt{2E_1E_2(1-cos(\theta_{12}))}$, where $E_i$ (i = 1,2) is the cluster energy and $\theta_{12}$ is defined as the angle formed by the two approximately straight lines, extending from the vertex to the positions of the cluster. The straight-line approximation is valid under the known masses of the electron/positron and $\Upsilon$, the decay kinematics of the $\Upsilon$ daughters, and the known STAR magnetic field strength.

In STAR 2006 (Run-6) p+p data, the total $\Upsilon$(1S+2S+3S) total yield is N = 67 $\pm$ 22 (stat). Integrated from 7 to 11 GeV, the background-subtracted $m_{ee}$ distribution results in $3\sigma$ significance [14]. Figure 2 shows the invariant mass spectrum for unlike-sign (blue) and like-sign (red).

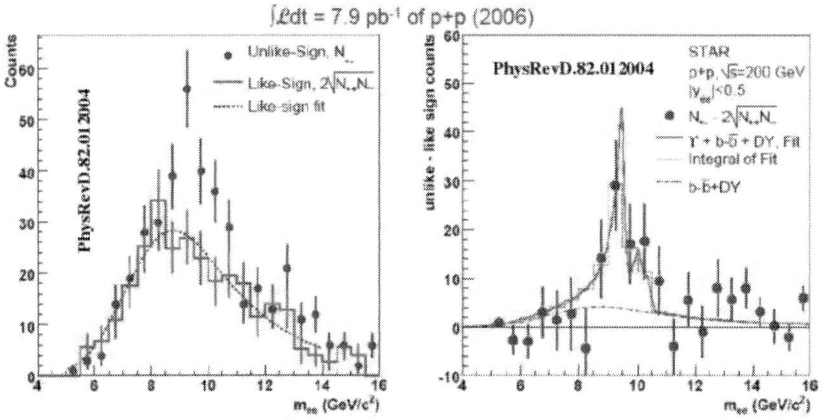


**Figure 2.** $\Upsilon$ invariant mass in Run-6 p+p data at $\sqrt{s}$ = 200 GeV: Unlike-sign raw yield at $y_{ee} \leq |0.5|$ and like-sign combinatorial background (left panel). The $e^+e^-$ signal on the Upsilon mass-region after subtraction of like-sign combinatorial background (right panel).

The STAR cross-section data point is consistent with the world data trend as a function of $\sqrt{s}$. The results of Next-to-Leading-Order perturbative Quantum Chromodynamics (NLO pQCD) calculations agree with the STAR data point within $1\sigma$ [15] in the Color Evaporation Model (CEM), while the Color Singlet Model (CSM) prediction underestimates the STAR data point by $2\sigma$ [16]. Figure 3 shows the STAR data point relative to the world data trend.

The Run-6 p+p data set is subject to a high rate of material conversion pairs, reducing the $\Upsilon$ mass peak signal to background (S/B). The reduced material in STAR Run-8 d+Au (and Run-9 p+p) data sets was the removal of the inner silicon detector systems (SVT + SSD). The final integrated luminosity for Run-8 was 32 $nb^{-1} \approx 12.5\ pb^{-1}$ (p+p equivalent).

In d+Au collisions, the signal + background pairings of unlike-sign electron pairs and background pairings of like-sign electron pairs results in a signal of $8\sigma$ significance [17]. Figure 4

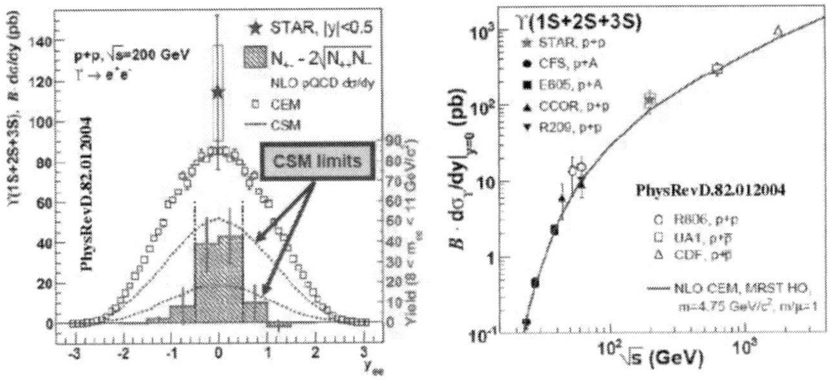


**Figure 3.** Run-6 p+p data at $\sqrt{s}$ = 200 GeV: $\Upsilon$ cross-section times branching ratio into electrons, with statistical error bars and a systematic uncertainty represented as a box (left panel). STAR cross-section data point relative to the world trend as a function of $\sqrt{s}$ (right panel).

shows the STAR $\Upsilon$(1S+2S+3S) unlike-sign (blue) and like-sign (black) mass spectrum (left panel), and the invariant mass peak after background subtraction (center panel) with an integrated (7 - 11 GeV/$c^2$) raw yield: 172 $\pm$ 20 (stat.).

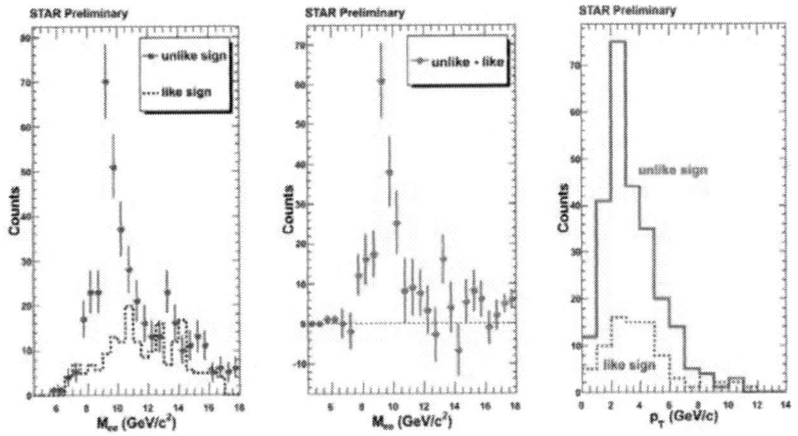


**Figure 4.** $\Upsilon$ invariant mass (left and center panels) and the $p_T$-spectrum (right panel) in Run-8 d+Au data at $\sqrt{s}$ = 200 GeV.

Figure 5 shows the cross-section results for Run-8 d+Au (left) and Run-6 p+p (right). The STAR cross-section data point for d+Au was found to be $B_{ee}(d\sigma/dy)_{y=0}^{\Upsilon^{1S}+\Upsilon^{2S}+\Upsilon^{3S}}$ = 35 $\pm$ 4 (stat) $\pm$ 5 (syst) nb, and is consistent with the CEM prediction with anti-shadowing effects + no absorption [18]. The measurement of the nuclear modification factor indicates that $\Upsilon$(1S+2S+3S) production is described by binary-collision scaling, within the current uncertainties, and that cold nuclear matter effects are not large. The most recent measured values for $R_{dAu}$ = 0.78 $\pm$ 0.28 (stat) $\pm$ 0.20 (syst).

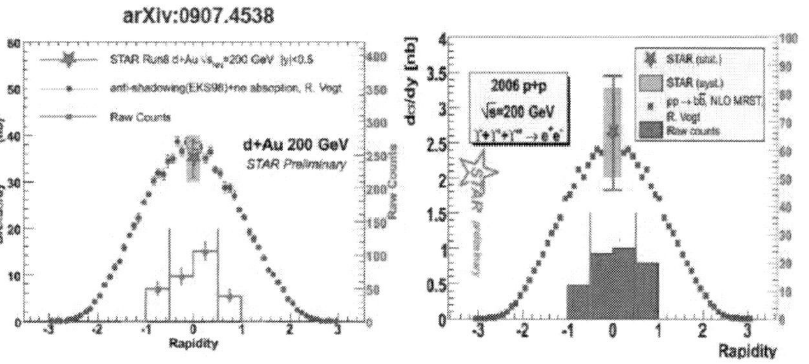


**Figure 5.** $R_{dAu}$ using $\Upsilon$ cross-sections at $\sqrt{s}$ = 200 GeV: Run-6 p+p (left) and Run-8 d+Au (right) data.

The reduced material budget and improved statistics in Run-9 p+p data (integrated luminosity $\approx$ 3 times greater than Run-6 p+p integrated luminosity) will help reduce the statistical uncertainty in $R_{dAu}$. Figure 6 shows the reconstructed mass from unlike-sign ($e^+e^-$) pairs (black) and like-sign ($e^+e^+ + e^-e^-$) pairs (red) in p+p data. There is very little background observed, even without the combinatorial background subtraction using like-sign pairs. The combinatorial background is normalized by the geometrical mean, following the prescription: $N = N_{+-} - 2\sqrt{N_{++}N_{--}}$, where the yield is extracted by integration of the invariant mass distribution of $\Upsilon$(1S+2S+3S) after the subtraction of the normalized like-sign pair distribution. The $\Upsilon$(1S+2S+3S) mass-analysis intends to separate the 'line shape' of the mass spectrum into the $\Upsilon$(1S), $\Upsilon$(2S) and $\Upsilon$(3S) states from the Drell-Yan and $b\bar{b} \rightarrow e^+e^-$ continuum contributions.

It is clear that the S/B region is higher in the Run-9 p+p measurement than the Run-6 p+p measurement. The high S/B ratio found in the reconstruction of $\Upsilon$ in the Run-9 p+p data enables us to attempt an analysis of Upsilon + Hadron azimuthal correlations. Prior to calculating the Upsilon + Hadron azimuthal correlation we cut tightly on the invariant mass-window from 8.0 to 11.5 GeV/$c^2$, for the purpose of optimizing our purity in the Upsilon's S/B ratio (blue vertical lines). Increased hadronic activity (measured in space coordinates around the $\Upsilon$) directly around the quarkonium has been proposed as an experimental observable to measure the radiation emitted off the colored quark pair during production [19]. This measured activity may provide insight into the production mechanism of heavy quarkonium. Figure 7 shows an $\Upsilon$ produced with the single hard-gluon emission expected in CSM production vs. the multiple soft-gluon emissions expected in COM production.

In d+Au data, we see that the correlation is not significant relative to the underlying-event. Figure 8 shows the $\Delta\Phi$ - correlation shape in d+Au data corrected for background contributions (red). The results in data are then compared to PYTHIA [20] simulations (blue).

In p+p data, it is notable that PYTHIA does have a similar underlying-event. However, the correlation has not yet been corrected for acceptance and track efficiencies. Figure 9 shows the $\Delta\Phi$ - correlation shape in p+p data, corrected for background contributions (red). The results in data are then compared to PYTHIA (blue). We will attempt to decompose the observed correlation shape found in data, into the predicted contributions from gluon activity in both COM and CSM. Interpretation of the $\Delta\Phi$ - correlation shape is still in progress.

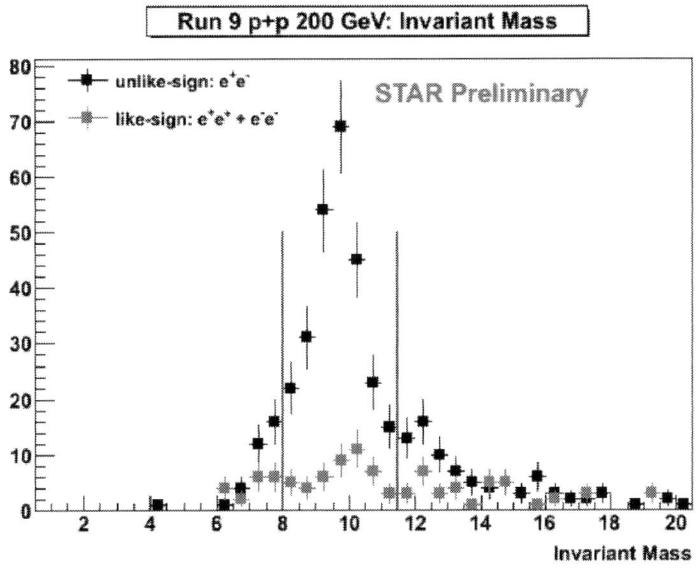


**Figure 6.** $\Upsilon$ invariant mass in Run-9 p+p data at $\sqrt{s} = 200$ GeV. Unlike-sign $e^+e^-$) pairs (black) and like-sign ($e^+e^+$ + $e^-e^-$) pairs (red). Upsilon + Hadron correlation cuts on the invariant mass from 8.0 to 11.5 GeV/c$^2$, to optimize the purity in the Upsilon S/B ratio (blue vertical lines).

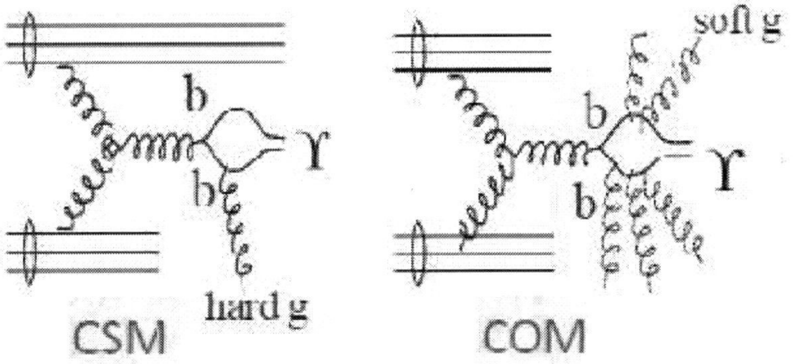


**Figure 7.** CSM vs. COM gluon radiation.

## 4. Future Measurements

The STAR program is expanding its lepton detection capability to include muons. Large mid-rapidity coverage of the muon telescope detector (MTD) allows for single muon detection from semi-leptonic decay of heavy flavor hadrons, as well as di-muon pairs from QGP thermal radiation and quarkonia decay. Muon detection expands the ability to study Drell-Yan production, light vector mesons, and constructing correlations of quarks and gluons as key signatures of resonances in the QGP.

The use of muons for analyses sidesteps the issue of $\gamma$-conversion pairs, minimizes the contributions from Dalitz decay, suffers less from the radiative losses in detector materials, and retains the ability to design muon-triggers in Au+Au collisions. The MTD triggering scenarios

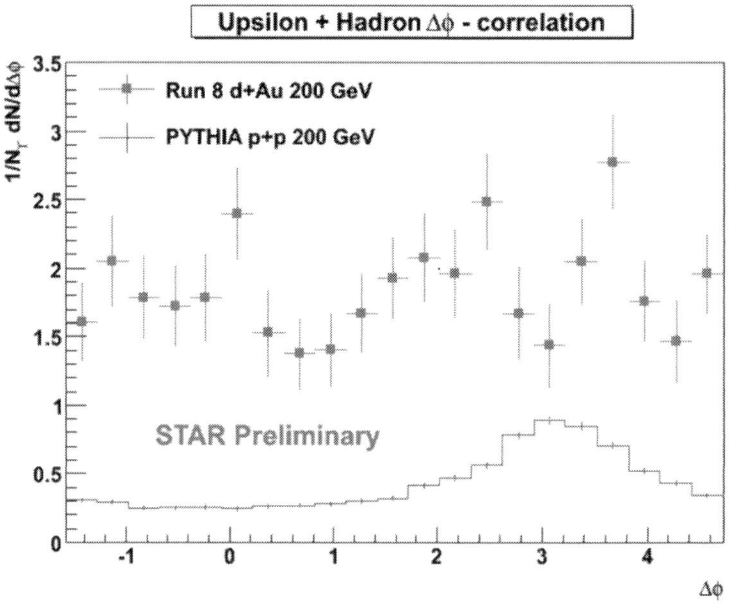


**Figure 8.** $\Delta\Phi$ - correlation: $\Upsilon$ + hadron at $\sqrt{s} = 200$ GeV in Run-8 d+Au data.

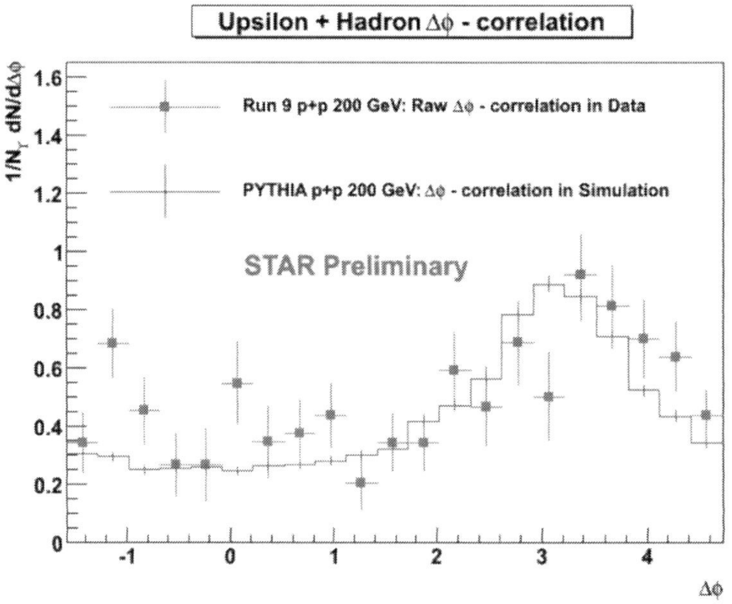


**Figure 9.** $\Delta\Phi$ - correlation: $\Upsilon$ + hadron at $\sqrt{s} = 200$ GeV in Run-9 p+p data.

extend from low to high-$p_T$ $J/\psi$ in central Au+Au collsions, provide excellent mass resolution for attempts at separating the different $\Upsilon$ states, and they open the door for analyses of electron-muon correlations, which aim to differentiate heavy flavor production from initial lepton pair production. The prototype of the MTD at STAR is operational from Run-7 to Run-10 [21-22].

## 5. Conclusions

STAR has measured the total $\Upsilon$(1S+2S+3S) yield to a $3\sigma$ level of significance in Run-6. The cross-section data point is consistent with the world data trend as a function of $\sqrt{s}$ and is within $1\sigma$ of the NLO pQCD CEM calculations, but is underestimated by $2\sigma$ in NLO pQCD CSM calculations. In STAR Run-8 d+Au data, the results from a reduction in material budget and improved statistics leads to a measured total $\Upsilon$(1S+2S+3S) yield at an $8\sigma$ level of significance. The cross-section data point was found to be consistent with the CEM prediction with anti-shadowing effects, and the nuclear modification factor indicates the $\Upsilon$(1S+2S+3S) production as being described by binary-collision scaling, within the current uncertainties, leading to the conclusion that the effect from cold nuclear matter is not large. Run-9 p+p data saw a final integrated luminosity $\approx$ 3 times greater than Run-6 p+p integrated luminosity. This will help reduce the statistical uncertainty in $R_{dAu}$. There is very little background observed in Run-9 $\Upsilon$ reconstruction, even without the combinatorial background subtraction using like-sign pairs. This reduction and increased luminosity makes it a real possibility for separating the $\Upsilon$(1S+2S+3S) mass states. These STAR measurements not only establish a baseline for future $\Upsilon$ measurements in Au+Au collision systems, but they also provide insight into the $\Upsilon$ production mechanism. In fact, the high Upsilon S/B ratio in Run-9 p+p data leads to an Upsilon + Hadron correlation in data with an underlying-event which is similar to that seen in PYTHIA simulations. The $\Delta\Phi$ - correlation analysis is not yet corrected for acceptance and track efficiencies and awaits an attempt to decompose the observed $\Delta\Phi$ - correlation shape from Run-9 p+p into possible contributions from the CSM and COM gluon activity. The STAR program is also expanding its event reconstruction capability to include processes involving muons. The MTD prototype at STAR is operational from Run-7 to Run-10, allowing for muon detection and the opportunity to minimize well-known issues that detract from electron-based analyses.

## 6. References

[1] T. Matsui and H. Satz, Phys. Lett. B 178, 416 (1986).
[2] S. Gavin, M. Gyulassy, and A. Jackson, Phys. Lett. B 207, 257 (1988).
[3] J.-P. Blaizot and J.-Y. Ollitrault, Phys. Rev. D 39, 232 (1989).
[4] L. Grandchamp et al., Phys. Rev. Lett. 92, 212301 (2004).
[5] R. L. Thews et al., Eur. Phys. J. C 43, 97 (2005).
[6] A. Andronic et al., Nucl.Phys. A789: 334-356 (2007).
[7] M. Cacciari, P. Nason and R. Vogt, Phys. Rev. Lett. 95, 60 122001 (2005); [arXiv:hep-ph/0502203].
[8] Z.w. Lin and C. M. Ko, Phys. Lett. B 503, 104 (2001); [arXiv:nucl-th/0007027].
[9] L. Grandchamp et al., Phys. Rev. C 73, 064906 (2006).
[10] J.P. Lansberg, Eur.Phys.J. C60: 693-703 (2009).
[11] CDF Collaboration, A. Abulencia et al., Phys. Rev. Lett. 99, 132001 (2007).
[12] J. Rafelski and B. Muller, Phys. Rev. Lett. 48 (1982) 1066.
[13] STAR Collaboration, B. I. Abelev et al., Phys. Rev. C 80, 041902(R) (2009).
[14] STAR Collaboration, B. I. Abelev et al., Phys. Rev. D 82, 012004 (2010).
[15] STAR Collaboration, A. D. Frawley, T. Ullrich, R. Vogt, Phys. Rept. 462, 125 (2008).
[16] S.J. Brodsky and J-P Lansberg, Phys. Rev. D 81, 051502 (2010).
[17] STAR Collaboration, H. Liu, Nucl.Phys. A830: 235c-238c (2009).
[18] K.Eskola, V.Kolhinen and C.Salgado, Eur. Phys. J. C 9 (1999) 61.
[19] A.C. Kraan, arXiv:0807.3123v1 [hep-ex] 19 Jul 2008.
[20] T. Sjostrand et al., Comput.Phys.Commun.178:852-867,2008.
[21] L. Ruan et al., Journal of Physics G: Nucl. Part. Phys. 36 (2009)095001; 0904.3774.
[22] Y. Sun et al., NIMA 593 (2008) 430

# Dimuons in pp and PbPb collisions with the CMS experiment

**Jorge A. Robles**
One Shields Ave. Physics Department, Davis CA, 95616
E-mail: Robles@physics.ucdavis.edu

**Abstract.** The unprecedented centre-of-mass energy available at the LHC offers unique opportunities for studying the properties of the strongly-interacting QCD matter created in PbPb collisions at extreme temperatures and very low parton momentum fractions. With its high precision, large acceptance for tracking and calorimetry, and a trigger scheme that allows analysis of each minimum bias PbPb event, CMS is fully equipped to measure dimuons in the high multiplicity environment of nucleus-nucleus collisions. Such probes are especially relevant for these studies since they are produced at early times and propagate through the medium, mapping its evolution. Inclusive and differential measurements of the Z boson yield will be shown, establishing that no modification is observed with respect to theoretical next-to-leading order pQCD calculations. The validity of the Glauber scaling for perturbative cross sections in nucleus-nucleus collisions at the LHC is confirmed. In addition, the status of the quarkonia measurements (primary and secondary $J/\psi$, and the 3 Upsilon states) will be reviewed, both in pp and in PbPb collisions.

## 1. Introduction

The strongly interacting state of matter created in heavy-ion collisions can be cleanly probed via processes that decay in the dilepton channel. Given the fact that leptons do not interact via the strong force, these can traverse the medium unaffected. When studying dileptons experimentally, muons offer some advantages over electrons. The latter suffer energy losses from Bremsstrahlung radiation, while muons are much less susceptible to these effects. In the CMS experiment dimuon reconstruction offers a very good mass resolution for quarkonia and electro-weak processes. In pp collisions quarkonia can be studied to test production models, while precision NNLO calculations can be tested with electroweak processes. The observables in pp act as a baseline to study the modification of probes seen in PbPb collisions. The study of quarkonia in heavy-ion collisions serves as a probe of the temperature of the Quark Gluon Plasma (QGP), which is expected to melt bound quarkonia states as the temperature rises. The study of Z bosons, since they are unaffected by the medium, are a very important tool to understand the cold nuclear effects. The energy density of the medium can also be measured via energy loss in Z+jet events, in which the Z balances the momentum (at tree level) of the jet on the opposite side.

## 2. Muon reconstruction

The CMS experiment is specially suited for muon reconstruction spanning a wide mass, with very good mass resolution. The reconstruction algorithm makes use of various detectors, combining the good muon identification and fast muon triggering from the muon chambers with the fine

$p_T$ resolution from the inner tracker. The tracker, made up from pixels and silicon strips, has a coverage of $|\eta| < 2.5$ with a $p_T$ resolution of 1-2% up to 100 GeV/c. The muon chambers, made up of 3 technologies, has a coverage of $|\eta| < 2.4$. A full description of the CMS detector can be found in [1].

## 3. pp Results

CMS has accumulated enough statistics at $\sqrt{s} = 7$ TeV to measure the cross sections and kinematic distributions for J/$\psi$, $\Upsilon$ and Z to compare to different available models. The J/$\psi$, $\Upsilon$ and Z results in pp are explained in more detail in [2], [3] and [4] respectively.

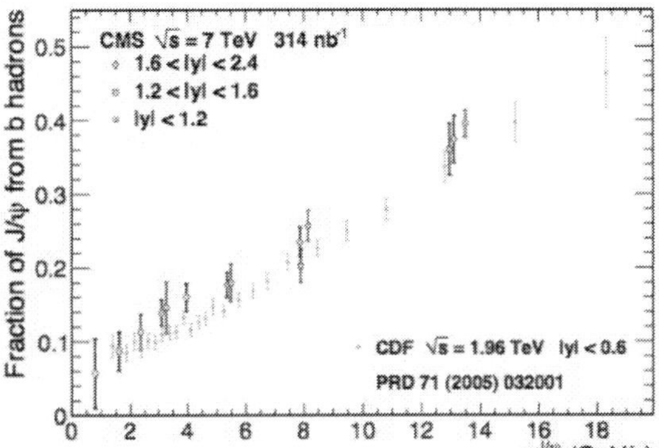


**Figure 1.** Fraction of J/$\psi$ from b-hadrons with data points from CMS at $\sqrt{s} = 7$ TeV and CDF points at $\sqrt{s}$= 1.96 TeV.

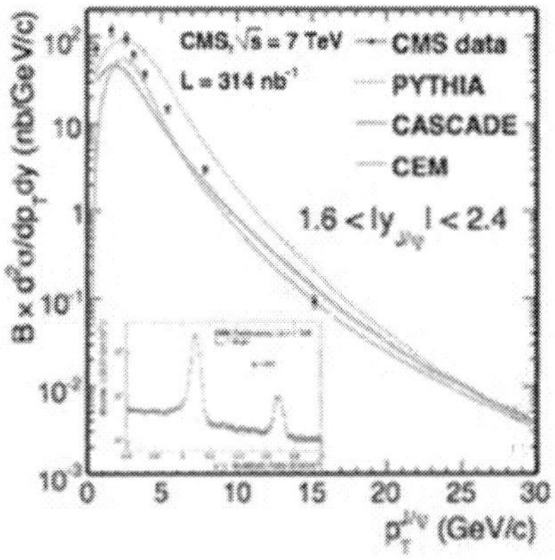


**Figure 2.** J/$\psi$ $p_T$ distribution at forward rapidity, compared to models. Lower-left corner, J/$\psi$ and $\psi$' mass distribution

### 3.1. J/$\psi$ in pp collisions

The J/$\psi$ measurement requires proper identification of prompt J/$\psi$ and decay-in-flight contributions coming from b-hadrons. Measuring the fraction of prompt to non-prompt relies in the discrimination of J/$\psi$ mesons produced away from the collision vertex. Fig. 1 shows the fraction of J/$\psi$ from $b$-hadron in three rapidity ranges as a function of transverse momentum. The measured fraction of J/$\psi$ from $b$-hadron obtained by CMS lies on top of the data points obtained from the CDF experiment, regardless of the different center-of-mass energies. The lower-left plot on Fig. 2 shows the invariant mass of the J/$\psi$ and the $\psi$' in the dimuon channel. The mass resolution observed is $\sigma = 20$ MeV/$c^2$ in the rapidity range $|y| <0.5$. The $p_T$ spectra for prompt and non-prompt J/$\psi$ is also measured in three rapidity bins and compared to available models (Color Evaporation Model, Pythia and CASCADE). In the main frame of Fig 2 the $p_T$ spectra of J/$\psi$ at high rapidity is shown. A reasonable agreement is found between data and theory in all bins except at low $p_T$ at forward rapidity for the prompt J/$\psi$. The non-prompt J/$\psi$ is divided also in the same rapidity bins and compared to theory (CASCADE, Pythia and a B$\rightarrow$ J/$\psi$ FONLL calculation) in all bins a good agreement is found.

### 3.2. $\Upsilon$ in pp collisions

The superb mass resolution achieved by CMS allows for a clear separation of the$\Upsilon$ ,$\Upsilon$ ' and$\Upsilon$" states. The reconstructed mass resolution obtained over the full CMS acceptance, $|\eta^{\mu}| <2.4$, is $\sigma = 100$ MeV/$c^2$. This resolution can be further improved to $\sigma = 67$ MeV/$c^2$ by restricting to muons reconstructed in the barrel region, $|\eta^{\mu}|$ <1.0 shown in Fig 3. The differential cross-section down to zero $p_T$ for the 3 $\Upsilon$ states at $\sqrt{s} = 7$ TeV was obtained. In Fig. 4 the$\Upsilon$(1S) $p_T$ spectra is shown and compared to Pythia in the $|y^{\Upsilon}| < 2$ rapidity range assuming unpolarized$\Upsilon$(ns) production. For the 3 $\Upsilon$ states a good agreement is found within statistical uncertainties. The $\Upsilon$(1s) and$\Upsilon$ (2s) include feed-down from $\chi_b$ and$\Upsilon$(3s).

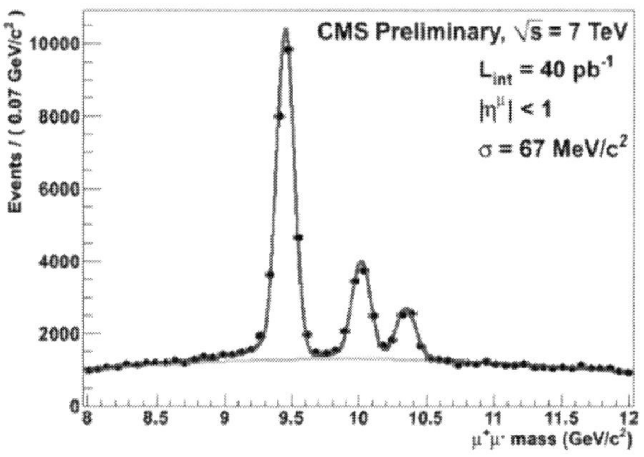


**Figure 3.** $\Upsilon$ states reconstructed in the CMS barrel $|\eta^{\mu}|$ <1.0

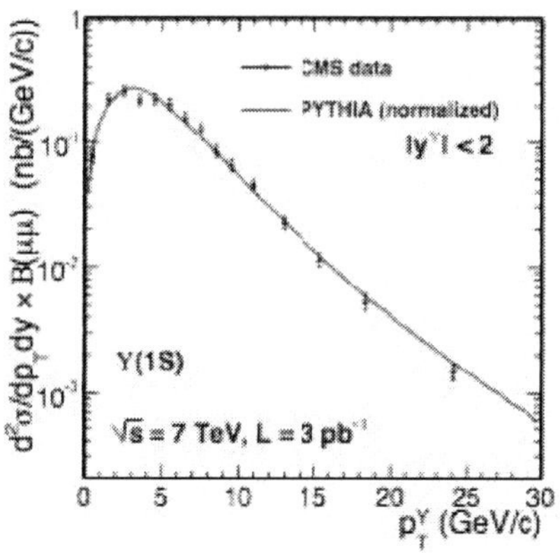


**Figure 4.** $\Upsilon$1S $p_T$ spectra, compared to PYTHIA in the $|y^{\Upsilon}| < 2.0$ range

*3.3. Z in pp collisions*
Z boson production has been studied to a high degree of precision in proton-proton experiments. Simulation is able to accurately describe the data obtained by CMS in the Z$\rightarrow \mu^+\mu-$ channel at $\sqrt{s} = 7$ TeV. Next-to-Next Leading Order(NNLO) calculations are in superb agreement with Z measurements in CMS. The Z cross-section measured in the di-electron and dimuon channel are compared to a NNLO calculation shown in Fig 6, the experimental uncertainties are $< 4$ %. The comparison uncertainties are dominated by the luminosity error bars ($\approx$11 %), which are expected to diminish in the future.

## 4. Heavy Ions in CMS
The Heavy Ion program at the LHC represents an increase in the center of mass energies of a factor of 14 compared to RHIC. This allows us to study the QGP with probes that were previously inaccessible. The production rate at the LHC allows for precision bottomonia studies, jets with energy above 100 GeV are be available to probe the medium, and is the first time electroweak processes are available in heavy ion collisions. Overall, the high-$p_T$ regime plays an important role in measurements that aim to quantify medium effects. CMS is ideal to carry out measurements in the dimuon channel, taking a advantage of the high resolution of muon and tracker detectors that have a performance that is just as good as in pp collisions. The results shown in PbPb collisions were obtained with the CMS experiment during the first HI run at the LHC at $\sqrt{s} = 2.76$ TeV, with data collected over November-December 2010.

Preliminary results obtained by a quick analysis are shown in Fig. 7 and Fig. 8, in which the invariant mass distribution around the J/$\psi$ and$\Upsilon$ , respectively, are obtained in the dimuon channel. In the case of the J/$\psi$, the reconstructed muons are restricted to have a $p_T^{\mu} > 6.5$ GeV/c, and the dimuon is restricted to be in the rapidity range $|y^{J/\psi}| < 2.4$. A total of $\approx 200$ J/$\psi$ counts is found in a peak of opposite sign pairs over the same-sign flat background. Fig 8

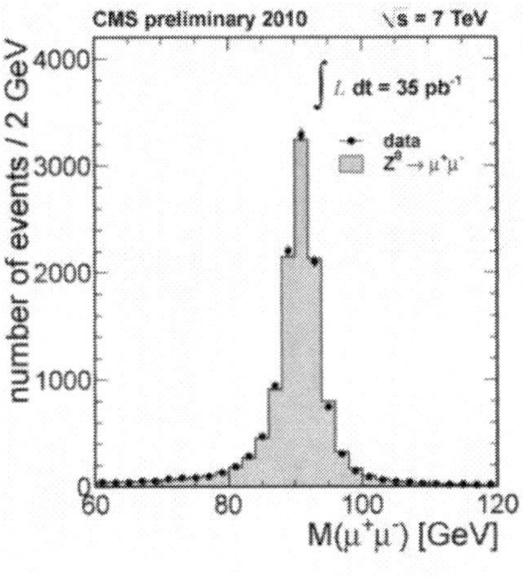


**Figure 5.** Z Invariant mass at $\sqrt{s} =$ 7 TeV data points with MC histogram in yellow.

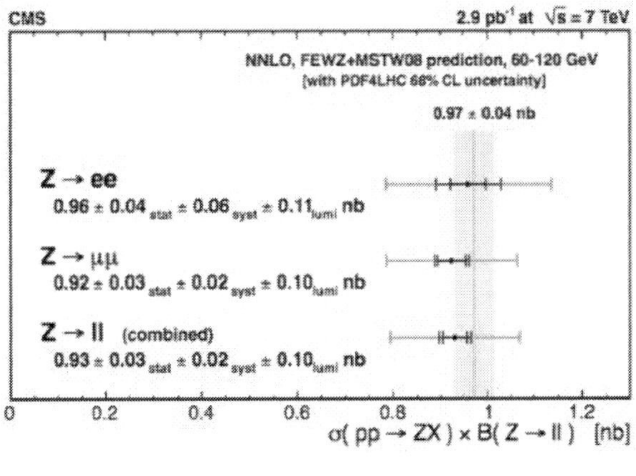


**Figure 6.** Ratio of Z cross-section measurement in the di-lepton channels with NNLO calculations.

shows the reconstructed invariant mass peak that arises from muons with a $p_T^{\mu} > 4$ GeV/c cut, and in the dimuon rapidity region $|y^{\Upsilon}| < 2.4$. These plots were obtained from a subset of heavy ion events available at the time.

### *4.1. Quarkonia in PbPb collisions*

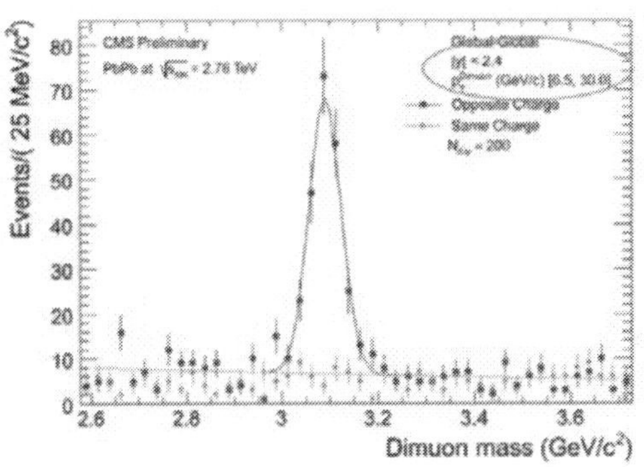


**Figure 7.** J/$\psi$ Invariant mass t $\sqrt{s} = 2.76$ TeV in dimuon channel. Opposite sign pairs in blue and same-sign pairs in red.

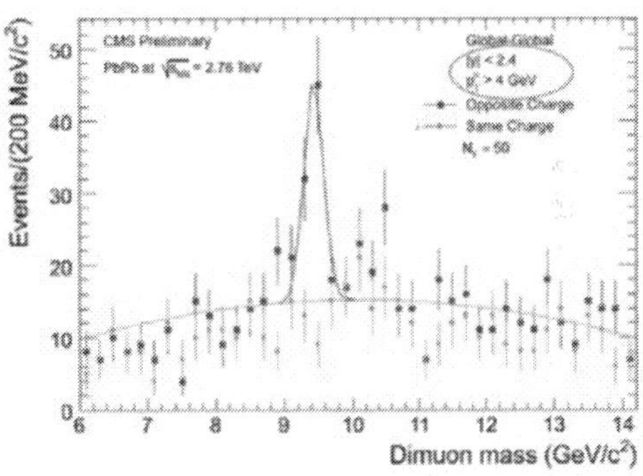


**Figure 8.** $\Upsilon$ Invariant mass t $\sqrt{s} = 2.76$ TeV in dimuon channel. Opposite sign pairs in blue and same-sign pairs in red.

### *4.2. Z Results in PbPb collsions*

The Z $\rightarrow \mu^+\mu^-$ process in PbPb collisions offers a benchmark probe of the initial quark pdf in the Pb nucleus that can be cleanly extracted. The Z is produced from hard collisions via $q\bar{q} \rightarrow Z$ process, once produced it decays within the medium on a time scale of $\approx 10^{-24}$ seconds. The dimuon decay channel is of special interest because the Z, as well as the muons, traverse the medium unaffected by the strongly interacting Quark Gluon Plasma. Given that the entire process is virtually untouched by the medium it offers a probe to measure initial state effects.

Predictions[5, 6, 7, 8, 9] indicate that the effects that can affect the Z boson production in heavy ion collisions are rather small. From these effects, shadowing is expected to have the biggest contribution ≈10 - 20% [8]. The isospin effect arises from the difference of quark content of protons and neutrons that collide in the PbPb case as opposed to pp, and is expected to contribute less tha 3% [8]. Finally, scattering and energy loss of the initial partons should have an effect of the order of 2 %[7].

The first measurement of Z production in PbPb collisions in CMS is further detailed in [10]. The reconstruction of the Z boson in the dimuon channel is done my requiring two opposite-charge muons with $p_T^{\mu} > 10$ GeV/c and $|\eta^{\mu}| < 2.4$. A loose set of quality cuts applied in each of the muons is enough to extract a clear peak in the 30 - 120 GeV/$c^2$ mass regime. Fig. 9 shows invariant mass from opposite-charge muon pairs in blue squares and only one same-sign pair in the above mentioned mass regime. The data points are overlaid with the empty histogram from pp collision data normalized to an integral of 39 counts. It is evident that $Z \to \mu^+\mu^-$ process can be extracted almost background free in a wide mass range around the Z mass. The structure found in the 30 - 50 GeV/$c^2$, is due to the dimuon continuum from other physics processes, mainly $b\bar{b}$ production. A visible agreement between the PbPb data points and the pp histogram data points to a comparable detector performance in pp and heavy ion collisions.

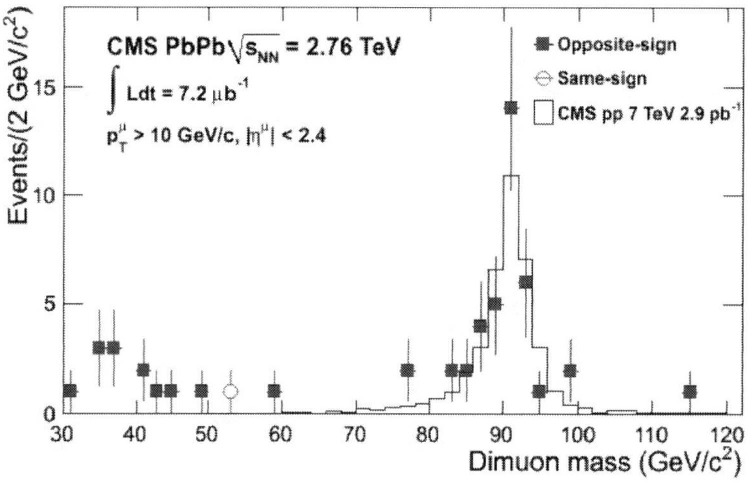


**Figure 9.** Z invariant mass in PbPb collisions at $\sqrt{s} = 2.76$ TeV in dimuon channel

Given the number of candidate events obtained from the run, the data were subdivided into transverse momentum($p_T^Z$), rapidity($y^Z$) and event-centrality bins. In absence of in-medium modification Z boson production is expected to scale by the number of binary collisions with respect to the pp measurement. Figure 10 shows the Z differential yield as a function of transverse momentum. The statistical error bars are shown in black and the systematic uncertainties are shown in orange boxes. The red data points in the rapidity region $|y^Z| < 2.0$ are compared to a next-to-leading order POWHEG calculation scaled by the nuclear geometry $(A^2/\sigma_{PbPb})$. Within statistical uncertainties an agreement between data and theory is found.

Figure 11 shows the Z differential yield as a function of rapidity in three bins. The red data points, with statistical error bars and systematic uncertainties in boxes, are compared to various theoretical predictions scaled by the nuclear geometry. The black line is a POWHEG calculation interfaced with Pythia as a baseline. To further discern between initial state effects a few more models are included. With the CT10 parameterization, the green dashed line shows isospin effects, the blue dot-and-dash line shows isospin and shadowing effects,. With the MSTW parameterization, the dotted brown line shows isospin effects ,while the red dot-and-dash line shows isospin and energy loss effects. Within the measurement uncertainties none of these subtler effects can be excluded.

The differential yield divided by nuclear overlap function, $T_{AB}$, as a function of $N_{part}$ is shown in Fig 12 in three centrality bins and a minbias point. The $N_{part}$ value for each of the bins is obtained by the average number of participants in the given bin. The CMS data points, in the rapidity range $|y^Z| < 2.0$, are compared to the same theoretical models described in the previous paragraph. The quantity plotted can be related to the nuclear modification factor, $R_{AA}$, by the following equation:

$$\frac{dN_{AA}}{dy}/T_{AB} = R_{AA} \times \frac{d\sigma_{pp}}{dy} \quad (1)$$

The main feature is a flat distribution as a function of centrality. This is expected for a probe that is unmodified by the presence of the dense QGP produced in central collisions. The data is also observed to match the POWHEG pp simulation scaled by the nuclear geometry.

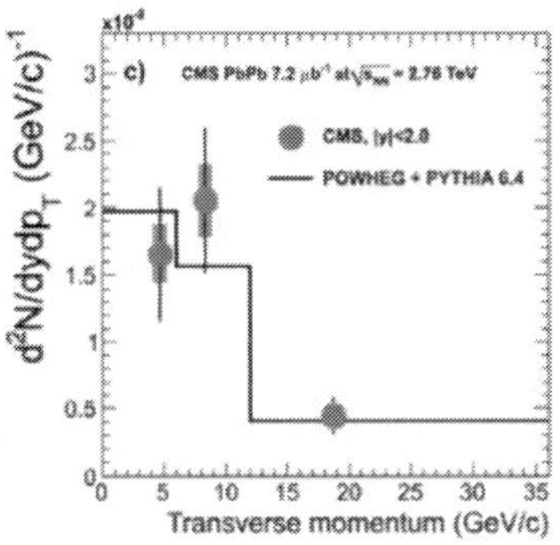


**Figure 10.** Z differential yield as a function of $p_T$

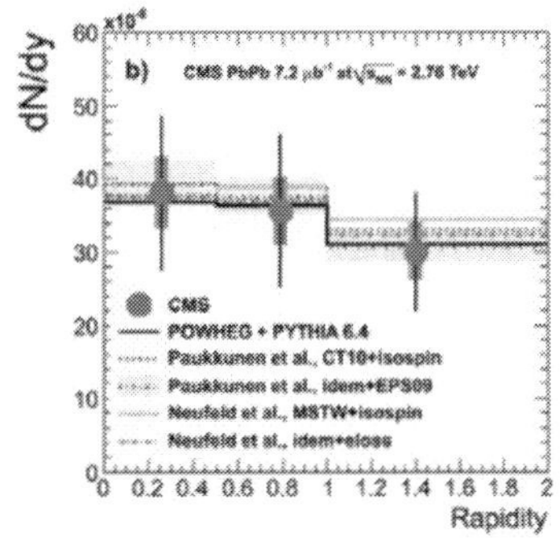


**Figure 11.** Z differential yield as a function of rapidity

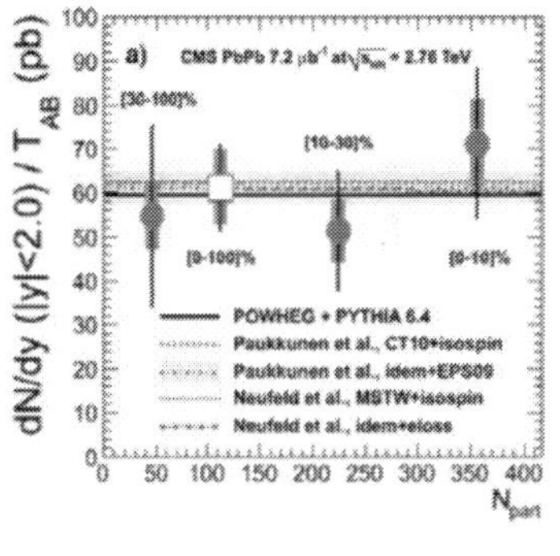


**Figure 12.** Z differential yield divided by the nuclear overlap function as a function of $N_{part}$

## 5. Conclusions

An overall agreement between theory and data for the J/$\psi$, $\Upsilon$ production is observed in pp collisions. A high level of agreement found for Z production measured in the dilepton channel and NNLO calculations. The J/$\psi$ and $\Upsilon$ were observed by CMS in PbPb collisions. The Z boson differential cross-sections, as a function of $y^Z$ and $p_T^Z$, were measured and observed to scale with the nuclear geometry with respect to theoretical NLO pQCD pp calculations. The Z yield as a function of centrality was found be flat, consistent an unmodified probe in central collisions.

## References

[1] R. Adolphi et al. (CMS), JINST 3, S08004 (2008)
[2] CMS collaboration, Eur. Phys. J. C (2011) 71: 1575, arXiv:1011.4193 [hep-ex].
[3] CMS collaboration, arXiv:1012.5545v1 [hep-ex]
[4] CMS collaboration, JHEP01 (2011) 080, arXiv1012.5545 [hep-ex].
[5] R. Vogt, Phys. Rev. C64, 044901 (2001), arXiv:hep-ph/0011242.
[6] V. Kartvelishvili, R. Kvatadze, and R. Shanidze, Phys.Lett. B356, 589 (1995), arXiv:hep-ph/9505418.
[7] R. B. Neufeld, I. Vitev, and B. W. Zhang, (2010),arXiv:1006.2389 [hep-ph].
[8] H. Paukkunen and C. A. Salgado, (2010),arXiv:1010.5392 [hep-ph].

[9] X.-F. Zhang and G. I. Fai, Phys. Lett. B545, 91 (2002),arXiv:hep-ph/0205155.
[10] CMS collaboration, PRL106, 212301 (2011), arXiv: 1102:5435 [nucl-ex].

Journal of Physics: Conference Series **316** (2011) 012025 doi:10.1088/1742-6596/316/1/012025

# Heavy Flavour Physics with the ALICE detector at the CERN-LHC

**Alessandro Grelli for the ALICE Collaboration**

ERC-Research Group QGP-ALICE, Utrecht University, Princeton Plein 5, 3584 CC Utrecht, The Netherlands

E-mail: a.grelli@uu.nl

**Abstract.** Heavy quarks have been proposed as powerful probes to study hot quark matter produced in high-energy heavy-ion collisions. The measurement of the charm and beauty production cross sections in pp collisions provides a reference for the studies in heavy ion collisions. Moreover, the pp measurements at LHC will allow to test the perturbative QCD calculations in a totally new energy domain. In this contribution we present the first results from the analysis of charm and beauty production in pp collisions at $\sqrt{s}$ = 7 TeV with the ALICE experiment, using the hadronic D meson and semi-leptonic D and B meson decays. In this talk I will present the first cross section measurement of muons from heavy flavour decays at forward rapidity and the heavy flavour electron spectra at central rapidity. We report on the first D meson cross section at central rapidity and the $J/\psi$ production cross section at central and forward rapidity.

## 1. Introduction

The main goal of the ALICE [1, 2] experiment is to study the strongly interacting matter in the conditions of high density and temperature. In such conditions lattice QCD calculations predict quark de-confinement and the formation of the so called Quark-Gluon Plasma (QGP) [3]. Heavy flavour hadrons are excellent probes for the investigation of the QGP. Open charm and beauty hadrons are expected to be sensitive to the energy density through the in-medium energy loss of the heavy quarks. Quarkonium states are sensitive to the initial temperature of the system via their dissociation due to color screening. A detailed discussion on physics motivation for heavy flavour measurements in Pb-Pb collisions can be found in Refs. [1, 2]. In this paper we discuss the preliminary ALICE results on heavy flavour production in pp collisions at $\sqrt{s} = 7$ TeV. The proton-proton results provide the baseline for the study of medium-induced effects in nucleus-nucleus collisions. Moreover they offer the possibility to test $p$QCD in the new energy domain of the Large Hadron Collider. Perturbative QCD predictions extracted using FONLL [4] (Fixed-Order-plus-Next-to-Leading-Log) describe well the beauty production cross section measured in pp collisions at $\sqrt{s} = 1.96$ TeV by CDF at the Tevatron [5]. Additionally, the CMS experiment found that at the LHC energy scale ($\sqrt{s} = 7$ TeV) the beauty production is well described by the theory [6]. In the case of charm, FONLL prediction agrees with data within errors. Nevertheless, the data points are systematically on the upper edge of the theory error band suggesting that charm production could be underestimated in the calculation as also seen at RHIC energies ($\sqrt{s}$ = 0.2 TeV) [7, 8]. Concerning the quarkonium production, several models have attempted to describe the hadroproduction of quarkonium states at Tevatron energies but failed to reproduce simultaneously the production cross section, transverse momentum distribution and polarization.

## 2. The ALICE detector and running conditions

The ALICE apparatus [1] has excellent capabilities for heavy-flavour measurements, for both open heavy-flavoured hadrons and quarkonia. It is composed of a central barrel and a forward muon arm. The central barrel covers the pseudo-rapidity region $-0.9 < \eta < 0.9$ and is equipped with tracking detectors and particle identification systems embedded in a magnetic field B=0.5 T along the $z$-axis. The Inner Tracking System (ITS) is the innermost central barrel detector and it is composed of six cylindrical layers of silicon detectors. The two layers closest to the beam pipe, positioned at radii of $\sim$ 4 and 7 cm, are equipped with pixel detectors. The two intermediate layers, at radii $\sim$ 15 and 24 cm, are made of drift detectors. The two outermost layers, at radii $\sim$ 39 and 44 cm, use strip detectors. Using ITS the track impact parameter (i.e. the distance of the closest approach of the track to the primary vertex) can be measured with a resolution better than 70 $\mu$m for $p_T \simeq 1$ GeV/c, thus providing the capability to detect the secondary vertices originating from heavy-flavour decays ($c\tau \sim$ 100-300 $\mu$m for D mesons). As standalone tracker the ITS allows to track particles down to $p_T \sim 70$ MeV/c. The Inner Tracking System is surrounded by the main ALICE tracking detector, a 510 cm long cylindrical Time Projection Chamber (TPC). It provides track reconstruction with up to 160 three-dimensional space-points as well as particle identication via specific energy deposit $dE/dx$. The TPC hadron identication capability is complemented by the Time-Of-Flight detector (TOF) and the Transition Radiation Detector (TRD). All mentioned detectors have full azimuthal coverage. The TRD is surrounded by a large acceptance (110 degrees azimuthal and $-0.7 < |\eta| < 0.7$) electromagnetic calorimeter (EMCal) providing and efficient and fast trigger for electron measurements.
The ALICE muon spectrometer is located at large rapidity $-4 < \eta < -2.5$. It is made of five tracking stations with two planes of multi-wire proportional chambers each, having a spatial resolution of about 100 $\mu$m. The muon spectrometer is completed by a dipole magnet with an integral field of 3 Tm and two trigger stations of resistive plate chambers placed behind an iron-wall muon filter with a thickness of about 7 interaction lengths. The system is completed by a frontal absorber of composite material. The spectrometer can detect quarkonia down to $p_T = 0$. The results presented in this paper are based on the 2010 minimum bias (MB) data sample [9]. It is defined as the logical "OR" of at least one of the pixel tracker layers and two rings of forward scintillator detectors ($V0A$ and $V0C$, placed asymmetrically with respect to the interaction point) is required. At least one charged particle over eight rapidity units is requested and all the ALICE detector is read out. This trigger configuration is sensitive to about 87% of the pp inelastic cross section. A muon trigger was also used during the 2010 data taking. In addition to the MB conditions, it requires at least one trigger signal ($p_T > 0.5$ GeV/c).

## 3. Electrons from heavy flavours decays

The semi-leptonic decay of D and B mesons, with a branching ratio of $\sim$ 10 %, gives access to the investigation of open charm and beauty production. The identification of the electrons is assured by the excellent particle identification of the ALICE central barrel detectors, namely TPC, TOF, TRD and EMCal. The results presented in this paper are obtained using TPC and TOF as PID detectors. The TRD and EMCal will allow us to extend the $p_T$ reach.
The extraction of the production cross section of the heavy flavour decay electrons can be performed in two different ways. A well establish method [11] consists of subtraction of cocktail of background electrons from the inclusive electron spectrum. The cocktail is obtained by summing up contributions of electrons from light hadrons decays ($\pi^0$ Dalitz, $\rho$, $\eta$, $\omega$ and $\phi$ decays), photon conversion, direct radiation and electrons from decays of heavy quarkonia. The determination of the cocktail is obtained on the basis of the measured $\pi^0$ cross section. A second

method is based on the displacement of electrons from beauty with respect to the primary vertex thanks to the relatively large decay length of B mesons ($c\tau \sim 500$ $\mu$m). A cut on the track impact parameter allows to select mainly decay electrons from beauty. The left panel of figure 1 shows

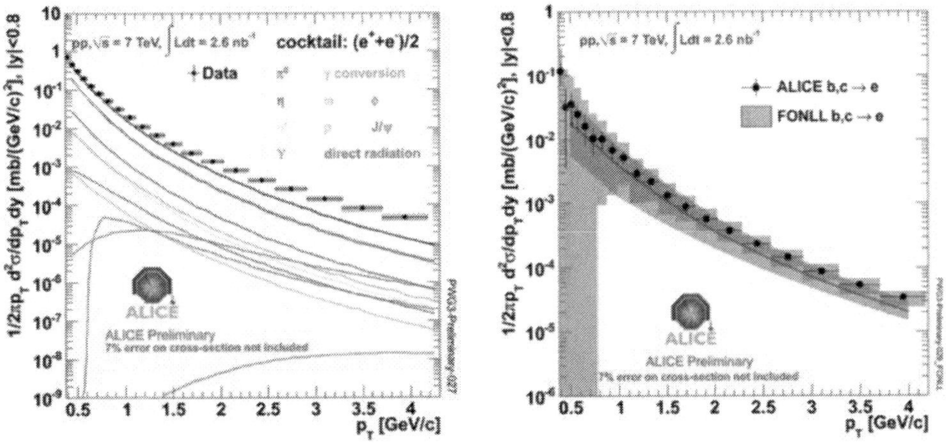


**Figure 1.** Left: Inclusive electron spectrum overlaid on the electron cocktail. Right: The heavy flavour electron spectrum is compared to the FONLL prediction.

the inclusive electron spectra after PID selection in the region 0.4< $p_T$ <4 GeV/$c$, corrected for efficiency and acceptance. Unfolding was performed to account for $p_T$ smearing due to Bremsstrahlung effects. The cocktail is shown in the same figure and is obtained adopting as input the $\pi^0$ cross section, measured by ALICE via the reconstruction of photon conversions and mt scaling for other light mesons ($\eta$, $\omega$, $\phi$, $\rho$). On the right panel, the heavy flavour spectrum is compared with FONLL predictions for charm and beauty summed together.

## 4. Cross section of forward single muons from heavy flavour decay

At forward rapidity heavy flavour production is studied using the single-muon $p_T$ distribution measured with the ALICE muon spectrometer. The main sources of background for this analysis are: muons coming from in-flight decays of light hadrons (decay muons), muons from decay of hadrons produced in the absorber (secondary muons) and punch-through hadrons. Due to the lower mass of the parent particles, background muons have softer transverse momentum than muons from heavy flavour decays and dominate in the low $p_T$ region. Punch-through hadrons are efficiently rejected by requiring that the track candidate in the tracking system has to be matched with the track reconstructed in the trigger system. In the region $2 < p_T < 6.5$ GeV/$c$, simulation studies indicated that the contamination of secondary muons is about 3% and is therefore small. The contribution of decay muons at 25% level is the main background source and it is subtracted using Monte Carlo simulations using Pythia and PHOJET as event generators. The systematic error introduced by this procedure is found to range from 20 to 30% from high to low $p_T$. After background subtraction, the muon $p_T$ spectrum is corrected for efficiency and normalized using the minimum bias pp production cross section found to be $\sigma_{MB} = 62.3 \pm 4.3$ mb. Figure 2 shows the $p_T$ differential cross section of single muons from heavy flavour decays in the $p_T$ range $0.4 < p_T < 6.5$ GeV/$c$ and -4< $\eta$ < -2.5. The corresponding FONLL calculations agree, within the errors, with the experimental data points. With the improved alignment of the muon spectrometer and the full 2010 data sample, the $p_T$ reach is expected to be extended to about 20 GeV/$c$. These improvements will allow the study of inclusive single muon production

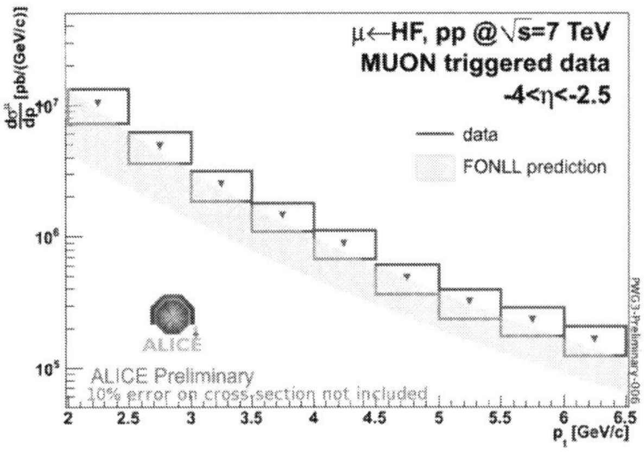


**Figure 2.** Comparison between the measured $d\sigma/dp_T$ distribution of single muons from heavy flavour decays in the rapidity range $-4 < \eta < -2.5$ and the FONLL predictions. The result corresponds to $\mathscr{L}= 3.49$ nb$^{-1}$.

cross section in the region 10< $p_T$ <20 GeV/$c$, where the dominant contribution is coming from beauty decay. This will establish a reference for the study of the beauty quenching in Pb-Pb collisions at forward rapidities.

## 5. Cross section of D mesons in the central rapidity region

The study of open charm production via D meson hadronic decays is based on displaced vertex reconstruction and invariant mass analysis. The displaced vertex reconstruction, as stated in section 2, is achieved thanks to the excellent tracking system. The resolution on single track impact parameter with the 2010 data sample is close to the design value, being 70 $\mu$m at $p_T \sim$ 1 GeV/$c$, allowing selection of decay vertices of open charm mesons ($c\tau \sim 120 - 310$ $\mu$m) in the whole accessible $p_T$ region. Several D meson measurements are under study and among them, the most advanced are the $D^0$, the $D^{*+}$ and the $D^+$. The reconstruction of $D^0$ meson is

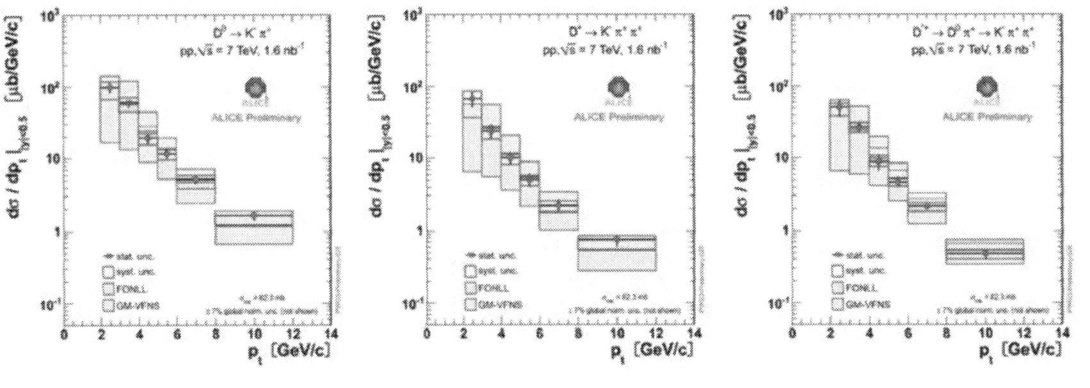


**Figure 3.** From left to right measured $d\sigma/dp_T$ for $D^0$, $D^+$ and $D^{*+}$ mesons in the central rapidity region ($|y| < 0.5$). Data are compared with predictions from FONLL and GM-VFNS. Results correspond to $\mathscr{L} = 1.6$ nb$^{-1}$.

performed using the hadronic channel $D^0 \to K^-\pi^+$ (BR = $(3.91\pm0.05)\%$) while the $D^{*+}$ meson

Journal of Physics: Conference Series **316** (2011) 012025 doi:10.1088/1742-6596/316/1/012025

is reconstructed in the channel $D^{*+} \rightarrow D^0\pi_s^+$ (BR = (67.7±0.5)%), where $D^0$ is reconstructed using the $K^-\pi^+$ channel. The $D^+$ is reconstructed in the channel $D^+ \rightarrow K^-\pi^+\pi^+$ (BR = (9.22±0.21)%). The selection strategy is based on three steps. At the beginning a series of quality requirements on single tracks (i.e. number of ITS clusters and TPC points) are applied. The acceptance for single tracks is $|\eta| < 0.8$. In the second step selections are applied to the topology of the decay taking advantage of the reconstruction on the secondary vertex. In the case of the $D^{*+}$, the secondary vertex cannot be reconstructed (being strong decay) and the selections are applied to the daughter $D^0$. To further suppress the combinatorial background a selection based on PID information from TPC and TOF detectors is applied on the daughter tracks. The PID strategy is developed to retain close to 100% of the signal so that the systematic on PID is kept under control. The raw yield extracted from the invariant mass analysis is then corrected for acceptance and efficiency. The efficiencies for prompt and secondary D mesons are accounted for. The contamination of D mesons from B decays is subtracted using a method based on FONLL expectation value for the $b$ cross section. A systematic error ($p_T$ dependent), of the order of 15%, is arising from such subtraction. The differential $p_T$ production cross section in the central rapidity region is evaluated for each of the three mesons in the $p_T$ range [2-12] GeV/$c$ and it is shown in figure 3. The results are obtained with ~ 20% of the 2010 data sample. As clear from the invariant mass analysis in figure 4 there are good prospects to extend the $p_T$ reach in the low and high region with the full 2010 data sample. Besides the aforementioned

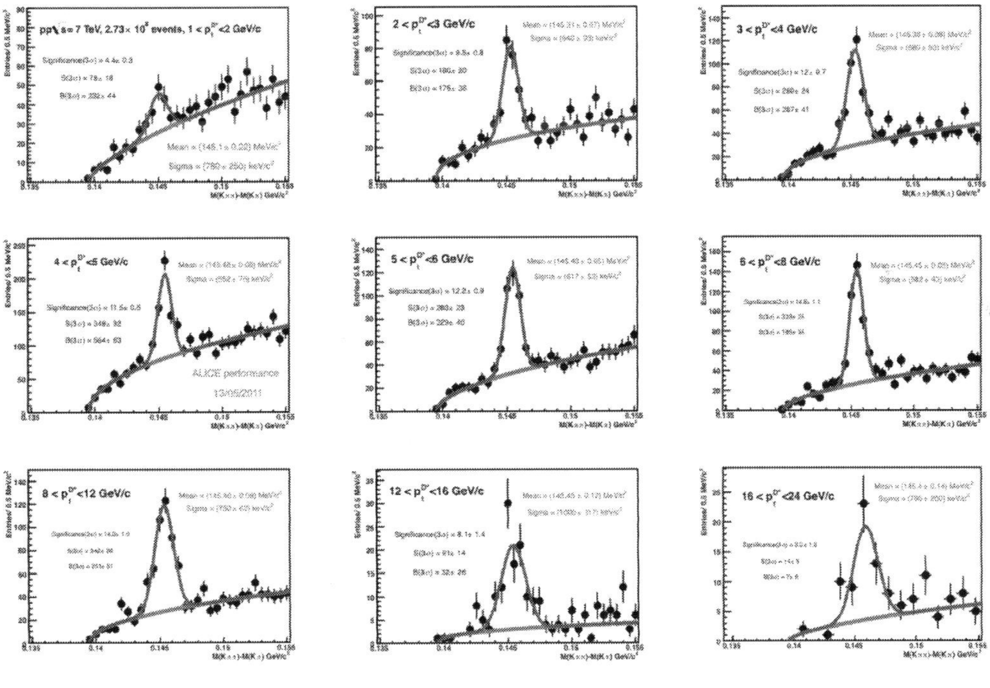


**Figure 4.** $D^{*+}$ signal in the $p_T$ range [1,24] GeV/$c$ extracted from ~ 4.8 nb$^{-1}$.

analysis, the $D^0 \rightarrow K^-\pi^+\pi^-\pi^+$ and the $D_s \rightarrow K^-K^+\pi^+$ analyses have allowed to extract promising signals in 4(5) $p_T$ regions. Efficiency and acceptance correction studies are on going. Finally, the $\Lambda_c \rightarrow pK^-\pi^+$ analysis already shows signals in different $p_T$ bins. In this case the displaced vertex reconstruction is less effective than for D mesons since its proper decay length is only 59 $\mu$m, although in the low $p_T$ (< 3 GeV/$c$) region it benefits from the excellent proton identification assured by TPC and TOF.

## 6. Cross section of inclusive J/$\psi$ production

The ALICE detector has the capability to measure the $J/\psi$ production in the central barrel and in the forward rapidity region with a $p_T$ acceptance down to 0 . The analysis in the central barrel ($|y| < 0.9$) is performed in the di-electron channel ($e^+e^-$) while in the forward region ($-4 < y < -2.5$), it is performed in the dimuon channel. The $J/\psi \to e^+e^-$ analysis is carried

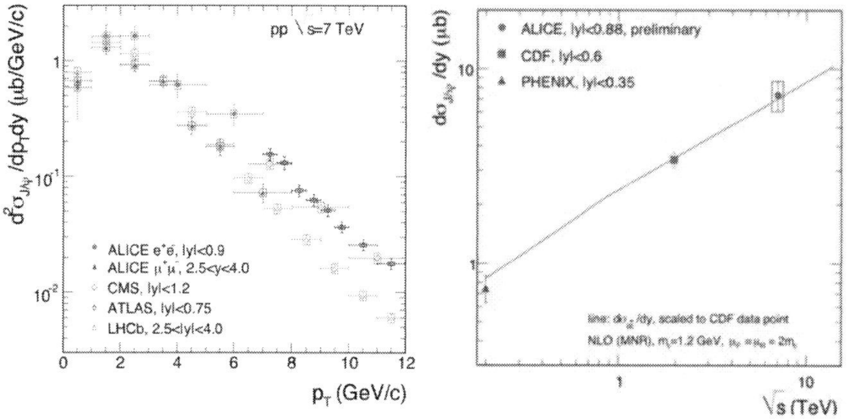


**Figure 5.** Left: $d^2\sigma_{J/\psi}/dp_Tdy$ for the mid-rapidity range and for the forward rapidity data, compared with results from the other LHC experiments [12]. Right: ALICE preliminary $J/psi$ production cross section compared to the result obtained by previous experiments at lower $\sqrt{s}$ [12]. See text for more details

out with a data sample of $2.4\times10^8$ MB events corresponding to an integrated luminosity of 3.9 nb$^{-1}$. Tracks were required to have a minimum $p_T$ of 1 GeV/$c$, at least one associated hit in the pixel detector and a minimum of 70 points in the TPC. Particle identification via specific energy loss in the TPC was applied. A total of $249\pm27(stat) \pm 20(syst)$ $J/\psi \to e^+e^-$ were detected. The left panel of figure 5 shows the $d^2\sigma_{J/\psi}/dp_Tdy$ for the mid-rapidity range and for the forward rapidity data compared with results from the other LHC experiments. In the right panel, the $p_T$ integrated $J/\psi$ cross section is plotted together with the results of CDF and PHENIX experiments at lower energies. In the di-muon channel, a total of $1.0\times10^7$ of events, satisfying the muon MB conditions were analyzed, corresponding to a total integrated luminosity of 15.6 nb$^{-1}$. A series of selection requirements were applied to the candidate di-muon. Among them at least an interaction vertex reconstructed by the pixel detector and at least one of the two reconstructed muon candidates has to match the corresponding track in the trigger chambers. The signal was extracted by fitting the di-muon invariant mass distribution with two Crystal Ball functions (to account for $J/\psi$ and $\psi^{'}$) and two exponentials [12]. The number of reconstructed and selected $J/\psi$ mesons is $1924\pm77(stat) \pm 144(syst)$. In order to extract the cross section the raw yield was corrected for the acceptance and for trigger and reconstruction efficiency. The $d\sigma_{J/\psi}/dy$, compared with results from other LHC experiments, is shown in figure 6. The $p_T$-integrated inclusive $J/\psi$ production cross sections in pp collisions at $\sqrt{s} = 7$ TeV are: $\sigma_{J/\psi}(|y| < 0.9) = 10.8 \pm 1.2(stat) \pm 1.7(syst)^{+1.6}_{-2.3}(pol)$ $\mu$b and $\sigma_{J/\psi}(2.5 < |y| < 4) = 6.31 \pm 0.25(stat) \pm 0.72(syst)^{+0.95}_{-1.96}(pol)$ $\mu$b. The systematic error named *pol* accounts for the unknown polarization of the $J/\psi$. Considering both the analyses ALICE has a broad coverage in rapidity and allows to measure $J/\psi$ down to $p_T = 0$ even in the central rapidity region.

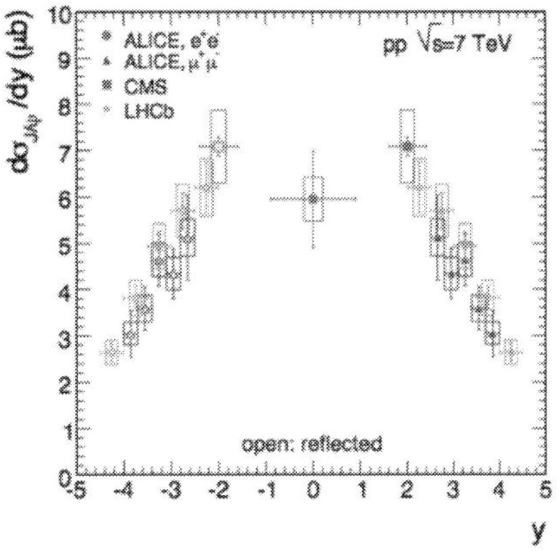


**Figure 6.** $d\sigma_{J/\psi}/dy$ compared with results from other LHC experiments [12]. The error bars are the sum in quadrature of the statistical and systematic errors. The box represent the systematic uncertainty on luminosity.

## 7. Conclusions

The recent ALICE results on heavy flavour measurements in pp collisions at $\sqrt{s} = 7$ TeV have been reviewed in this paper. The production cross section of muons and electrons from heavy flavour decay, D mesons and $J/\psi$ were presented. Results agree, within uncertainties, with FONLL pQCD calculations. It was shown that the D mesons analysis, using the full 2010 data sample, can extend the $p_T$ reach of the measurement down to 1 GeV/$c$ or below.

## 8. References

[1] K. Aamodt *et al.* [ALICE Coll.], J. Instrum. **3**, S08002 (2008)
[2] B. Alessandro *et al.* [ALICE Coll.], J. Phys. G: Nucl. Part. Phys. **32**, 1295 (2006)
[3] F. Karsch, Nucl. Phys. **A698**, 199 (2002)
[4] M. Cacciari, M. Greco and P. Nason, JHEP **9805** 007 (1998)
[5] M. Cacciari *et al.*, JHEP **0407**, 33 (2007)
[6] V. Khachatryan *et al.* [CMS Coll.], ArXiv:1011.4193 (2010)
[7] A. Adare *et al.* [PHENIX Coll.], Phys. Rev. Lett. **97**, 252002 (2006)
[8] W. Xie *et al.* [STAR Coll.], Nucl. Phys. A. **855** (2011)
[9] K. Aamodt *et al.* [ALICE Coll.], JINST 5 P03003 (2010)
[10] Jenkins D W and Dow J D *Phys. Rev.* B **39**, 3317 (1989)
[11] S. S Adler *et al.*, Phys. Rev. Lett. **96**, 032001 (2006)
[12] K. Aamodt *et al.* [ALICE Coll.], arXiv:1105.0380 (2011)

# Direct photons in Au+Au collisions measured with the PHENIX detector at RHIC

**Richard Petti**

Department of Physics and Astronomy, Stony Brook University, SUNY, Stony Brook, New York 11794, USA

E-mail: richard.petti@gmail.com

**Abstract.** A major goal of experiments in heavy-ion physics is the characterization of the quark gluon plasma (QGP) produced in the collision of heavy ions at high energy. Direct photons are a particularly good probe of the produced medium because they do not interact strongly and so can escape the medium unmodified, carrying information about when the photon was produced. It is expected that direct photon contributions from different sources (QGP radiation, hard scattering, hadron gas radiation) dominate at different transverse momentum ranges. Low momentum direct photons are dominated by thermal radiation (both from the QGP and hadron gas), while high momentum direct photons dominantly come from hard parton scatterings in the initial collision. We present a summary of techniques to measure direct photons with the PHENIX detector, with a focus on low momentum direct photons through their external conversion to dilepton pairs.

## 1. Introduction

Experiments at the Relativistic Heavy Ion Collider (RHIC) at BNL have established evidence that a quark-gluon plasma has been produced in heavy-ion collisions at $\sqrt{s_{NN}} = 200$ GeV [1]. One signature amongst many is that the system appears to behave like a liquid with partonic degrees of freedom, as evidenced by elliptic flow measurements and number of constituent quark scaling [2]. Another observation was the suppression of high momentum probes, indicating that the medium is not completely transparant to colored objects [3]. Recent PHENIX measurements of the temperature of the plamsa also indicate a temperature well above the expected critical temperature needed to form a QGP predicted by lattice QCD calculations [4]. A major goal of the RHIC program is to now study and characterize this extreme state of matter.

In this paper we focus on photons as probes of the QGP. Photons do not interact strongly with the medium and so escape virtually unmodfied, carrying information about each stage of the collisions where the photon was created. We are interested in direct photons, which are defined as all photons that do not originate from hadronic decays. This includes thermal and prompt photons, along with bremsstrahlung and jet-conversion photons. We mention two important pieces of information (although there are many others) that can be extracted from the study of direct photons. One is the temperature of the plasma, which can be accessed by the shape of the photon distribution. Another is the elliptic flow of direct photons. Theory calculations [5] indicate that the elliptic flow of thermal photons is quite sensitive to the thermalization time, $\tau_0$, of the plasma. It is expected that a small thermalization time leads to a small $v_2$, as the proportion of photons coming from the QGP phase is larger for earlier thermalization times.

The shape of the $v_2$ as a function of $p_T$ also changes drastically depending on the assumed $\tau_0$. Thermal photons are expected to dominate at low momentum, and so we focus on measuring direct photons below 5 GeV.

The measurement of low momentum direct photons is notoriously difficult due to very large background from hadron (mostly $\pi^0$) decays. This fact, combined with the limited energy resolution of the electromagnetic calorimeter (EMCal) at low energy, requires an alternate method to simply measuring photons that directly deposit their energy in the EMCal. Thus it is advantageous to measure dilepton pairs originating from the conversion of photons. The momentum resolution in the tracking systems for charged particles improves as $p_T$ decreases. We consider two sources of dilepton pairs from photons, internal and external conversions. Internal conversions occur from processes that happen to produce a virtual photon, rather than a real photon, which will decay into a low mass $e^+e^-$ pair [4]. External conversions occur when a photon interacts with material in the experimental apperature and converts into the dilepton pair. The external conversion analysis is complementary to the virtual (internal) conversion analysis and is a good cross-check of the validy of the procedure and is one of the main motivations for undertaking the external conversion analysis.

## 2. Experimental Apparatus

A detailed description of the PHENIX detector subsystems can be found in the reference [6]. Photons are measured in the central arms of the PHENIX spectrometer, each of which cover $\pi/2$ in azimuth and total cover a psuedorapidity range of $\eta < |0.35|$. Charged particles are tracked in layers of multi-wire proportional chambers, calculating the bend in the magnetic field and hence the momentum of the track [7]. A combination of the Ring-Imaging Cherenkov Dector (RICH) [8] and the electromagnetic calorimeter (EMCal) [9] assist in electron identification. There are forward and backward beam-beam counters (BBCs) at $3.1 < |\eta| < 3.9$, utilized as global detectors to determine both centrality and the vertex of the collision along the beamline (the z direction). A dedicated reaction plane detector consisting of plastic scintillator paddles was also installed for the 2007 run [10], which reside at $1.0 < |\eta| < 2.8$.

## 3. Analysis Techniques

The details of the internal conversion analysis can be found in the reference [4]. The basic idea of this technique is that low mass $e^+e^-$ pairs are formed when a virtual direct photon is produced in the system. Many of these pairings exist from pure combinatorics and have no physical correlation. These are removed by studying pairs formed in mixed events. Further sources of correlated background pairs must also be removed (such as external conversions, cross pairs, jet pairs). These can be removed by studying like-sign foreground distributions. Once all background is removed, the yield of pairs is compared to a cocktail of expected hadronic sources. An excess of pairs is observed in Au+Au collisions and is interpreted as originating from virtual photons. The yield of real photons can be calculated from the pair rate observed. Upon calculating the cross-section of direct photons, an excess above the expected rate is seen below $p_T < 3$ GeV. This excess is exponential in shape and is fit to extract a temperature parameter. The temperature measured in the paper for central Au+Au collisions is $T = 221 \pm 19(stat) \pm 19(sys)$ MeV, which is above the expected critical temperature needed for the phase transition. This is an important result and it is worthwhile to seek independent measurements of the same quantity as a cross check. The external conversion channel provides us with this cross check.

The details of the external conversion analysis are described here. We start by identifying single electrons, which will be paired later. The electron id detectors used are the RICH and EMCal using common PHENIX electron id cuts. It is required that the track is associated with

$> 2$ phototubes in the RICH. It is also required that the energy to momentum ratio, $E/p > 0.5$. The reconstructed momentum of the track must be $0.2 < p_T < 20$ GeV.

We need to identify pairs that actually come from an external photon conversion. During the 2007 RHIC run, the Hadron Blind Detector (HBD) [11] was installed for its engineering run. The backplane of the detector is fairly thick (about $4\% X_0$) and so is a significant localized source of conversions. Hence we choose to focus on conversions occurring in this backplane. It is located at a radius of about 60 cm from the interaction point. This significant conversion radius is the key to indentifying these photon conversions.

PHENIX does all of its tracking outside the magnetic field. This requires an assumption about the origin of the track to determine the bend in the field and calculate the momentum of the particle. A natural assumption is that tracks originate from the event vertex. But if we consider electrons coming from a radius of 60 cm, then this assumption is wrong and leads to a misreconstruction of the electron track. This misreconstruction leads to an artificial opening angle of the pair, which leads to an apparent mass of the pair. This apparent mass is proportional to the radius at which the conversion occurs (the controlling parameter is how much field the particle missed, even though it was assumed to pass through the entire field), causing a slight over-estimate of the momentum. The apparent mass can therefore be used to help identify conversions at the HBD backplane. Simualtion studies have shown that we expect an invariant mass peak of the dilepton pairs to be at 12 MeV with the Run 7 magnetic field configuration, see the open symbols in Fig. 1. This same peak can be seen in the data, Fig. 2.

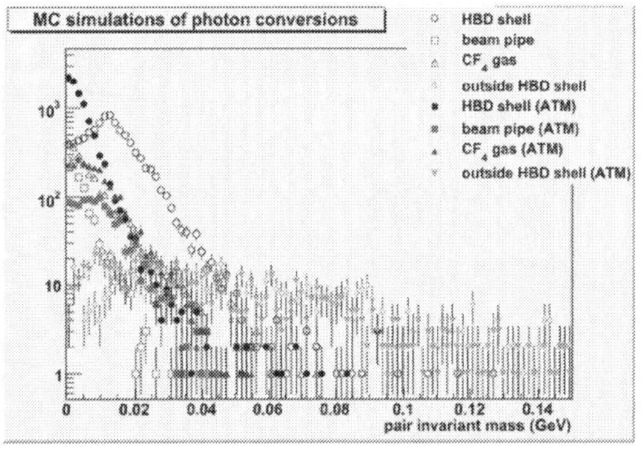


**Figure 1.** The $e^+e^-$ pair invariant mass distriubtion for photon conversions at different radii is plotted from GEANT based Monte Carlo simulations. The source of the conversion is represented by various shapes as shown in the legend. Open symbols represent the normally reconstructed mass, with the closed symbols as the ATM calculated mass.

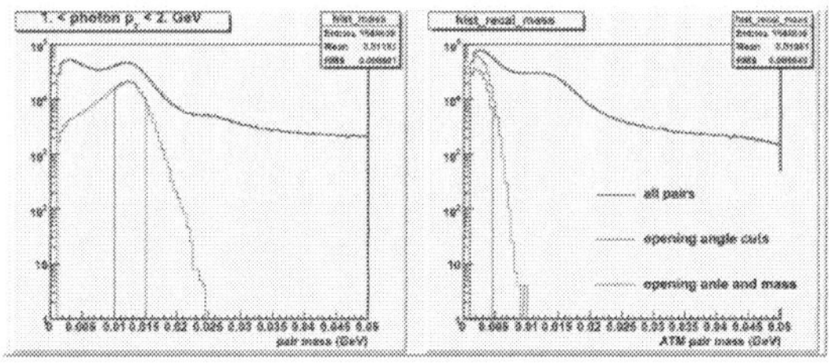


**Figure 2.** The application of the pair cuts on the raw pair invariant mass distribution shown for one particular $p_T$ bin in real data. The left side shows the normal track model and the right side the ATM. Each color represents the application of more cuts as shown in the legend.

Of course it is desirable to have the correct momentum of the converted photon, not a misreconstructed momentum. Therefore we correct for this mis-measurement with an alternate track model (ATM) assumption. Rather than assuming that tracks originate from the event vertex, it is assumed that all the tracks originate from a radius of 60 cm. The momentum of the reconstructed track (under the normal track model assumption) is reparameterized for the ATM assumption. This is studied through Monte Carlo simulations. The conversion pairs are then reconstructed properly under this ATM assumption, and the invariant mass moves from peaking at 12 MeV towards having a peak at 0 MeV. All other particles originating from the event vertex will now be misreconstructed, but this is a desired effect. Pairs from particles originating from the event vertex will be shifted up in mass. This is especially important for Dalitz decays of $\pi^0$'s, since the HBD conversion peak sits on the side of the Dalitz peak (see the left hand side of Fig. 2, where the first peak seen in the black historam is the Dalitz peak and the second peak from conversions at the HBD). We cut on how the mass moves under the two track models, allowing us to indentify converted photons. It is required that the ATM calculated mass is less than 4.5 MeV and the normally calculated mass is between $10 < mass < 15$ MeV.

In addition to these mass cuts, we also cut on the separation of the tracks at the HBD backplane calculated with the corrected ATM momenta. Conversion pairs should have no opening angle at the point of conversion. Therefore we require that the tracks are close in $\phi$ and $z$. These cuts allow us to get a very clean photon id at low momentum. The effect of these pair cuts on the raw $e^+e^-$ pair invariant mass distribution is shown in Fig. 2. An unfortunate detector problem offers a good opportunity for an internal cross-check of the analysis, namely the removal of half of the HBD for repair during the run. And so we can check the misidentification rate of the photons by analyzing the pairs we id in the arms serparately and comparing the number of photons that we find to convert in the arms. It is observed that the misidentification of converted photons is less than 3%. The combinatorial background is insignificant in this very low invarant mass range (mass less than 20 MeV).

The final goal of the measurement is the direct photon $v_2$ (at the time of writing this analysis is still a work in progress and so only an analysis strategy and current status is presented). We need three basic pieces of information to derive the direct photon $v_2$. We measure the $v_2$ of all photons (we cannot a priori distinguish the origin of a photon) and statistically subtract background sources to the $v_2$. Therefore, we measure the inclusive photon $v_2$. In addition to the inclusive photon $v_2$, we need to also know the $\pi^0$ decay photon $v_2$ (since most of the inclusive photons come from $\pi^0$ decays). This is the only hadronic $v_2$ source that we measure. Contributions to the $v_2$ from other (less significant) sources of hadrons will be calculated from $KE_T$ scaling, where $KE_T = m_T - m_0$. We also need to know the relative fraction of direct photons compared to hadronic sources or $R_\gamma = N^{incl}/N^{hadron}$, the ratio of inclusive photons to hadronic decay photons. Then the direct photon $v_2$ can be calculated from Eqn. 1.

$$v_2^{dir} = \frac{N^{incl}v_2^{inc} - N^{BG}v_2^{BG}}{N^{inc} - N^{BG}} = \frac{R_\gamma v_2^{incl} - v_2^{BG}}{R_\gamma - 1} \quad (1)$$

The inclusive photon $v_2$ is measured by taking the identified external photon conversions using the aforementioned method and correlating the angle of emission, $\phi$, relative to the reaction plane angle, $\Psi$, in azimuth. This $\phi - \Psi$ distribution is binned in $p_T$. Correlation functions (C.F.) are constructed from these distributions. We perform a Fourier decomposition of the correlation functions, following Eqn. 2. The correlation functions are fit with Eqn. 2 and the elliptic flow strength, $v_2$, is extracted in each $p_T$ bin. Higher order Fourier terms in Eqn. 2 are assumed to be negligible and are not included. Sine terms are zero by symmetry. We also calculate $v_2$ as $\langle cos(2(\phi - \Psi))\rangle$, the mean projection method. This yields consistent results with the fit and the average of the two is taken for the result presented here. The extracted $v_2$ is corrected by the reaction plane resolution, determined by comparing correlations between the north and south detectors. The major source of systematic error on the points is from the reaction plane determination, which is roughly 10%.

$$C.F. = B[1 + 2v_2cos(2(\phi - \Psi))] \quad (2)$$

The ratio $R_\gamma$ is measured using the double ratio shown in Eqn. 3. It is defined as the ratio of the yield of inclusive photons to the yield of hadronic decay photons. A ratio above one indicates a signal of direct photons. The terms that go into the numerator of the double ratio, along with the correction factors, are shown in Eqns. 4 and 5. Note that all the terms are measured as a function of the converted photon $p_T$. The numerator of Eqn. 3 is purely from data and consists of the ratio of the inclusive photon yield (the converted photons) to the yield of those photons that we tag as coming from $\pi^0$ decays. It is necessary to measure the $\pi^0$ contribution since that is the largest source of hadronic decay photons. This is done by reconstructing $\pi^0$'s by combining the converted photons with photons we measure in the EMCal. To identify a photon in the EMCal we cut on the shower shape, comparing the shower shape of the photon candidate with the expected shower shape for a photon, and require a minimum energy deposition of 500 MeV. The combinatorial background is removed with a mixed event technique and after this subtraction, we integrate the mass distrubution around $2\sigma$ of the $\pi^0$ peak. In this way we (statistically) tag some of our inclusive photons as coming from a $\pi^0$ decay.

$$R_\gamma = \frac{\gamma^{incl}(p_T)}{\gamma^{hadr}} = \frac{\epsilon_\gamma(p_T) f(p_T) \cdot \left( \frac{N_\gamma^{incl}(p_T)}{N^{\pi^0 tag}(p_T)} \right)_{Data}}{\left( \frac{N_\gamma^{hadr}(p_T)}{N^{\pi^0}(p_T)} \right)_{Sim}} \tag{3}$$

$$N_\gamma^{incl}(p_T) = c\epsilon_{pair} a_{pair} \gamma^{incl}(p_T) \tag{4}$$

$$N_\gamma^{\pi^0 tag}(p_T) = c\epsilon_{pair} a_{pair} \epsilon_\gamma f \gamma^{\pi^0}(p_T) \tag{5}$$

In the Eqns. 3, 4, and 5, $\gamma$ represents the yield of photons that nature produces, N represents the value we actually measure. The correction factors for the inclusive photon yield are shown in Eqn. 4. $\epsilon_{pair}$ and $a_{pair}$ are the $e^+e^-$ pair efficiency and acceptance in the PHENIX detector respectively. There is also some factor corresponding to the probability of the photon externally converting, denoted by c. These corrections apply to both the inclusive and the $\pi^0$ tagged photon samples. For the $\pi^0$ tagged sample we additionaly have to correct for the reconstruction efficiency, $\epsilon_\gamma$, and apply an acceptance correction, f, for reconstructing the unconverted decay photon in the EMCal. The f factor is a conditional acceptance correction and is defined as the probability of getting the unconverted photon in the EMCal, given that we already have the $e^+e^-$ pair from the other photon. The $\epsilon_\gamma$ accounts for occupancy effects in the detector by embedding simulated $\pi^0$ decays into real events. These corrections are in progress and are being studied with full GEANT3 based Monte Carlo simulations. We account for sources of decay photons other than the $\pi^0$ in the denominator of Eqn. 3 determined from simulations. Main sources of decay photons are decays of mesons such as $\pi^0$, $\eta$, and $\eta'$, and are included in a photon cocktail. The photon distributions from these decays are calculated in a decay generator, which realistically handles the kinematics of the decay. The measured $\pi^0$ $p_T$ distribution from published data is fit with a Hagedorn function as is used as input to the photon cocktail for the $\pi^0$ $p_T$ shape. The shape of the $p_T$ distributions for the other hadrons is determined by $m_T$ scaling, where the $p_T$ is replaced by $m_T = \sqrt{m + p_T}$, with m the mass of the hadron. The relative hadron/$\pi^0$ ratios are also taken from actual measurement and input into the cocktail. Then the denominator of the double ratio can be calculated, which is the ratio of the yield of all hadronic decay photons to the yield of decay photons coming from $\pi^0$s.

The $\pi^0$ tagging technique will also be extended, allowing us to calculate the $\pi^0$ $v_2$ in a manner consistent with the rest of the analysis.

## 4. Results and Conclusions

This analysis is still a work in progress at the time of writing, therefore only the methodology and current status have been discussed. A preliminary result has been obtained for the inclusive photon $v_2$, shown in Fig 3, for three centrality bins in a range $0.4 < p_T < 3.5$ GeV. From left to right in the figure are the $0-20\%$, $20-40\%$, and $40-60\%$ centrality bins. The major source of systematic error on the points comes from the reaction plane determination (about 10%), with photon id and calculation method contributing only a few percent. The systematic error associated with the reaction plane resolution is common to both the inclusive photon $v_2$ and the $\pi^0$ $v_2$, thus when the final systematic error is propagated to the full direct photon $v_2$ result, the reaction plane systematic will only come in once. These results are compared to published data [12] in which photons are directly measured in the EMCal. The agreement is good, indicating the validity of this method.

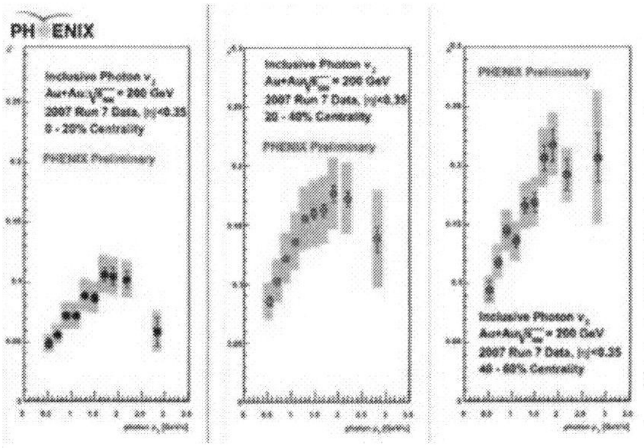

**Figure 3.** The preliminary measured inclusive photon $v_2$ via the external conversion method. This is shown in three centrality bins. From left to right the bins are $0-20\%$, $20-40\%$, and $40-60\%$ centrality.

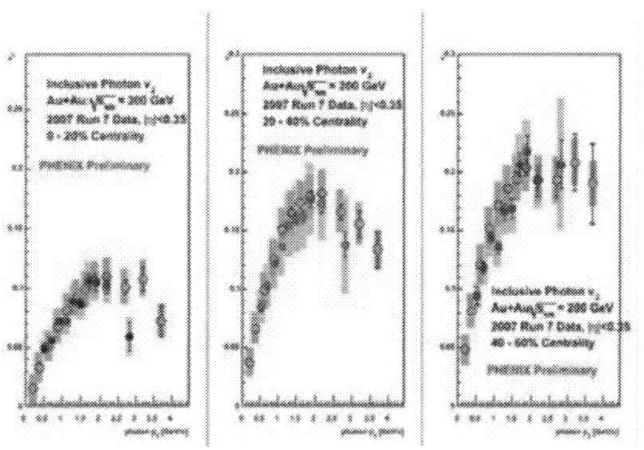

**Figure 4.** The same points as in 3, but with published data overlayed in the open black cirles with the red systematic error boxes [12]

In summary, we have presented the current status and the analysis method for measuring the elliptic flow of direct photons via their external conversion to dilepton pairs in the detector material. A comparison of the measured inclusive photon $v_2$ with an earlier PHENIX result obtained using the EMCal shows the reliability of this method. In a next step, we will subtract the $v_2$ of decay photons and calculate the direct photon $v_2$, using the external conversion method.

## References

[1] Adcox K et al 2005 *Nucl. Phys.* A **757** 184-283
[2] Adare A et al 2007 *Phys. Rev. Lett.* **98** 162301
[3] Adcox K et al 2002 *Phys. Rev. Lett.* **88** 022301
[4] Adare A et al 2010 *Phys. Rev. Lett.* **104** 132301
[5] Chatterjee R and Srivastava D 2009 *Phys. Rev.* C **79** 021901
[6] Adcox K et al 2003 *Nucl. Instrum. Meth.* A **499** 469-79
[7] Adcox K et al 2003 *Nucl. Instrum. Meth.* A **499** 489-507
[8] Aizawa M et al 2003 *Nucl. Instrum. Meth.* A **499** 508-20
[9] Aphecetche A et al 2003 *Nucl. Instrum. Meth.* A **499** 521-36
[10] Richardson E et al 2011 *Nucl. Instrum. Meth.* A **636** 99-107
[11] Anderson W et al 2011 arXiv:1103.4277
[12] Adler S et al 2006 *Phys. Rev. Lett.* **96** 032302

# Properties of the partonic phase at RHIC within PHSD

**E. L. Bratkovskaya**
Institute for Theoretical Physics, University of Frankfurt, Frankfurt, Germany
Frankfurt Institute for Advanced Study, Frankfurt, Germany
E-mail: Elena.Bratkovskaya@th.physik.uni-frankfurt.de

**W. Cassing**
Institute for Theoretical Physics, University of Giessen, Giessen, Germany

**V. P. Konchakovski**
Institute for Theoretical Physics, University of Giessen, Giessen, Germany
Bogolyubov Institute for Theoretical Physics, Kiev, Ukraine

**O. Linnyk**
Institute for Theoretical Physics, University of Giessen, Giessen, Germany

**V. Ozvenchuk**
Frankfurt Institute for Advanced Study, Frankfurt, Germany

**V. Voronyuk**
Joint Institute for Nuclear Research, Dubna, Russia
Frankfurt Institute for Advanced Study, Frankfurt, Germany
Bogolyubov Institute for Theoretical Physics, Kiev, Ukraine

**Abstract.** The dynamics of partons, hadrons and strings in relativistic nucleus-nucleus collisions is analyzed within the novel Parton-Hadron-String Dynamics (PHSD) transport approach, which is based on a dynamical quasiparticle model for partons (DQPM) matched to reproduce recent lattice-QCD results – including the partonic equation of state – in thermodynamic equilibrium. The transition from partonic to hadronic degrees of freedom is described by covariant transition rates for the fusion of quark-antiquark pairs or three quarks (antiquarks), respectively, obeying flavor current-conservation, color neutrality as well as energy-momentum conservation. In order to explore the space-time regions of 'partonic matter' the PHSD approach is applied to nucleus-nucleus collisions from SPS to RHIC energies. Detailed comparisons are presented for hadronic rapidity spectra and transverse mass distributions. The traces of partonic interactions are found in particular in the elliptic flow of hadrons as well as in an approximate quark-number scaling at the top RHIC energy.

## 1. Introduction

The 'Big Bang' scenario implies that in the first micro-seconds of the universe the entire state has emerged from a partonic system of quarks, antiquarks and gluons – a quark-gluon plasma (QGP) – to color neutral hadronic matter consisting of interacting hadronic states (and resonances) in which the partonic degrees of freedom are confined. The nature of confinement and the dynamics of this phase transition has motivated a large community for several decades and is still an outstanding question of todays physics. Early concepts of the QGP were guided by the idea of a weakly interacting system of partons which might be described by perturbative QCD (pQCD). However, experimental observations at the Relativistic Heavy Ion Collider (RHIC) indicated that the new medium created in ultrarelativistic Au+Au collisions is interacting more strongly than hadronic matter and consequently this concept had to be severely questioned. Moreover, in line with theoretical studies in Refs. [1, 2, 3] the medium showed phenomena of an almost perfect liquid of partons [4, 5] as extracted from the strong radial expansion and the scaling of elliptic flow $v_2(p_T)$ of mesons and baryons with the number of constituent quarks and antiquarks [4].

The question about the properties of this (nonperturbative) QGP liquid is discussed controversially in the literature and dynamical concepts describing the formation of color neutral hadrons from colored partons are scarce. A fundamental issue for hadronization models is the conservation of 4-momentum as well as the entropy problem because by fusion/coalescence of massless (or low constituent mass) partons to color neutral bound states of low invariant mass (e.g. pions) the number of degrees of freedom and thus the total entropy is reduced in the hadronization process [6, 7, 8]. This problem - a violation of the second law of thermodynamics as well as the conservation of four-momentum and flavor currents - has been addressed in Ref. [9] on the basis of the DQPM employing covariant transition rates for the fusion of 'massive' quarks and antiquarks to color neutral hadronic resonances or strings. In fact, the dynamical studies for an expanding partonic fireball in Ref. [9] suggest that the latter problems have come to a practical solution.

A consistent dynamical approach - valid also for strongly interacting systems - can be formulated on the basis of Kadanoff-Baym (KB) equations [10] or off-shell transport equations in phase-space representation, respectively [10, 11, 12]. In the KB theory the field quanta are described in terms of dressed propagators with complex selfenergies. Whereas the real part of the selfenergies can be related to mean-field potentials (of Lorentz scalar, vector or tensor type), the imaginary parts provide information about the lifetime and/or reaction rates of time-like 'particles' [13]. Once the proper (complex) selfenergies of the degrees of freedom are known the time evolution of the system is fully governed by off-shell transport equations (as described in Refs. [10, 13]). The determination/extraction of complex selfenergies for the partonic degrees of freedom has been performed before in Refs. [14, 15] by fitting lattice QCD (lQCD) 'data' within the Dynamical QuasiParticle Model (DQPM). In fact, the DQPM allows for a simple and transparent interpretation of lattice QCD results for thermodynamic quantities as well as correlators and leads to effective strongly interacting partonic quasiparticles with broad spectral functions. For a review on off-shell transport theory and results from the DQPM in comparison to lQCD we refer the reader to Ref. [13].

The actual implementations in the PHSD transport approach have been presented in detail in Refs. [16, 17]. Here we report again on the actual description of hadronization (Section 2). We also present results for rapidity distributions, transverse mass spectra and elliptic flow for heavy ion collisions at SPS (Section 3) and RHIC energies (Section 4) in comparison to data from the experimental collaborations.

Journal of Physics: Conference Series **316** (2011) 012027 doi:10.1088/1742-6596/316/1/012027

## 2. Hadronization in PHSD

The hadronization, i.e. the transition from partonic to hadronic degrees of freedom, is described in PHSD by local covariant transition rates as introduced in Ref. [9] e.g. for $q + \bar{q}$ fusion to a meson $m$ of four-momentum $p = (\omega, \mathbf{p})$ at space-time point $x = (t, \mathbf{x})$:

$$\frac{dN_m(x,p)}{d^4x d^4p} = Tr_q Tr_{\bar{q}}\ \delta^4(p - p_q - p_{\bar{q}})\ \delta^4\left(\frac{x_q + x_{\bar{q}}}{2} - x\right)$$
$$\times \omega_q\ \rho_q(p_q)\ \omega_{\bar{q}}\ \rho_{\bar{q}}(p_{\bar{q}})\ |v_{q\bar{q}}|^2\ W_m(x_q - x_{\bar{q}}, (p_q - p_{\bar{q}})/2)$$
$$\times N_q(x_q, p_q)\ N_{\bar{q}}(x_{\bar{q}}, p_{\bar{q}})\ \delta(\text{flavor, color}). \tag{1}$$

In Eq. (1) we have introduced the shorthand notation,

$$Tr_j = \sum_j \int d^4x_j \int \frac{d^4p_j}{(2\pi)^4}\ , \tag{2}$$

where $\sum_j$ denotes a summation over discrete quantum numbers (spin, flavor, color); $N_j(x,p)$ is the phase-space density of parton $j$ at space-time position $x$ and four-momentum $p$. In Eq. (1) $\delta$(flavor, color) stands symbolically for the conservation of flavor quantum numbers as well as color neutrality of the formed hadron $m$ which can be viewed as a color-dipole or 'pre-hadron'. Furthermore, $v_{q\bar{q}}(\rho_p)$ is the effective quark-antiquark interaction from the DQPM (displayed in Fig. 10 of Ref. [15]) as a function of the local parton ($q + \bar{q} + g$) density $\rho_p$ (or energy density). Furthermore, $W_m(x,p)$ is the dimensionless phase-space distribution of the formed 'pre-hadron', i.e.

$$W_m(\xi, p_\xi) = \exp\left(\frac{\xi^2}{2b^2}\right)\ \exp\left(2b^2(p_\xi^2 - (M_q - M_{\bar{q}})^2/4)\right) \tag{3}$$

with $\xi = x_1 - x_2 = x_q - x_{\bar{q}}$ and $p_\xi = (p_1 - p_2)/2 = (p_q - p_{\bar{q}})/2$. The width parameter $b$ is fixed by $\sqrt{\langle r^2 \rangle} = b = 0.66$ fm (in the rest frame) which corresponds to an average rms radius of mesons. We note that the expression (3) corresponds to the limit of independent harmonic oscillator states and that the final hadron-formation rates are approximately independent of the parameter $b$ within reasonable variations. By construction the quantity (3) is Lorentz invariant; in the limit of instantaneous 'hadron formation', i.e. $\xi^0 = 0$, it provides a Gaussian dropping in the relative distance squared $(\mathbf{r}_1 - \mathbf{r}_2)^2$. The four-momentum dependence reads explicitly (except for a factor 1/2)

$$(E_1 - E_2)^2 - (\mathbf{p}_1 - \mathbf{p}_2)^2 - (M_1 - M_2)^2 \leq 0 \tag{4}$$

and leads to a negative argument of the second exponential in (3) favoring the fusion of partons with low relative momenta $p_q - p_{\bar{q}} = p_1 - p_2$.

Related transition rates (to Eq. (1)) are defined for the fusion of three off-shell quarks ($q_1 + q_2 + q_3 \leftrightarrow B$) to color neutral baryonic ($B$ or $\bar{B}$) resonances of finite width (or strings) fulfilling energy and momentum conservation as well as flavor current conservation using Jacobi coordinates (cf. Ref. [16]).

On the hadronic side the PHSD transport approach includes explicitly the baryon octet and decouplet, the $0^-$- and $1^-$-meson nonets as well as selected higher resonances as in HSD [18]. Hadrons of higher masses (> 1.5 GeV in case of baryons and > 1.3 GeV in case of mesons) are treated as 'strings' (color-dipoles) that decay to the known (low-mass) hadrons according to the JETSET algorithm [19]. We discard an explicit recapitulation of the string decay and refer the reader to the original work [19] or Ref. [20].

## 3. Application to nucleus-nucleus collisions at SPS energies

In this Section we employ the PHSD approach to nucleus-nucleus collisions at moderate relativistic energies. It is of interest, how the PHSD approach compares to the HSD [18] model (without explicit partonic degrees-of-freedom) as well as to experimental data. In Fig. 1 we show the transverse mass spectra of $\pi^-$, $K^+$ and $K^-$ mesons for 7% central Pb+Pb collisions at 40 and 80 A·GeV and 5% central collisions at 158 A·GeV in comparison to the data of the NA49 Collaboration [21]. Here the slope of the $\pi^-$ spectra is only slightly enhanced in PHSD relative to HSD which demonstrates that the pion transverse motion shows no sizeable sensitivity to the partonic phase. However, the $K^\pm$ transverse mass spectra are substantially hardened with respect to the HSD calculations at all bombarding energies - i.e. PHSD is more in line with the data - and thus suggests that partonic effects are better visible in the strangeness-degrees of freedom. The hardening of the kaon spectra can be traced back to parton-parton scattering as well as a larger collective acceleration of the partons in the transverse direction due to the presence of repulsive vector fields for the partons. The enhancement of the spectral slope for kaons and antikaons in PHSD due to collective partonic flow shows up much clearer for the kaons due to their significantly larger mass (relative to pions). We recall that in Refs. [22] the underestimation of the $K^\pm$ slope by HSD (and also UrQMD) had been suggested to be a signature for missing partonic degrees of freedom; the present PHSD calculations support this early suggestion. Moreover, the PHSD calculations for RHIC energies show a very similar trend - the inverse slope increases by including the partonic phase.

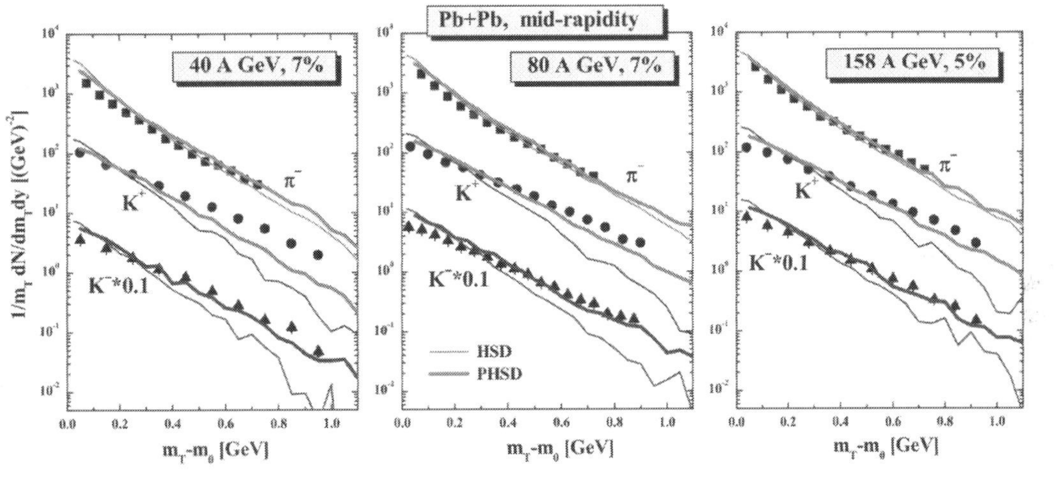


**Figure 1.** The $\pi^-$, $K^+$ and $K^-$ transverse mass spectra for central Pb+Pb collisions at 40, 80 and 158 A·GeV from PHSD (thick solid lines) in comparison to the distributions from HSD (thin solid lines) and the experimental data from the NA49 Collaboration [21].

The strange antibaryon sector is of further interest since here the HSD calculations have always underestimated the yield [23]. Our detailed studies in Ref. [16] show that the HSD and PHSD calculations both give a reasonable description of the $\Lambda + \Sigma^0$ yield of the NA49 Collaboration [24]; both models underestimate the NA57 data [25] by about 30%. An even larger discrepancy in the data from the NA49 and NA57 Collaborations is seen for $(\bar{\Lambda} + \bar{\Sigma}^0)/N_{wound}$; here the PHSD calculations give results which are in between the NA49 data and the NA57 data whereas HSD underestimates the $(\bar{\Lambda} + \bar{\Sigma}^0)$ midrapidity yield at all centralities.

The latter result suggests that the partonic phase does not show up explicitly in an enhanced production of strangeness (or in particular strange mesons and baryons) but leads to a different redistribution of antistrange quarks between mesons and antibaryons. In fact, as demonstrated in Ref. [16], we find no sizeable differences in the double strange baryons from HSD and PHSD – in a good agreement with the NA49 data – but observe a large enhancement in the double strange antibaryons for PHSD relative to HSD.

## 4. Application to nucleus-nucleus collisions at RHIC energies

In this section we continue the comparison of the PHSD transport approach to the experimental data from the RHIC collaborations as well as to the correspondent HSD results [17]. We find the rapidity distributions of the charged mesons to be slightly narrower in PHSD than those from HSD and actually closer to the experimental data. Also note that there is slightly more production of $K^{\pm}$ mesons in PHSD than in HSD while the number of charged pions is slightly lower. The actual deviations between the PHSD and HSD spectra are not dramatic but more clearly visible than at SPS energies (cf. Ref. [16]). Nevertheless, it becomes clear from Fig. 2 that the energy transfer in the nucleus-nucleus collision from initial nucleons to produced hadrons – reflected dominantly in the light meson spectra – is rather accurately described by PHSD. Fig. 2 also demonstrates that the longitudinal motion is well understood within the PHSD approach.

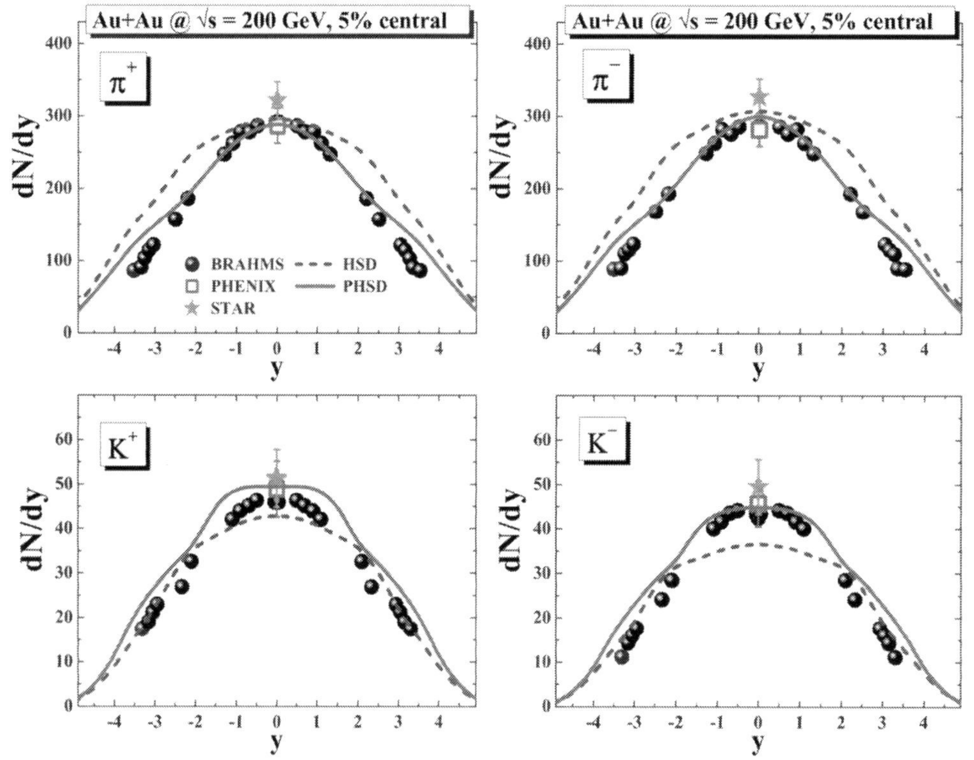


**Figure 2.** The rapidity distribution of $\pi^+$ (upper part, l.h.s.), $K^+$ (lower part, l.h.s.), $\pi^-$ (upper part, r.h.s.) and $K^-$ (lower part, r.h.s.) for 5% central Au+Au collisions at $\sqrt{s} = 200$ GeV from PHSD (solid red lines) in comparison to the distribution from HSD (dashed blue lines) and the experimental data from the RHIC Collaborations [26, 27, 28].

Independent information on the active degrees of freedom is provided by transverse mass spectra of the hadrons especially in central collisions. The actual results for RHIC energies are displayed in Fig. 3 where we show the transverse mass spectra of $\pi^-$, $K^+$ and $K^-$ mesons for 5% central Au+Au collisions at $\sqrt{s} = 200$ GeV in comparison to the data of the RHIC

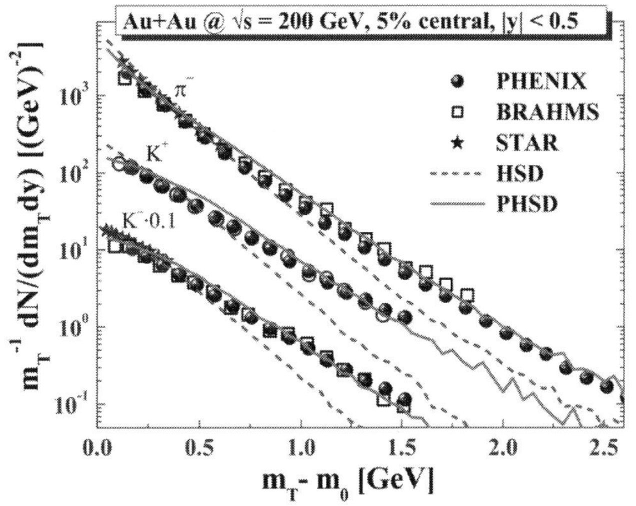


**Figure 3.** The $\pi^-$, $K^+$ and $K^-$ transverse mass spectra for 5% central Au+Au collisions at $\sqrt{s}$ = 200 GeV from PHSD (solid red lines) in comparison to the distributions from HSD (dashed blue lines) and the experimental data from the BRAHMS, PHENIX and STAR Collaborations [26, 27, 28] at midrapidity.

Collaborations [26, 27, 28]. Here the slope of the $\pi^-$ spectra is slightly enhanced in PHSD (solid red lines) relative to HSD (dashed blue lines) which demonstrates that the pion transverse mass spectra also show some sensitivity to the partonic phase (contrary to the SPS energy regime). The $K^\pm$ transverse mass spectra are substantially hardened with respect to the HSD calculations – i.e. PHSD is more in line with the data – and thus suggest that partonic effects are better visible in the strangeness degrees-of-freedom. The hardening of the kaon spectra can be traced back to parton-parton scattering as well as a larger collective acceleration of the partons in the transverse direction due to the presence of the repulsive mean-field for the partons. The enhancement of the spectral slopes for kaons and antikaons in PHSD due to collective partonic flow shows up much clearer for the kaons due to their significantly larger mass (relative to pions).

Of additional interest are the collective properties of the strongly interacting system which are explored experimentally via the elliptic flow

$$v_2(p_T, y) = \left\langle (p_x^2 - p_y^2)/(p_x^2 + p_y^2) \right\rangle |_{p_T, y} \tag{5}$$

of hadrons as a function of centrality, rapidity $y$, transverse momentum $p_T$ or transverse kinetic energy per participating quarks and antiquarks. We note that the reaction plane in PHSD is given by the $x - z$ plane with the $z$-axis in beam direction.

We start in Fig. 4 with the elliptic flow $v_2$ (for Au+Au collisions at the top RHIC energy) as a function of the centrality of the reaction measured by the number of participating nucleons $N_{part}$. The solid (red) line stands for the results from PHSD which is compared to the data for charged particles from the PHOBOS Collaboration [30]. The dashed blue line refers to the corresponding results for $v_2$ from HSD (taken from Ref. [29]). The momentum integrated results in the pseudo-rapidity window $|\eta| \leq 1$ from PHSD compare well to the data from Ref. [30] whereas the HSD results clearly underestimate the elliptic flow as pointed out before in Ref. [29]. The relative enhancement of $v_2$ in PHSD with respect to HSD can be traced back to the high interaction rate in the partonic phase and to the repulsive mean field for partons. We note in

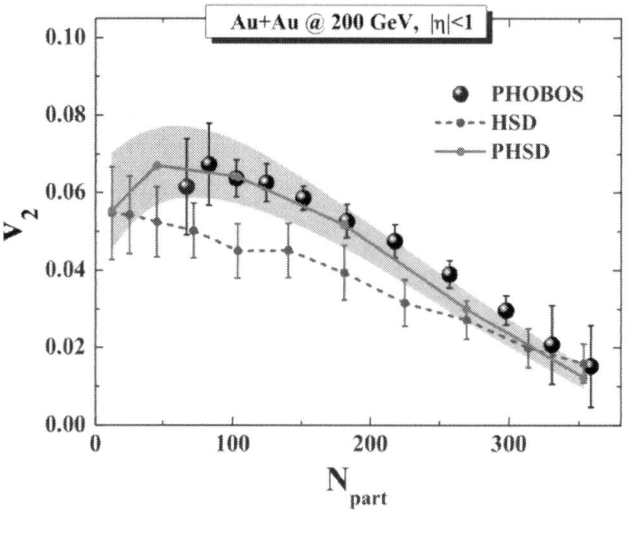


**Figure 4.** The elliptic flow $v_2$ for Au+Au collisions at the top RHIC energy $\sqrt{s} = 200$ GeV as a function of the centrality measured by the number of participating nucleons $N_{part}$. The solid (red) line stands for the results from PHSD whereas the dashed (blue) line represents the results from HSD (from Ref. [29]). The data are taken from the PHOBOS Collaboration [30] and correspond to momentum integrated events in the pseudo-rapidity window $|\eta| \leq 1$ for charged particles. The shaded band signals the statistical uncertainties of the PHSD calculations.

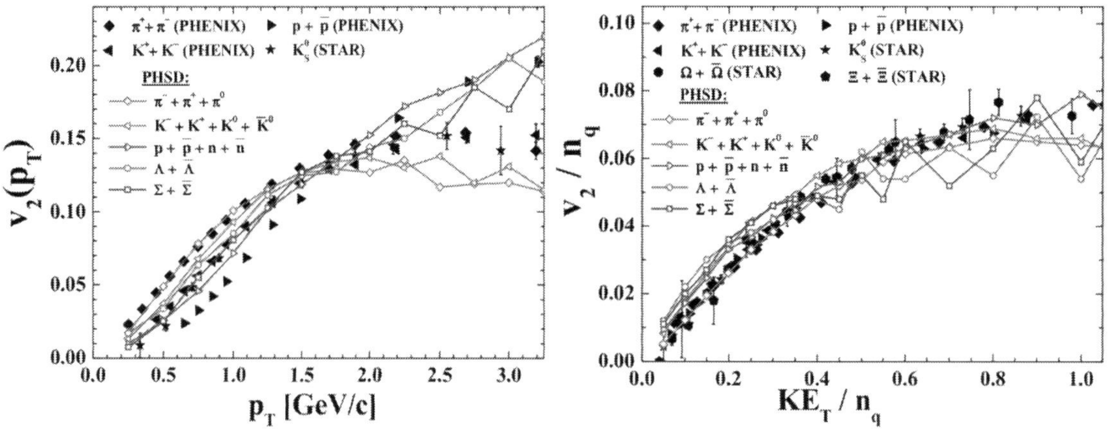


**Figure 5.** *Left:* The hadron elliptic flow $v_2$ for inclusive Au+Au collisions as a function of the transverse momentum $p_T$ (in GeV) for different hadrons in comparison to the data from the STAR [31] and PHENIX Collaborations [32] within the same rapidity cuts. *Right:* The elliptic flow $v_2$ - scaled by the number of constituent quarks $n_q$ - versus the transverse kinetic energy (6) divided by $n_q$ for different hadron species in comparison to the data from the STAR [31] and PHENIX [32] Collaborations.

passing that PHSD calculations without mean fields only give a tiny enhancement for the elliptic flow relative to HSD.

Fig. 5 (l.h.s.) shows the final hadron $v_2$ versus the transverse momentum $p_T$ for different particle species in comparison to the data from the STAR [31] and PHENIX Collaborations [32]. We observe a mass separation in $p_T$ as well as a separation in mesons and baryons for $p_T >$ 2 GeV roughly in line with data. The elliptic flow of mesons is slightly underestimated for $p_T > 2$ GeV in PHSD which is opposite to ideal hydrodynamics which overestimates $v_2$ at high transverse momenta. On the other hand, the proton (and antiproton) elliptic flow is slightly overestimated at low transverse momenta $p_T < 1.5$ GeV.

A further test of the PHSD hadronization approach is provided by the 'constituent quark number scaling' of the elliptic flow $v_2$ which has been observed experimentally in central Au+Au collisions at RHIC [4, 32, 33]. In this respect we plot $v_2/n_q$ versus the transverse kinetic energy per constituent parton,

$$KE_T = \frac{m_T - m}{n_q} , \tag{6}$$

with $m_T$ and $m$ denoting the transverse mass and actual hadron mass, respectively. For mesons we have $n_q = 2$ and for baryons/antibaryons $n_q = 3$. The results for the scaled elliptic flow are shown in Fig. 5 (r.h.s.) in comparison to the data from the STAR [31] and PHENIX Collaborations [32] for different hadrons and suggest an approximate scaling. For $KE_T > 0.5$ GeV there is a tendency to underestimate the experimental measurements for $\Lambda, \Sigma, \bar{\Lambda}, \bar{\Sigma}$ baryons which we attribute to an underestimation of interaction terms in PHSD for high momentum hadrons. In this respect we recall that the momentum independence of the quasiparticle width $\gamma$ and mass $M$ is presently a rough approximation and has to be refined. Due to the limited statistics especially in the baryonic sector with increasing $p_T$ this issue will have to be re-addressed with high statistics in future.

## 5. Summary

In this contribution we have addressed relativistic collisions of Pb+Pb at SPS energies and Au+Au collisions at RHIC energies in the PHSD approach which includes explicit partonic degrees of freedom as well as dynamical local transition rates from partons to hadrons (1). The hadronization process conserves four-momentum and all flavor currents and slightly increases the total entropy since the 'fusion' of rather massive partons dominantly leads to the formation of color neutral strings or resonances that decay microcanonically to lower mass hadrons. Since this dynamical hadronization process increases the total entropy the second law of thermodynamics is not violated (as is the case for simple coalescence models incorporating massless partons).

The PHSD approach has been also applied to nucleus-nucleus collisions from 40 to 160 A·GeV as well as for RHIC energies in order to explore the space-time regions of 'partonic matter' [16]. We have found that even central collisions at the top SPS energy of ~158 A· GeV show a large fraction of non-partonic, i.e. hadronic or string-like matter, which can be viewed as a 'hadronic corona' [34]. This finding implies that neither purely hadronic nor purely partonic 'models' can be employed to extract physical conclusions in comparing model results with data. On the other hand - studying in detail Pb+Pb reactions at SPS energies in comparison to the data [16] - it is found that the partonic phase has only a very low impact on the longitudinal rapidity distributions of hadrons but a sizeable influence on the transverse-mass distribution of final kaons due to the partonic interactions. The most pronounced effect is seen on the production of multi-strange antibaryons due to a slightly enhanced $s\bar{s}$ pair production in the partonic phase from massive time-like gluon decay and a more abundant formation of strange antibaryons in the hadronization process. This enhanced formation of strange antibaryons in central Pb+Pb collisions at SPS energies by hadronization supports the early suggestion by Braun-Munzinger and Stachel [35, 36] in the statistical hadronization model - which describes well particle ratios from AGS to RHIC energies.

At RHIC energies the PHSD calculations show also a good reproduction of the hadron transverse mass and rapidity spectra. Furthermore, the elliptic flow $v_2$ is well described for

Au+Au reactions at $\sqrt{s} = 200$ GeV as a function of centrality as well as of transverse momenta up to $p_T \simeq 1.5$ GeV/$c$. Due to the local transition rates from partons to hadrons (1) the PHSD approach also gives approximate quark number scaling of the elliptic flow as found experimentally by the RHIC Collaborations.

**Acknowledgements**

Work supported in part by the HIC for FAIR framework of the LOEWE program and by DFG.

[1] E. Shuryak, Prog. Part. Nucl. Phys. 53 (2004) 273.
[2] M. H. Thoma, J. Phys. G 31 (2005) L7; Nucl. Phys. A 774 (2006) 307.
[3] A. Peshier and W. Cassing, Phys. Rev. Lett. 94 (2005) 172301.
[4] I. Arsene *et al.*, Nucl. Phys. A 757 (2005) 1; B. B. Back *et al.*, Nucl. Phys. A 757 (2005) 28; J. Adams *et al.*, Nucl. Phys. A 757 (2005) 102; K. Adcox *et al.*, Nucl. Phys. A 757 (2005) 184.
[5] T. Hirano and M. Gyulassy, Nucl. Phys. A 769 (2006) 71.
[6] R. C. Hwa and C. B. Yang, Phys. Rev. C 67 (2003) 034902; V. Greco, C. M. Ko and P. Levai, Phys. Rev. Lett. 90 (2003) 202302.
[7] R. J. Fries, B. Müller, C. Nonaka and S. A. Bass, Phys. Rev. Lett. 90 (2003) 202303.
[8] Z.-W. Lin *et al.*, Phys. Rev. C 72 (2005) 064901.
[9] W. Cassing and E. L. Bratkovskaya, Phys. Rev. C 78 (2008) 034919.
[10] S. Juchem, W. Cassing and C. Greiner, Phys. Rev. D 69 (2004) 025006; Nucl. Phys. A 743 (2004) 92.
[11] W. Cassing and S. Juchem, Nucl. Phys. A 665 (2000) 377; *ibid* A 672 (2000) 417.
[12] Y. B. Ivanov, J. Knoll and D. N. Voskresensky, Nucl. Phys. A 672 (2000) 313.
[13] W. Cassing, E. Phys. J. ST 168 (2009) 3.
[14] W. Cassing, Nucl. Phys. A 791 (2007) 365.
[15] W. Cassing, Nucl. Phys. A 795 (2007) 70.
[16] W. Cassing and E. L. Bratkovskaya, Nucl. Phys. A 831 (2009) 215.
[17] E. L. Bratkovskaya, W. Cassing, V. P. Konchakovski, O. Linnyk, Nucl. Phys. A 856 (2011) 162.
[18] W. Cassing and E. L. Bratkovskaya, Phys. Rep. 308 (1999) 65.
[19] H.-U. Bengtsson and T. Sjöstrand, Comp. Phys. Commun. 46 (1987) 43.
[20] T. Falter, W. Cassing, K. Gallmeister, and U. Mosel, Phys. Rev. C 70 (2004) 054609.
[21] C. Alt et al., NA49 Collaboration, Phys. Rev. C 66 (2002) 054902; Phys. Rev. C 77 (2008) 024903.
[22] E. L. Bratkovskaya, S. Soff, H. Stöcker, M. van Leeuwen, and W. Cassing, Phys. Rev. Lett. 92 (2004) 032302.
[23] J. Geiss, W. Cassing and C. Greiner, Nucl. Phys. A 644 (1998) 107.
[24] T. Anticic *et al.*, Phys. Rev. C 80 (2009) 034906.
[25] F. Antinori *et al.*, Phys. Lett. B 595 (2004) 68; J. Phys. G: Nucl. Phys. 32 (2006) 427.
[26] S. S. Adler *et al.*, Phys. Rev. C 69 (2004) 034909.
[27] J. Adams *et al.*, Phys. Rev. Lett. 92 (2004) 112301.
[28] I. G. Bearden *et al.*, Phys. Rev. Lett. 94 (2005) 162301.
[29] E. L. Bratkovskaya, W. Cassing, and H. Stöcker, Phys. Rev. C 67 (2003) 054905.
[30] B. B. Back *et al.*, Phys. Rev. C 72 (2005) 051901.
[31] J. Adams *et al.*, Phys. Rev. Lett. 92 (2004) 052302; Phys. Rev. Lett. 95 (2005) 122301.
[32] A. Adare *et al.*, Phys. Rev. Lett. 98 (2007) 162301.
[33] S. Afanasiev *et al.*, Phys. Rev. C 80 (2009) 024909.
[34] J. Aichelin and K. Werner, Phys. Rev. C79 (2009) 064907.
[35] P. Braun-Munzinger *et al.*, Phys. Lett. B 365 (1996) 1; *ibid.* B 465 (1999) 15; *ibid.* B 518 (2001) 41.
[36] A. Andronic, P. Braun-Munzinger and J. Stachel, Nucl. Phys. A 772 (2006) 167.

Journal of Physics: Conference Series **316** (2011) 012028 doi:10.1088/1742-6596/316/1/012028

# Dilepton production at SPS and RHIC energies within the PHSD off-shell transport approach

**O. Linnyk**
Institut für Theoretische Physik, Universität Gießen, 35392 Gießen, Germany
E-mail: Olena.Linnyk@theo.physik.uni-giessen.de

**E. L. Bratkovskaya**
Institut für Theoretische Physik, Universität Frankfurt, 60438 Frankfurt am Main, Germany

**W. Cassing**
Institut für Theoretische Physik, Universität Gießen, 35392 Gießen, Germany

**Abstract.** The microscopic parton-hadron-string dynamics (PHSD) transport approach describes the full evolution of relativistic heavy-ion collisions: from the initial hard scatterings to the partonic phase in the early hot and dense reaction region (when the local energy density is above $\approx$1 GeV/fm$^3$) followed by hadronization and off-shell hadron propagation and interactions. The description of quarks and gluons in PHSD is based on the dynamical quasiparticle model (DQPM) matched to reproduce lattice QCD results in thermodynamic equilibrium. The DQPM describes QCD properties in terms of single-particle Green's functions and leads to the notion of effective quasiparticles which are massive and have broad spectral functions (due to large interaction rates). Dressing the quark and gluon lines with effective propagators provided by the DQPM, we obtain cross sections of dilepton production in the reactions $q+\bar{q}\to l^+l^-$ (Born mechanism), $q+\bar{q}\to l^+l^-+g$ (quark annihilation with the gluon Bremsstrahlung in the final state), and $q/\bar{q}+g\to l^+l^-+q/\bar{q}$ (gluon Compton scattering) by off-shell quarks and gluons in the strongly interacting partonic phase. After implementing the off-shell cross sections for the partonic processes into the PHSD transport approach, we address the dilepton production from the parton interactions in the early stage of relativistic heavy-ion collisions at SPS and RHIC energies. Dilepton production in $In+In$ collisions at 158 AGeV and in $Au+Au$ at $\sqrt{s}=200$ GeV is calculated. By comparing to the data of the NA60 and PHENIX Collaborations, we study the relative importance of different dilepton production mechanisms and point out the regions in phase space where partonic channels are dominant.

## 1. The PHSD approach to dilepton radiation

Early on, dileptons have been suggested as probes of the quark-gluon plasma (QGP) that is expected to be produced during the early stage of heavy-ion collisions at Super-Proton-Synchrotron (SPS) energies [1, 2, 3, 4, 5]. To address the dilepton production in a hot and dense medium – as created in heavy-ion collisions – we employ an up-to-date relativistic transport model, i.e. the Parton Hadron String Dynamics [9] (PHSD) that incorporates the explicit partonic phase in the early reaction region. Within PHSD, one solves generalized transport equations on the basis of the off-shell Kadanoff-Baym equations for Greens functions

Journal of Physics: Conference Series **316** (2011) 012028 doi:10.1088/1742-6596/316/1/012028

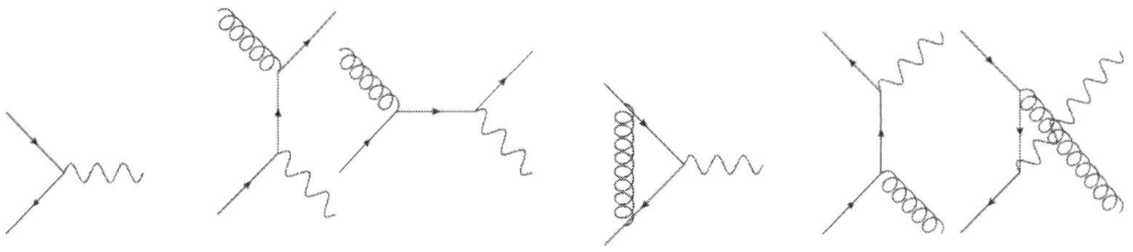


**Figure 1.** Diagrams contributing to the dilepton production from a QGP: (a) Drell-Yan mechanism, (b) gluon-Compton scattering (GCS), (c) vertex correction, (d) gluon Bremsstrahlung (NLODY), where virtual photons (wavy lines) split into lepton pairs, spiral lines denote gluons, arrows denote quarks. In each diagram the time runs from left to right.

in phase-space representation (in the first order gradient expansion, beyond the quasiparticle approximation). The approach consistently describes the full evolution of a relativistic heavy-ion collision from the initial hard scatterings and string formation through the dynamical deconfinement phase transition to the quark-gluon plasma (QGP) as well as hadronization and to the subsequent interactions in the hadronic phase.

In the hadronic sector PHSD is equivalent to the Hadron-String-Dynamics (HSD) transport approach [10, 11, 12] that has been used for the description of $pA$ and $AA$ collisions from SIS to RHIC energies and has lead to a fair reproduction of hadron abundances, rapidity distributions and transverse momentum spectra. In particular, HSD incorporates off-shell dynamics for vector mesons – according to Refs. [13] – and a set of vector-meson spectral functions [14] that covers possible scenarios for their in-medium modification. Various models predict that hadrons change in the (hot and dense) nuclear medium; in particular, a broadening of the spectral function or a mass shift of the vector mesons has been expected. Furthermore, QCD sum rules indicated that a mass shift may lead to a broadening and vice versa [15]; therefore both modifications should be studied simultaneously. In the off-shell transport, hadron spectral functions change dynamically during the propagation through the medium and evolve towards the on-shell spectral function in the vacuum.

The description of quarks and gluons in PHSD is based on a dynamical quasiparticle model for partons matched to reproduce lattice QCD results in thermodynamic equilibrium (DQPM). The DQPM describes QCD properties in terms of single-particle Green's functions (in the sense of a two-particle irreducible approach) and leads to the notion of the constituents of the sQGP being effective quasiparticles, which are massive and have broad spectral functions (due to large interaction rates). The transition from partonic to hadronic degrees of freedom in PHSD is described by covariant transition rates for the fusion of quark-antiquark pairs to mesonic resonanses or three quarks (antiquarks) to baryonic states.

Dilepton radiation by the constituents of the QGP proceeds via the elementary processes illustrated in Fig. 1: the Born (Drell-Yan) $q + \bar{q}$ annihilation mechanism, Gluon Compton scattering ($q + g \to \gamma^* + q$ and $\bar{q} + g \to \gamma^* + \bar{q}$), and quark + anti-quark annihilation with gluon Bremsstrahlung in the final state ($q + \bar{q} \to g + \gamma^*$). In the on-shell approximation, one uses perturbative QCD cross sections for the processes listed above. However, in order to make quantitative predictions at experimentally relevant low dilepton mass and strong coupling, we have to take into account the non-perturbative spectral functions and self-energies of the quarks, anti-quarks and gluons thus going beyond the on-shell approximation. For this purpose, off-shell cross sections were derived in [16] for dilepton production in the reactions $q + \bar{q} \to l^+l^-$ (Drell-Yan mechanism), $q + \bar{q} \to g + l^+l^-$ (quark annihilation with the gluon Bremsstrahlung in the final state), $q(\bar{q}) + g \to q(\bar{q}) + l^+l^-$ (gluon Compton scattering), $g \to q + \bar{q} + l^+l^-$ and

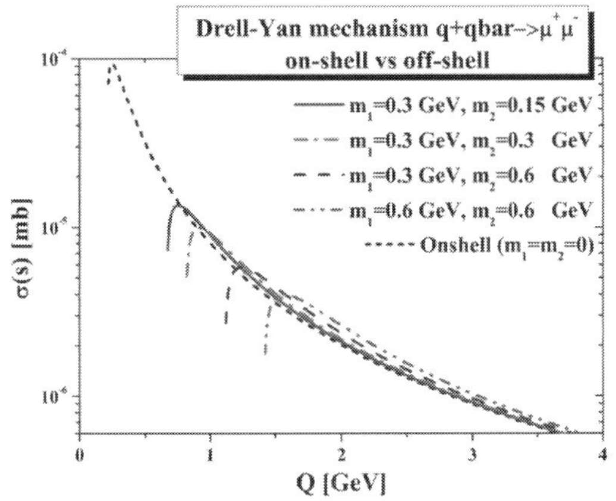


**Figure 2.** Dilepton production cross sections in the Born channel ($q + \bar{q} \to \mu^+ + \mu^-$). The short dashes (black) line shows the on-shell, i.e. the standard pQCD result. The other lines show the off-shell cross section, in which the annihilating quark and antiquark have finite masses $m_1$ and $m_2$ with different values (see legend).

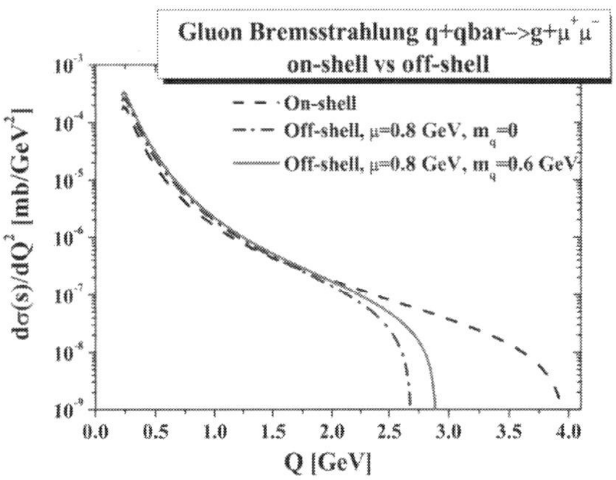


**Figure 3.** Comparison of off-shell and on-shell cross sections for dilepton production in the gluon Bremsstrahlung $q+\bar{q} \to g+\mu^+\mu^-$ channel at $\sqrt{s} = 4$ GeV. Dashed black line shows the on-shell (pQCD) cross section regularized by the cut on gluon mass $\mu_{cut} = 0.206$ GeV, blue dash-dotted and red solit lines present the off-shell cross section for the finite parton masses.

$q(\bar{q}) \to q(\bar{q}) + g + l^+l^-$ (virtual gluon decay, virtual quark decay) in the sQGP in effective perturbation theory by dressing the quark and gluon lines with the DQPM propagators for quarks and gluons. The obtained off-shell elementary cross sections then are implemented into the PHSD transport code, where the masses of quarks and gluons are distributed according to the DQPM spectral functions.

For illustration, we plot the dilepton production cross sections in the Born mechanism on the l.h.s. of Fig. 2. The short dashes (black) line shows the on-shell, i.e. the standard perturbative result. The other lines show the off-shell cross section, in which the annihilating quark and antiquark have finite masses $m_1$ and $m_2$ with different values: $m_1 = 0.3$ GeV, $m_2 = 0.15$ GeV (solid magenta line), $m_1 = 0.3$ GeV, $m_2 = 0.3$ GeV (dash-dotted red line), $m_1 = 0.3$ GeV, $m_2 = 0.6$ GeV (dashed blue line), $m_1 = 0.6$ GeV, $m_2 = 0.6$ GeV (dash-dot-dot green line).

In Fig. 3, the off-shell cross sections for the quark annihilation with gluon bremsstrahlung channel at various values of quark and gluon off-shellnesses (masses) are compared to the on-shell (pQCD) result. Dashed black line shows the on-shell cross section regularized by a cut on gluon mass $\mu_{cut} = 0.206$ GeV, red solid line presents the off-shell cross section for the gluon mass fixed to $\mu = 0.8$ GeV and on-shell quark and anti-quark ($m_1 = m_2 = m_3 = 0$). Blue dash-dotted line gives the off-shell result for $\mu = 0.8$ GeV, $m_1 = m_2 = m_3 = m_q = 0.6$ GeV. One readily notices the shift of the maximum allowed mass of the lepton pair to a lower value (in order to produce a massive gluon in the final state).

## 2. Comparison to data at SPS energies

By employing the HSD approach to the low mass dilepton production in relativistic heavy-ion collisions, it was shown in [17] that the NA60 Collaboration data for the invariant mass spectra for $\mu^+\mu^-$ pairs from In+In collisions at 158 A·GeV favored the 'melting $\rho$' scenario [19]. Also the data from the CERES Collaboration [20] showed a preference for the 'melting $\rho$' picture. On the other hand, the dilepton spectrum from In+In collisions at 158 A·GeV for $M > 1$ GeV

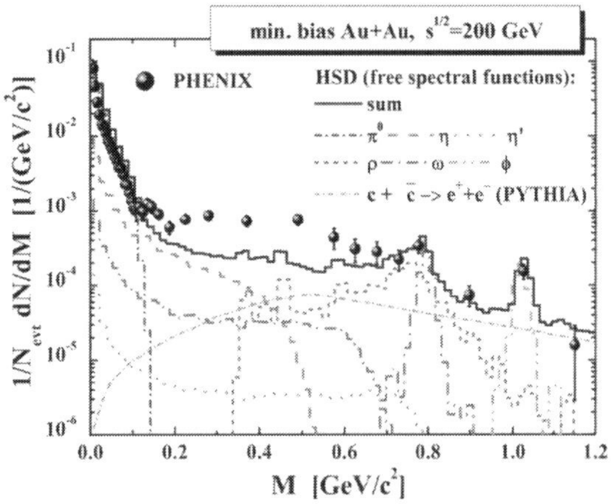


**Figure 4.** The HSD results for the mass differential dilepton spectra in case of inclusive $Au + Au$ collisions at $\sqrt{s}$ = 200 GeV in comparison to the data from PHENIX [24]. The results are shown for vacuum spectral functions (for $\rho, \omega, \phi$) including the channel decompositions.

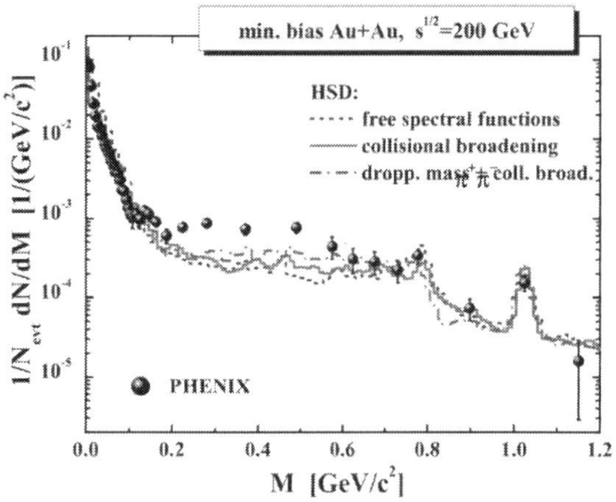


**Figure 5.** The HSD results for the mass differential dilepton spectra in case of inclusive $Au + Au$ collisions at $\sqrt{s}$ = 200 GeV in comparison to the data from PHENIX [24] in the 'free' scenario, the 'collisional broadening' picture as well as the 'dropping mass + collisional broadening' model.

could not be accounted for by the known hadronic sources (see Fig.2 of [17]). Also, hadronic models do not reproduce the softening of the $m_T$ distribution of dileptons at $M > 1$ GeV [19]. This observation pointed towards a partonic origin of the spectra at $M > 1$ GeV.

In Ref. [21] we presented PHSD results for the dilepton spectrum as produced in $In + In$ reactions at 158 AGeV compared to the NA60 data [19, 22]. We found that the spectrum at invariant masses below 1 GeV was well reproduced by the $\rho$ meson yield, if a broadening of the meson spectral function in the medium was assumed. On the other hand, the spectrum at $M > 1$ GeV was shown to be dominated by the partonic sources. Moreover, accounting for partonic dilepton sources allowed to reproduce in PHSD the effective temperature of the dileptons (slope parameters) in the intermediate mass range. The softening of the transverse mass spectrum with growing invariant mass implies that the partonic channels occur dominantly before the collective radial flow has developed.

## 3. Comparison to data at RHIC energies

The PHENIX Collaboration has presented dilepton data from $pp$ and $Au + Au$ collisions at Relativistic-Heavy-Ion-Collider (RHIC) energies of $\sqrt{s}$=200 GeV [23, 24] which show a large enhancement in $Au + Au$ reactions (relative to scaled $pp$ collisions) in the invariant mass regimes from 0.15 to 0.6 GeV and from 1 to 4 GeV [25]. We recall that HSD provides a reasonable description of hadron production in $Au + Au$ collisions at $\sqrt{s}$ = 200 GeV [26] and present in Figs 4 and 5 the HSD results for $e^+e^-$ pairs in inclusive $Au + Au$ collisions in comparison to the data from PHENIX [24] as calculated in [17]. Whereas the total yield is quite well described in the region of the pion Dalitz decay as well as around the $\omega$ and $\phi$ mass, HSD clearly underestimates the measured spectra in the regime from 0.2 to 0.6 GeV by an average factor of 3. When including the in-medium modification scenarios for the vector mesons, we achieve a sum spectrum which is only slightly enhanced compared to the 'free' scenario. The low mass dilepton spectra from $Au + Au$ collisions at RHIC (from the PHENIX Collaboration) are clearly underestimated in the invariant mass range from 0.2 to 0.6 GeV in the 'collisional

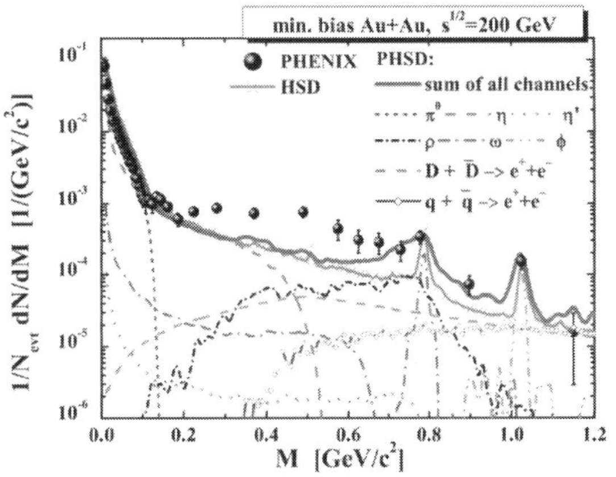


**Figure 6.** The PHSD results for the mass differential dilepton spectra in case of inclusive $Au + Au$ collisions at $\sqrt{s} = 200$ GeV in comparison to the data from PHENIX [24] in the low mass region ($M = 0-1.2$ GeV). For comparison, the HSD results are shown by the grey line with cross symbols.

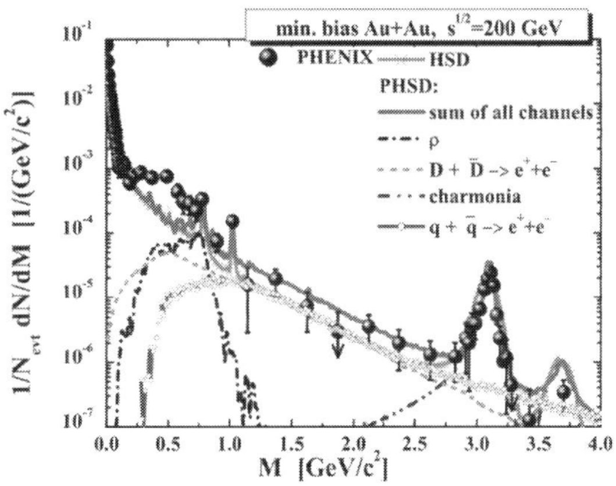


**Figure 7.** The HSD results for the mass differential dilepton spectra in case of inclusive $Au + Au$ collisions at $\sqrt{s} = 200$ GeV in comparison to the data from PHENIX [24] for $M = 0 - 4$ GeV. For comparison, the HSD results are shown by the grey line with cross symbols.

broadening' scenario as well as in the 'dropping mass + collisional broadening' model. We mention that HSD results for the low mass dileptons are very close to the calculated spectra from van Hees and Rapp as well as Dusling and Zahed [27] (cf. the comparison in Ref. [28]). At higher masses (from 1 to 4 GeV) the only hadronic sources of correlated lepton pairs are the charmed mesons: semi-leptonic decays of correlated D-mesons and the dilepton decays of charmonia. Between the $\phi$ and $J/\Psi$ peaks, the HSD results underestimate the PHENIX data by approximately a factor of two. In analogy to the analysis of the data at SPS energies in the previous Section, we attribute the missing yield to partonic contributions.

By implementing the off-shell partonic processes into the PHSD transport approach, we calculate the dilepton spectra in $Au + Au$ at $\sqrt{s}$=200 GeV and compare to the PHENIX data in Figs. 6 and 7. In Fig. 6 we present the low mass region ($M = 0 - 1.2$ GeV); in this region, the yield in PHSD is dominated by hadronic sources and essentially coincides with the HSD result. There is a discrepancy between the PHSD calculations and the data in the region of masses from 0.2 to 0.6 GeV. The discrepancy is not amended by accounting for the radiation from the QGP, since the latter is 'over-shone' by the radiation from hadrons integrated over the evolution of the collision. In contrary, the partonic radiation is visible in the mass region $M = 1 - 4$ GeV, as on sees in Fig. 7. Dileptons generated by the quark-antiquark annihilation in sQGP constitute about a half of the observed yield in the mass range between the masses of the $\phi$ and the $J/\Psi$ mesons. At $M > 2.5$ GeV, the partonic yield dominates over the D-meson contribution. Thus, accounting for partonic radiation in PHSD fills up the gap between the hadronic model results [17, 25] and the data at $M > 1$ GeV.

## 4. Summary

Parton Hadron String Dynamics [9] (PHSD) transport approach incorporates the relevant off-shell dynamics of the vector mesons as well as the explicit partonic phase in the early hot and dense reaction region. Comparison of the transport calculations to the data of the CERES and NA60 Collaborations points towards 'melting' modification of the $\rho$'-meson at high densities, i.e. the broadening of the vector meson's spectral function. On the other hand, the spectrum at $M > 1$ GeV is shown to be dominated by the partonic sources.

The low mass dilepton spectra from $Au + Au$ collisions at RHIC (from the PHENIX Collaboration) are clearly underestimated by the hadronic channels in the invariant mass range from 0.2 to 0.6 GeV. The discrepancy is not amended by accounting for the radiation from the QGP, since the latter is 'over-shone' by the radiation from hadrons integrated over the evolution of the collision.

In contrary, the partonic radiation is visible in the mass region $M = 1 - 4$ GeV. Dileptons generated by the quark-antiquark annihilation in sQGP constitute about a half of the observed yield in the mass range between the masses of the $\phi$ and the $J/\Psi$ mesons. At $M > 2.5$ GeV, the partonic yield dominates over the D-meson contribution. Thus, accounting for partonic radiation in PHSD fills up the gap between the hadronic model results [17, 25] and the data at $M > 1$ GeV.

A detailed calculation of the dilepton production in heavy ion collisions at RHIC energy versus collision centrality, dilepton mass $M$ and transverse momentum $p_T$ will be published in near future. The comparison to the dilepton data will open the possibility to study the relative importance of different processes in the dilepton production and guide us towards a better understanding of the properties of matter created in heavy-ion collisions.

## Acknowledgments

Work supported in part by the "HIC for FAIR framework of the "LOEWE program. We acknowledge stimulating discussions with S. Damjanovic, J.Manninen, A. Toia.

## References

[1] E. V. Shuryak, *Sov. Phys. JETP* **47** (1978) 212.
[2] E. V. Shuryak, *Phys. Lett.* B **78** (1978) 150, *Sov. J. Nucl. Phys.* **28** (1978) 408, *Yad. Fiz.* **28** (1978) 796.
[3] E. L. Feinberg, *Izv. Akad. Nauk Ser. Fiz.* **34** (1970) 1987.
[4] E. L. Feinberg, *Nuovo Cim.* A **34** (1976) 391.
[5] J. D. Bjorken and H. Weisberg, *Phys. Rev.* D **13** (1976) 1405.
[6] G. Agakichiev *et al.*, CERES Collaboration, *Phys. Rev. Lett.* **75** (1995) 1272; Th. Ullrich *et al.*, *Nucl. Phys.* A**610** (1996) 317c; A. Drees, *Nucl. Phys.* A**610** (1996) 536c.
[7] M. A. Mazzoni, HELIOS Collaboration, *Nucl. Phys.* A**566** (1994) 95c; M. Masera, *Nucl. Phys.* A**590** (1995) 93c; T. Åkesson et al., *Z. Phys.* C **68** (1995) 47.
[8] O. Linnyk, E. L. Bratkovskaya and W. Cassing, *Nucl. Phys.* A **830** (2009) 491C.
[9] W. Cassing and E. L. Bratkovskaya, *Phys. Rev.* C **78** (2008) 034919, *Nucl. Phys.* A **831** (2009) 215.
[10] W. Cassing, E. L. Bratkovskaya, *Phys. Rept.* **308** (1999) 65.
[11] E. L. Bratkovskaya, W. Cassing, *Nucl. Phys.* A**619** (1997) 413.
[12] W. Ehehalt, W. Cassing, *Nucl. Phys.* A **602** (1996) 449.
[13] W. Cassing, S. Juchem, *Nucl. Phys.* A **665** (2000) 377; *ibid.* A **672** (2000) 417.
[14] E. L. Bratkovskaya, W. Cassing, *Nucl. Phys.* A **807** (2008) 214.
[15] J. Ruppert, T. Renk and B. Müller, *Phys. Rev.* C **73** (2006) 034907.
[16] O. Linnyk, *J. Phys.* G **38** (2011) 025105.
[17] E. L. Bratkovskaya, W. Cassing and O. Linnyk, *Phys. Lett.* B **670** (2009) 428.
[18] W. Cassing, *Nucl. Phys.* A **791** (2007) 365; *ibid.* A **795** (2007) 70.
[19] R. Arnaldi *et al.*, NA60 Collaboration, *Phys. Rev. Lett.* **96** (2006) 162302, J. Seixas *et al.*, *J. Phys.* G **34** (2007) S1023, S. Damjanovic *et al.*, *Nucl. Phys.* A **783** (2007) 327c.
[20] D. Adamova *et al.* CERES Collaboration, *Nucl. Phys.* A **715** (2003) 262, *Phys. Rev. Lett.* **91** (2003) 042301, G. Agakichiev *et al.*, *Eur. Phys. J.* C **41** (2005) 475, D. Adamova *et al. Phys. Lett.* B **666** (2008) 425, A. Marin *et al.*, PoS C **POD07** (2007) 034.
[21] O. Linnyk, E. L. Bratkovskaya and W. Cassing, *Nucl. Phys.* A **830** (2009) 491C, O. Linnyk, E. L. Bratkovskaya and W. Cassing, *AIP Conf. Proc.* **1257** (2010) 700.
[22] R. Arnaldi *et al.*, NA60 Collaboration, *Eur. Phys. J.* C **59**, 607 (2009)
[23] A. Adare *et al.*, PHENIX Collaboration, *Phys. Lett.* B **670** (2009) 313.
[24] A. Toia *et al.*, PHENIX Collaboration, *Nucl. Phys.* A **774** (2006) 743, *Eur. Phys.* J **49** (2007) 243, S. Afanasiev

Journal of Physics: Conference Series **316** (2011) 012028 doi:10.1088/1742-6596/316/1/012028

*et al.*, PHENIX Collaboration, arXiv:0706.3034 [nucl-ex], A. Adare *et al.*, PHENIX Collaboration, *Phys. Rev.* C **81** (2010) 034911.

[25] J. Manninen, E. L. Bratkovskaya, W. Cassing and O. Linnyk, *Eur. Phys. J.* C **71** (2011) 1615.

[26] E. L. Bratkovskaya, W. Cassing, H. Stöcker, *Phys. Rev.* C **67** (2003) 054905, E. L. Bratkovskaya *et al. Phys. Rev.* C **69** (2004) 054907.

[27] K. Dusling and I. Zahed, *Nucl. Phys.* A **825** (2009) 212.

[28] A. Toia, *J. Phys.* G **35** (2008) 104037.

27th Winter Workshop on Nuclear Dynamics (WWND 2011) IOP Publishing
Journal of Physics: Conference Series **316** (2011) 012029 doi:10.1088/1742-6596/316/1/012029

# A mean field theory for the cold quark gluon plasma

**D.A. Fogaça[1] and F.S. Navarra[1]**

[1]Instituto de Física, Universidade de São Paulo
C.P. 66318, 05315-970 São Paulo, SP, Brazil

E-mail: navarra@if.usp.br

**Abstract.** A hot and strongly interacting quark gluon plasma (QGP) has been observed in heavy ion collisions at RHIC. Quarks and gluons in a deconfined phase might also exist in the core of compact stars. We shall assume here that the cold QGP is also a strongly interacting system. The strong interactions may be partly due to non-perturbative effects, which survive after the deconfinement transition and which can be related with the non-vanishing gluon condensates in the QGP. In this work, starting from the QCD Lagrangian we perform a gluon field decomposition in low ("soft") and high ("hard") momentum components, we make a mean field approximation with spatial and time dependence for the hard gluons and take the matrix elements of the soft gluon fields in the plasma. The latter are related to the gluon condensates. We derive an analytical expression for the equation of state, which is compared with the MIT bag model one. The effect of the condensates is to soften the equation of state whereas the hard gluons significantly increase the energy density and the pressure.

## 1. Introduction

From the study of heavy ion collisions we have learned one important lesson about the hot quark gluon plasma (QGP): it is a strongly interacting system. It is therefore natural to assume that the cold QGP is also strongly interacting. Presumably, this kind of system exists in the core of dense stars. Because of the limitations of lattice calculations in this domain and also because of the lack of experimental information, the cold QGP is less known than the hot QGP.

In [1] we have studied the non-perturbative effects in the cold QGP generated by the residual dimension two and dimension four gluon condensates. In the present work we extend the results of [1] and consider the inhomogeneities in the gluon fields, including their spatial and time derivatives.

In the vacuum, non-perturbative effects have been successfully understood in terms of the QCD condensates, i.e., vacuum expectation values of quark and gluon "soft" (low momentum) fields. These condensates can, in principle, be computed in lattice QCD. In practice, since they are vacuum properties and therefore universal, they can be extracted from phenomenological analyses of hadron masses, as it is customary done in QCD sum rules [2]. The condensates are expected to vanish in the limit of very high temperature or chemical potential. However, it has been suggested that they may survive after the deconfinement transition both in the high temperature [3, 4] and in the high chemical potential cases [5]. For our purposes the relevant gluon condensates are those of dimension four, $\langle 0|\frac{\alpha_s}{\pi}F^2|0\rangle$ $(=\langle F^2\rangle)$, and of dimension two [6, 7, 8, 9], $\langle 0|g^2A^2|0\rangle$ $(=\langle g^2A^2\rangle)$.

In what follows we shall repeat the procedure outlined in [1] and derive an equation of state for the QGP but, in contrast to our previous work, we will not neglect the field derivatives.

## 2. The equation of state

In this section we introduce a mean field approximation for QCD, extending previous works along the same line [10, 11]. The Lagrangian density of QCD is given by:

$$\mathcal{L}_{QCD} = -\frac{1}{4}F^a_{\mu\nu}F^{a\mu\nu} + \sum_{q=1}^{N_f} \bar{\psi}^q_i\Big[i\gamma^\mu(\delta_{ij}\partial_\mu - igT^a_{ij}G^a_\mu) - \delta_{ij}m_q\Big]\psi^q_j \tag{1}$$

where

$$F^{a\mu\nu} = \partial^\mu G^{a\nu} - \partial^\nu G^{a\mu} + gf^{abc}G^{b\mu}G^{c\nu} \tag{2}$$

The summation on $q$ runs over all quark flavors, $m_q$ is the mass of the quark of flavor $q$, $i$ and $j$ are the color indices of the quarks, $T^a$ are the SU(3) generators and $f^{abc}$ are the SU(3) antisymmetric structure constants. For simplicity we will consider only light quarks with the same mass $m$. Moreover, we will drop the summation and consider only one flavor. At the end of our calculation the number of flavors will be recovered. Following [10, 11], we shall start writting the gluon field as:

$$G^{a\mu} = A^{a\mu} + \alpha^{a\mu} \tag{3}$$

where $A^{a\mu}$ and $\alpha^{a\mu}$ are the low ("soft") and high ("hard") momentum components of the gluon field respectively. We will assume that $A^{a\mu}$ represents the soft modes which populate the vacuum and $\alpha^{a\mu}$ represents the modes for which the running coupling constant is small.

Inserting (3) into (2) we obtain:

$$F^{a\mu\nu} = (\partial^\mu A^{a\nu} - \partial^\nu A^{a\mu} + gf^{abc}A^{b\mu}A^{c\nu}) + (\partial^\mu \alpha^{a\nu} - \partial^\nu \alpha^{a\mu} + gf^{abc}\alpha^{b\mu}\alpha^{c\nu})$$
$$+gf^{abc}A^{b\mu}\alpha^{c\nu} + gf^{abc}\alpha^{b\mu}A^{c\nu} \tag{4}$$

### 2.1. *The mean field approximation*

In a cold quark gluon plasma the density is much larger than the ordinary nuclear matter density. These high densities imply a very large number of sources of the gluon field. Assuming that the coupling constant is not very small, the existence of intense sources implies that the bosonic fields tend to have large occupation numbers at all energy levels, and therefore they can be treated as classical fields. This is the famous approximation for bosonic fields used in relativistic mean field models of nuclear matter [12]. It has been applied to QCD in the past and amounts to assume that the "hard" gluon field, $\alpha^a_\mu$, is simply a function of the coordinates [12]:

$$\alpha^a_\mu(\vec{x},t) = \delta_{\mu 0}\,\alpha^a_0(\vec{x},t) \tag{5}$$

with $\partial_\nu\alpha^a_\mu \neq 0$. This space and time dependence goes beyond the standard mean field approximation, where $\alpha^a_\mu$ is constant in space and time [12] and consequently $\partial_\nu\alpha^a_\mu = 0$. We keep assuming, as in [1], that the soft gluon field $A^{a\mu}$ is independent of position and time and thus:

$$\partial^\nu A^{a\mu} = 0 \tag{6}$$

Substituting (5) and (6) into (4) and (1) we have:

$$\mathcal{L}'_{QCD} = -\frac{1}{4}\Gamma^a_{\mu\nu}\Gamma^{a\mu\nu} - \frac{f^{abc}}{2}\Gamma^a_{\mu\nu}(gA^{b\mu}\alpha^{c\nu} + g\alpha^{b\mu}A^{c\nu} + g_sA^{b\mu}A^{c\nu})$$
$$-\frac{f^{abc}f^{ade}}{4}\Big[g_s^2A^b_\mu A^c_\nu A^{d\mu}A^{e\nu}$$
$$+g^2A^b_\mu A^c_\nu A^{d\mu}\alpha^{e\nu} + g^2A^b_\mu A^c_\nu\alpha^{d\mu}A^{e\nu} + g^2A^b_\mu\alpha^c_\nu A^{d\mu}A^{e\nu} + g^2\alpha^b_\mu A^c_\nu A^{d\mu}A^{e\nu}$$

$$+g^2 A^b_\mu \alpha^c_\nu \alpha^{d\mu} A^{e\nu} + g^2 A^b_\mu \alpha^c_\nu A^{d\mu} \alpha^{e\nu} + g^2 \alpha^b_\mu A^c_\nu A^{d\mu} \alpha^{e\nu} + g^2 \alpha^b_\mu A^c_\nu \alpha^{d\mu} A^{e\nu}\Big]$$

$$+\bar{\psi}_i\Big\{i\gamma^\mu[\delta_{ij}\partial_\mu - iT^a_{ij}(g_s A^a_\mu + g_h \alpha^a_\mu)] - \delta_{ij} m\Big\}\psi_j \tag{7}$$

and

$$\Gamma^{a\mu\nu} = \partial^\mu \alpha^{a\nu} - \partial^\nu \alpha^{a\mu} + g_h f^{abc} \alpha^{b\mu} \alpha^{c\nu} \tag{8}$$

In the above expressions the coupling is running. When we have only soft fields $A^a_\mu$ the corresponding $g$ is therefore large. Accordingly, we shall call it $g = g_s$. The interactions of the hard fields among themselves and also the interactions between $\alpha^a_0(\vec{x}, t)$ and the quarks are dominated by high momenta and the hard coupling is small. We shall call it $g = g_h$. In the mixed terms containing soft and hard fields the couplings could be soft or hard.

We next replace in (7) all the terms that contain soft gluon fields by their corresponding expectation values in the cold QGP:

$$\mathcal{L}'_{QCD} = -\frac{1}{4}\Gamma^a_{\mu\nu}\Gamma^{a\mu\nu} - \frac{f^{abc}}{2}\Gamma^a_{\mu\nu}\Big(\langle gA^{b\mu}\rangle\alpha^{c\nu} + \alpha^{b\mu}\langle gA^{c\nu}\rangle + \langle g_s A^{b\mu} A^{c\nu}\rangle\Big)$$

$$-\frac{f^{abc} f^{ade}}{4}\Big[\langle g_s^2 A^b_\mu A^c_\nu A^{d\mu} A^{e\nu}\rangle$$

$$+\langle g^2 A^b_\mu A^c_\nu A^{d\mu}\rangle\alpha^{e\nu} + \langle g^2 A^b_\mu A^c_\nu A^{e\nu}\rangle\alpha^{d\mu} + \alpha^c_\nu\langle g^2 A^b_\mu A^{d\mu} A^{e\nu}\rangle + \alpha^b_\mu\langle g^2 A^c_\nu A^{d\mu} A^{e\nu}\rangle$$

$$+\alpha^c_\nu\langle g^2 A^b_\mu A^{e\nu}\rangle\alpha^{d\mu} + \alpha^c_\nu\langle g^2 A^b_\mu A^{d\mu}\rangle\alpha^{e\nu} + \alpha^b_\mu\langle g^2 A^c_\nu A^{d\mu}\rangle\alpha^{e\nu} + \alpha^b_\mu\langle g^2 A^c_\nu A^{e\nu}\rangle\alpha^{d\mu}\Big]$$

$$+\bar{\psi}_i\Big\{i\gamma^\mu[\delta_{ij}\partial_\mu - iT^a_{ij}(\langle g_s A^a_\mu\rangle + g_h\alpha^a_\mu)] - \delta_{ij} m\Big\}\psi_j \tag{9}$$

These expectation values have the following properties:

$$\langle A^{a\mu}\rangle = 0 \tag{10}$$

$$\langle A^{a\mu} A^{b\nu} A^{c\rho}\rangle = 0 \tag{11}$$

$$\langle {g_s}^2 A^{a\mu} A^a_\mu\rangle \equiv \langle A^2\rangle \neq 0 \tag{12}$$

$$\left\langle \frac{\alpha_S}{\pi} F^{a\mu\nu} F_{a\mu\nu}\right\rangle = \left\langle \frac{{g_s}^2}{4\pi^2} F^{a\mu\nu} F_{a\mu\nu}\right\rangle = \left\langle \frac{{g_s}^4}{4\pi^2} A^4\right\rangle \equiv \langle F^2\rangle \neq 0 \tag{13}$$

As in [10, 11] we have:

$$\langle g_s^2\ A^{a\mu} A^{b\nu}\rangle = -\frac{\delta^{ab}}{8}\frac{g^{\mu\nu}}{4}{\mu_0}^2 = -\frac{\delta^{ab} g^{\mu\nu}}{32}{\mu_0}^2 \tag{14}$$

As a consequence:

$$\langle g_s^2\ A^{a\mu} A^b_\mu\rangle = \langle g_s^2\ A^{a\mu} A^b_\nu g^\nu{}_\mu\rangle = \langle g_s^2\ A^{a\mu} A^b_\nu\rangle g^\nu{}_\mu =$$

$$-\frac{\delta^{ab}}{32} g^\mu{}_\nu g^\nu{}_\mu {\mu_0}^2 = -\frac{\delta^{ab}}{32} 4 \times {\mu_0}^2 = -\frac{\delta^{ab}}{8}{\mu_0}^2 \tag{15}$$

and also [11]:

$$\langle g_s^2\ A^a_\mu A^b_\nu A^{c\rho} A^{d\eta}\rangle = \frac{{\phi_0}^4}{(32)(34)}\Big[g_{\mu\nu} g^{\rho\eta}\delta^{ab}\delta^{cd} + g_\mu{}^\rho g_\nu{}^\eta\delta^{ac}\delta^{bd} + g_\mu{}^\eta g_\nu{}^\rho\delta^{ad}\delta^{bc}\Big] \tag{16}$$

The $\langle g_s^2 A^2\rangle$ condensate is associated with a dynamical gluon mass [10, 11] which is defined as:

$${m_G}^2 \equiv \frac{9}{32}{\mu_0}^2 \tag{17}$$

Journal of Physics: Conference Series **316** (2011) 012029 doi:10.1088/1742-6596/316/1/012029

In spite of the recent progress in the field, still very little is known about the parameter $\langle A^2 \rangle$ at finite (and high) density. In our approach, as in [13], we have $\langle A^2 \rangle < 0$ in (14) so ${m_G}^2$ is positive.

Inserting (10) to (17) into (8) and (9), using the fact that $-(g_h^2 f^{abc} f^{ade} \alpha_0^b \alpha_0^c \alpha_0^d \alpha_0^e)/4 = 0$, because of the color symmetry, we obtain the following effective Lagrangian:

$$\mathcal{L}_0 = -\frac{1}{2}\alpha_0^a(\vec{\nabla}^2\alpha_0^a) + \frac{{m_G}^2}{2}\alpha_0^a\alpha_0^a - b{\phi_0}^4 + \bar{\psi}_i\Big(i\delta_{ij}\gamma^\mu\partial_\mu + g_h\gamma^0 T^a_{ij}\alpha_0^a - \delta_{ij}m\Big)\psi_j \tag{18}$$

where the constant $b$ is defined as $b \equiv 9/(4\times 34)$ and

$$-\frac{1}{4}\langle F^{a\mu\nu}F_{a\mu\nu}\rangle = -\frac{\pi^2}{g_s^2}\langle\frac{\alpha_s}{\pi}F^{a\mu\nu}F^a_{\mu\nu}\rangle = -b\phi_0^4 \tag{19}$$

*2.2. Energy density and pressure*

The equations of motion [14] are given by:

$$\frac{\partial\mathcal{L}}{\partial\eta_i} - \partial_\mu\frac{\partial\mathcal{L}}{\partial(\partial_\mu\eta_i)} + \partial_\nu\partial_\mu\left[\frac{\partial\mathcal{L}}{\partial(\partial_\mu\partial_\nu\eta_i)}\right] = 0 \tag{20}$$

Inserting (18) into(20) with $\eta_1 = \alpha_0^a$ and $\eta_2 = \bar{\psi}$ we find:

$$-\vec{\nabla}^2\alpha_0^a + {m_G}^2\alpha_0^a = -g_h\rho^a \tag{21}$$

$$\Big(i\gamma^\mu\partial_\mu + g_h\gamma^0 T^a\alpha_0^a - m\Big)\psi = 0 \tag{22}$$

where $\rho^a$ is the temporal component of the color vector current given by $j^{a\nu} = \bar{\psi}_i\gamma^\nu T^a_{ij}\psi_j$ . From [14] the energy-momentum tensor is given by:

$$T^{\mu\nu} = \frac{\partial\mathcal{L}}{\partial(\partial_\mu\eta_i)}(\partial^\nu\eta_i) - g^{\mu\nu}\mathcal{L} - \left[\partial_\beta\frac{\partial\mathcal{L}}{\partial(\partial_\mu\partial_\beta\eta_i)}\right](\partial^\nu\eta_i) + \frac{\partial\mathcal{L}}{\partial(\partial_\mu\partial_\beta\eta_i)}(\partial_\beta\partial^\nu\eta_i) \tag{23}$$

The energy density is:

$$\varepsilon = < T_{00} > \tag{24}$$

which, with the use of (22), becomes:

$$\varepsilon = \frac{1}{2}\alpha_0^a(\vec{\nabla}^2\alpha_0^a) - \frac{{m_G}^2}{2}\alpha_0^a\alpha_0^a + b{\phi_0}^4 + i\bar{\psi}\gamma^0(\partial_0\psi) \tag{25}$$

Rewritting the last term of the above expression [1] we arrive at:

$$\varepsilon = \frac{1}{2}\alpha_0^a(\vec{\nabla}^2\alpha_0^a) - \frac{{m_G}^2}{2}\alpha_0^a\alpha_0^a + b{\phi_0}^4 - g_h\rho^a\alpha_0^a + 3\frac{\gamma_Q}{2\pi^2}\int_0^{k_F} dk\, k^2\sqrt{\vec{k}^2 + m^2} \tag{26}$$

where $\gamma_Q$ is the quark degeneracy factor $\gamma_Q = 2(\text{spin}) \times 3(\text{flavor})$. The sum over all the color states was already performed and resulted in the pre-factor 3 in the expression above. $k_F$ is the Fermi momentum defined by the quark number density $\rho$:

$$\rho = \langle N|\psi_i^\dagger\psi_i|N\rangle = \frac{3}{V}\sum_{\vec{k},\lambda}\langle N|N\rangle = 3\frac{\gamma_Q}{(2\pi)^3}\int d^3k = 3\frac{\gamma_Q}{2\pi^2}\int_0^{k_F} dk\, k^2 = \frac{\gamma_Q}{2\pi^2}{k_F}^3 \tag{27}$$

In the above expression $|N\rangle$ denotes a state with N quarks.

In a first approximation the field $\alpha_0^a$ may be estimated from (21). Neglecting the terms containing the derivatives ${\partial_0}^2\alpha_0^a$ and $\vec{\nabla}^2\alpha_0^a$ of (21) we have: [14]:

$$\alpha_0^a \cong -\frac{g_h}{{m_G}^2}\rho^a \tag{28}$$

Inserting (28) back on the left-hand-side of (21) and then solving it for $\alpha_0^a$ we find:

$$\alpha_0^a = -\frac{g_h}{{m_G}^2}\rho^a - \frac{g_h}{{m_G}^4}\vec{\nabla}^2\rho^a \tag{29}$$

We can write the color charge density $\rho^a$ in terms of the quark number density $\rho$ through:

$$\rho^a\rho^a = 3\rho^2 \tag{30}$$

Analogously we have

$$\rho^a\vec{\nabla}^2\rho^a = 3\rho\vec{\nabla}^2\rho\,,\quad \rho^a\vec{\nabla}^4\rho^a = 3\rho\vec{\nabla}^4\rho\,,\quad \vec{\nabla}^4\rho^a\,\frac{\partial^2\rho^a}{\partial t^2} = 3\vec{\nabla}^4\rho\,\frac{\partial^2\rho}{\partial t^2}\,,\quad \ldots \tag{31}$$

Inserting (29), (30) and (31) into (26) and performing the momentum integral we arrive at the final expression for the energy density:

$$\varepsilon = \left(\frac{3g_h^2}{2{m_G}^2}\right)\rho^2 + \left(\frac{3g_h^2}{2{m_G}^4}\right)\rho\,\vec{\nabla}^2\rho + \left(\frac{3g_h^2}{2{m_G}^6}\right)\rho\,\vec{\nabla}^4\rho + \left(\frac{3g_h^2}{2{m_G}^8}\right)\vec{\nabla}^2\rho\,\vec{\nabla}^4\rho + b\phi_0^4$$

$$+3\frac{\gamma_Q}{2\pi^2}\left\{\frac{{k_F}^3\sqrt{{k_F}^2+m^2}}{4}\right\} + \frac{3m^2}{8}\frac{\gamma_Q}{2\pi^2}\left[k_F\sqrt{{k_F}^2+m^2} - m^2 ln\left(\frac{k_F+\sqrt{{k_F}^2+m^2}}{m}\right)\right] \tag{32}$$

The pressure is given by

$$p = \frac{1}{3} < T_{ii} > \tag{33}$$

Repeating the same steps mentioned before we arrive at:

$$p = \left(\frac{3g_h^2}{2{m_G}^2}\right)\rho^2 + \left(\frac{2g_h^2}{{m_G}^4}\right)\rho\,\vec{\nabla}^2\rho - \left(\frac{g_h^2}{{m_G}^6}\right)\rho\,\vec{\nabla}^4\rho - \left(\frac{g_h^2}{2{m_G}^4}\right)\vec{\nabla}\rho\cdot\vec{\nabla}\rho$$

$$+\left(\frac{g_h^2}{2{m_G}^6}\right)\vec{\nabla}^2\rho\,\vec{\nabla}^2\rho - \left(\frac{g_h^2}{{m_G}^8}\right)\vec{\nabla}^2\rho\,\vec{\nabla}^4\rho$$

$$-\left(\frac{g_h^2}{2{m_G}^8}\right)\vec{\nabla}(\vec{\nabla}^2\rho)\cdot\vec{\nabla}(\vec{\nabla}^2\rho) - \left(\frac{g_h^2}{{m_G}^6}\right)\vec{\nabla}\rho\cdot\vec{\nabla}(\vec{\nabla}^2\rho) - b{\phi_0}^4$$

$$+\frac{\gamma_Q}{2\pi^2}\left\{\frac{{k_F}^3\sqrt{{k_F}^2+m^2}}{4}\right\} - \frac{3m^2}{8}\frac{\gamma_Q}{2\pi^2}\left[k_F\sqrt{{k_F}^2+m^2} - m^2 ln\left(\frac{k_F+\sqrt{{k_F}^2+m^2}}{m}\right)\right] \tag{34}$$

where the speed of sound $c_s$ is given by:

$${c_s}^2 = \frac{\partial p}{\partial\varepsilon} \tag{35}$$

Journal of Physics: Conference Series **316** (2011) 012029 doi:10.1088/1742-6596/316/1/012029

## 3. Numerical results and discussion

Including the inhomogeneities of the gluon field (and consequently of its derivatives) was an important step in the restoration of gauge invariance, which was broken in [1] by the assumption of constant fields. Moreover these derivatives play an important role in the hydrodynamics of the cold QGP. It was shown in [15] that these derivatives, coupled to derivatives already present in the equations of Euler and of the continuity lead to the Korteweg - de Vries (KdV) equation for perturbations in the baryon density. The existence of KdV solitons in the QGP is a remarkable phenomenon.

In spite of their conceptual relevance, the derivatives in (32) and (34) are not very important for the bulk properties of the system, since they are corrections to the results found in [1]. Therefore, in what follows we compare our results (32), (34) and (35) without the gradients, with the corresponding results obtained with the MIT bag model for a gas of quarks at zero temperature [12, 16]. We choose $\mathcal{B} = 110$ MeV fm$^{-3}$, which lies in the range ($50 < B < 200$ MeV fm$^{-3}$) used in calculations of stellar structure [17]. For the comparison we must rewrite our results as functions of the baryon density, which is $\rho_B = \frac{1}{3}\rho$.

We consider the MIT bag model with finite $\mathcal{B}$ and our model with massless quarks and soft gluons but no hard gluons. In this case we can identify our gluonic term with the gluonic component of the MIT bag model, represented by the bag constant. We then obtain an expression for the bag constant in terms of the gluon condensate:

$$\mathcal{B}_{QCD} = b\phi_0^4 = \langle \frac{1}{4} F^{a\mu\nu} F^a_{\mu\nu} \rangle \tag{36}$$

Fixing $\mathcal{B}$ and choosing a reasonable value of the coupling of the soft gluons, $g_s$, appearing in (19) we can infer the value of the dimension four condensate, $< F^2 >$, in the deconfined phase. For $\mathcal{B}_{QCD} = \mathcal{B} = 110$ MeV fm$^{-3}$ and $g_s = 2.7$ (which would correspond to $\alpha_{s(soft)} = g_s^2/4\pi = 0.6$) we find:

$$\langle F^2 \rangle = \langle \frac{\alpha_s}{\pi} F^{a\mu\nu} F^a_{\mu\nu} \rangle = \frac{g_s^2}{\pi^2} \mathcal{B}_{QCD} = 0.0006 \text{ GeV}^4 \tag{37}$$

In the lack of knowledge of the in-medium dimension two condensate, we use the factorization hypothesis, which, in the notation of Refs. [10] and [11], implies the choice ${\mu_0}^2 = g_s {\phi_0}^2$. As a consequence, (14), (16), (17) and (19) and are related and we obtain:

$$\langle g_s^2 A^2 \rangle = -\sqrt{\frac{(4)(34)\pi^2}{9} \langle F^2 \rangle} = -0.3 \text{ GeV}^2 \tag{38}$$

which corresponds to a dynamical mass of $m_G = 290$ MeV. This number is consistent with the values quoted in recent works [18, 19], which lie in the range $200 < m_G < 600$ MeV. Finally, the numerical evaluation of (32), (34) and (35) requires the choice of $g_h$, the coupling of the hard gluons, and of the quark mass, $m$. We choose them to be $g_h = 0.35$ (corresponding to $\alpha_{s(hard)} = g_h^2/4\pi = 0.01$) and $m = 0.02$ GeV.

For this set of parameters our EOS is harder than the MIT one. This can be seen in the Fig. 1 where we show the pressure as a function of the energy density, and also in Fig. 2 where we show the ratio of the speed of sound ${c_s}^2/{c_0}^2$, where ${c_s}^2$ is given by (35) and ${c_0}^2 = 1/3$ for MIT. In the same range of baryon densities, we have more energy, much more pressure and consequently a larger speed of sound. This behavior can be attributed to the first term of the equations (32) and (34), which comes from the hard gluons. This term is exactly the same both in (32) and (34) and in the limit of high densities becomes dominant yielding $p \simeq \varepsilon$ and hence $c_s \rightarrow 1$. Physically, this term represents the perturbative corrections to the MIT approach. Since the quark density is extremely large, even in the weak coupling regime (typical of the hard gluons) the field $\alpha_0^a$ is intense.

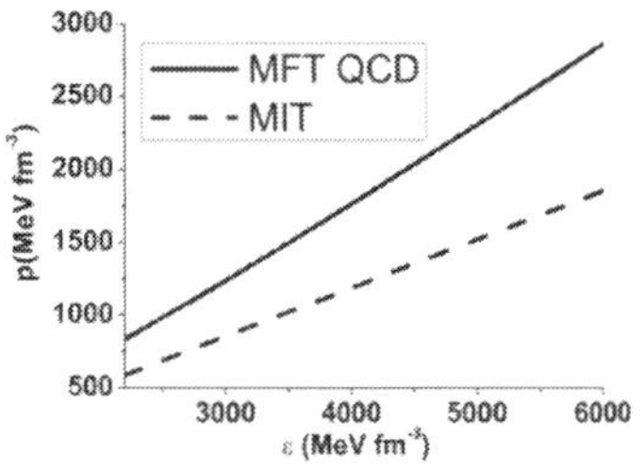


**Figure 1.** Pressure as a function of the energy density.

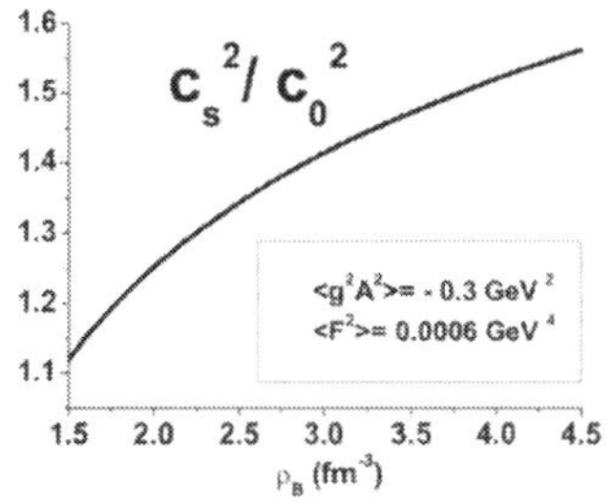


**Figure 2.** Sound speed ratio. ${c_0}^2 = 1/3$ for MIT.

We now show the EOS for different choices of the condensates, which are now treated as independent from each other. In Fig. 3 we fix $\langle F^2\rangle$ and vary $\langle g_s^2 A^2\rangle$, starting from the central value $-0.3$ GeV$^2$ and increasing its magnitude up to the lattice result $-2.56$ GeV$^2$ [20] (or $\langle g_s^2 A^2\rangle = -(1.6 \text{ GeV})^2$ which corresponds to $m_G = 848$ MeV). In Fig. 4 we perform the complementary study keeping $\langle g_s^2 A^2\rangle$ and increasing the magnitude of $\langle F^2\rangle$. As it can be seen, increasing the condensates reduces the pressure and, in the case of $\langle g_s^2 A^2\rangle$, softens the equation of state. This behavior could be anticipated from Eqs. (32), (34) and from equation of motion (21) without derivatives of the density. Indeed, keeping fixed the coupling and the quark density, when we increase the gluon mass, the field becomes weaker.

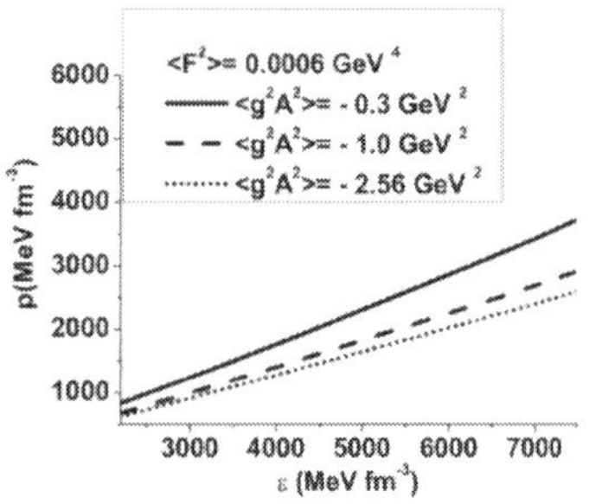


**Figure 3.** EOS for different values of dimension two gluon condensates.

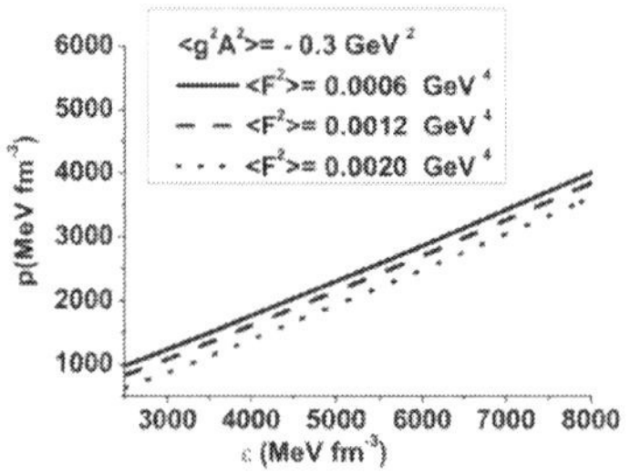


**Figure 4.** EOS for different values of dimension four gluon condensates.

To summarize, we have derived an equation of state for the cold QGP, which may be useful for calculations of stellar structure. The derivation is simple and based on three assumptions: i) decomposition of the gluon field into soft and hard components; ii) replacement of the soft gluon fields by their expectation values ("in-medium condensates") and iii) replacement of the hard gluon fields by their mean-field (classical) values. The effect of the condensates is to soften the EOS whereas the hard gluons significantly increase the energy density and the pressure.

**Acknowledgments**
We are deeply grateful to A. Natale, S. Sorella, R. Gavai, R. Gupta, R.B. Neufeld, I. Sagert and J.E. Horvath for fruitful discussions. We are especially grateful to D. Dudal for kindly and carefully answering our questions on the dimension two gluon condensate. We thank also

the referee of [1] for the comments and suggestions. This work was partially financed by the Brazilian funding agencies CAPES, CNPq and FAPESP.

## References

[1] D. A. Fogaça and F. S. Navarra, Phys. Lett. B **700**, 236 (2011).
[2] M. Nielsen, F. S. Navarra and S. H. Lee, Phys. Rep. **497**, 41 (2010) and references therein.
[3] D. E. Miller, Phys. Rep. **443**, 55 (2007).
[4] G. Boyd, J. Engles, F. Karsch, E. Laermann, C. Legeland, M. Lutgemeier and B. Petersson, Nucl. Phys. B **469**, 419 (1996).
[5] M. A. Metlitski and A. R. Zhitnitsky Nucl. Phys. B **731**, 309 (2005).
[6] M. J. Lavelle and M. Schaden, Phys. Lett. B **208**, 297 (1988);
[7] R. Akhoury and V. I. Zakharov, Phys. Lett. B **438**, 165 (1998).
[8] F. V. Gubarev, L. Stodolsky and V. I. Zakharov, Phys. Rev. Lett. **86**, 2220 (2001); F. V. Gubarev and V. I. Zakharov, Phys. Lett. B **501**, 28 (2001);
[9] D. Vercauteren, D. Dudal, J. Gracey, N. Vandersickel and H. Verschelde, Acta Phys. Polon. Supp. **3**, 829 (2010); PoS **LC2010**, 071 (2010); D. Dudal, O. Oliveira and N. Vandersickel, Phys. Rev. D **81**, 074505 (2010).
[10] L. S. Celenza and C. M. Shakin, Phys. Rev. D **34**, 1591 (1986).
[11] X. Li and C. M. Shakin, Phys. Rev. D **71**, 074007 (2005).
Phys. Lett. B **156**, 255 (1985).
[12] B.D. Serot and J.D. Walecka, Advances in Nuclear Physics **16**, 1 (1986).
[13] H. Verschelde, K. Knecht, K. Van Acoleyen and M. Vanderkelen, Phys. Lett. B **516**, 307 (2001).
[14] D. A. Fogaça, L. G. Ferreira Filho and F. S. Navarra, Nucl. Phys. A **819**, 150 (2009).
[15] D. A. Fogaça and F. S. Navarra, Phys. Lett. B **645**, 408 (2007); Phys. Lett. B **639**, 629 (2006).
[16] D. A. Fogaça, L. G. Ferreira Filho and F. S. Navarra, Phys. Rev. C **81**, 055211 (2010).
[17] M. Baldo, P. Castorina, D. Zappal, Nucl. Phys. A **743**, 3 (2004); G. F. Burgio, M. Baldo, P. K. Sahu and H. J. Schulze, Phys. Rev. C **66**, 025802 (2002).
[18] A. C. Aguilar and A. A. Natale, JHEP **0408**, 057 (2004).
[19] D. Dudal, S. P. Sorella, N. Vandersickel and H. Verschelde, Phys. Rev. D **77**, 071501 (2008); D. Dudal, J. A. Gracey, S. P. Sorella, N. Vandersickel and H. Verschelde, Phys. Rev. D **78**, 065047 (2008).
[20] P. Boucaud, et al., Phys. Rev. D **63**, 114003 (2001).

# Modeling Quark Gluon Plasma Using CHIMERA

**Betty Abelev**

Physics Division, Lawrence Livermore National Laboratory, 7000 East Avenue, L-211, Livermore, CA 94550-9698

E-mail: abelev@llnl.gov

**Abstract.** We attempt to model Quark Gluon Plasma (QGP) evolution from the initial Heavy Ion collision to the final hadronic gas state by combining the Glauber model initial state conditions with eccentricity fluctuations, pre-equilibrium flow, UVH2+1 viscous hydrodynamics with lattice QCD Equation of State (EoS), a modified Cooper-Frye freeze-out and the UrQMD hadronic cascade. We then evaluate the model parameters using a comprehensive analytical framework which together with the described model we call CHIMERA. Within our framework, the initial state parameters, such as the initial temperature ($T_{init}$), presence or absence of initial flow, viscosity over entropy density ($\eta/s$) and different Equations of State (EoS), are varied and then compared simultaneously to several experimental data observables: HBT radii, particle spectra and particle flow. $\chi^2$/nds values from comparison to the experimental data for each set of initial parameters will then used to find the optimal description of the QGP with parameters that are difficult to obtain experimentally, but are crucial to understanding of the matter produced.

## 1. Introduction

Recreating Quark Gluon Plasma (QGP), matter that existed in the first millisecond after the Big Bang, has been the goal of experiments at RHIC and now at the LHC. QGP is created during the course of a collision of massive relativistic ions, such as gold ions in the Brookhaven Laboratory or lead ions at CERN, and has been shown to exhibit the properties of a viscous liquid [1][2]. To understand the properties of the matter created in these collisions, the evolution of matter during the collision itself must be understood. Currently, we understand the basic stages of matter evolution, but lack the nuanced understanding of interdependencies and variation in parameters required for precise description of the collision.

Our present understanding of a heavy ion collision can be summarized as follows. After collision time $\tau \approx 1$ fm/c, a strongly interacting liquid of quarks and gluons is formed, which is described with various accuracy by hydrodynamical models. As the gas cools, mesons and hadrons "freeze out" and start interacting as colorless objects, until a phase of a relatively stable gas of hadrons is reached.

In this work we attempt to evaluate numerically the validity of models describing this picture by systematically varying model parameters and calculating $\chi^2$ distributions of proximity of fit to the experimental data. The evaluating algorithm used for these purposes is called CHIMERA, which stands for Comprehensive Hydrodynamical Integrated Modelling Evaluation and Reporting Algorithm, and will be described in detail in the next section.

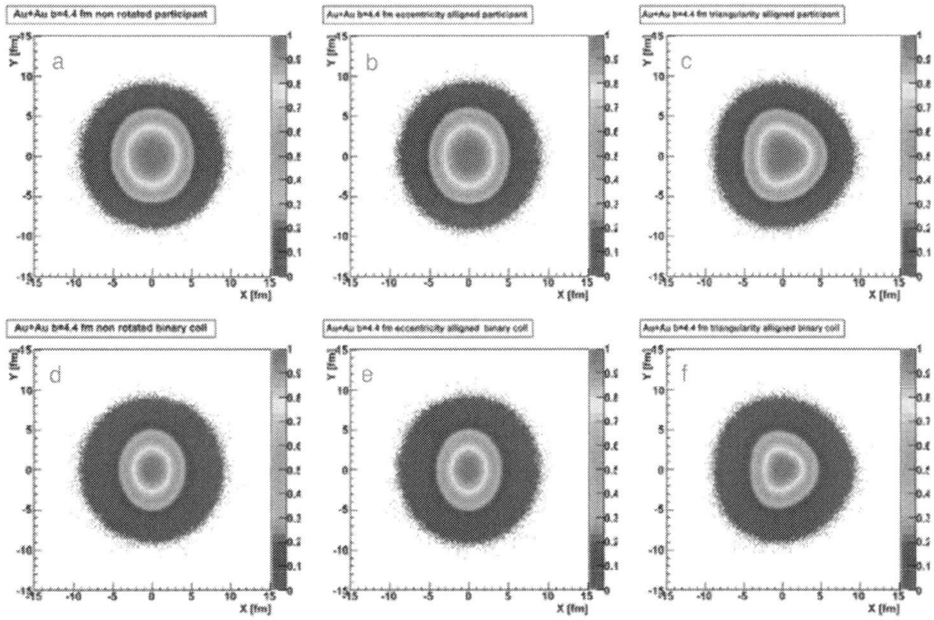


**Figure 1.** Hydrodynamical initial conditions energy density distributions for Au+Au collisions at $\sqrt{s_{NN}} = 200$ GeV with $b$ =4.4 fm, scaled as $\mathrm{N_{part}}$ $(a, b, c)$, or $\mathrm{N_{coll}}$ $(d, e, f)$. Panels $a$ and $d$ were produced using optical Glauber geometry, panels $b$ and $c$ using Monte Carlo Glauber, and panels $c$ and $f$ using Glauber geometry with event-by-event fluctuations.

## 2. Model description

CHIMERA is hydrodynamics-based framework, thus we first start by determining initial inputs and an initial energy distribution. In the analysis presented here, the initial conditions are computed using Monte Carlo Glauber model of the colliding nuclei [4] (e.g., Figure 1, panels $b$ and $e$). However, CHIMERA is capable of handling several initial condition implementations, including optical Glauber initial conditions (Figure 1, panels $a$ and $d$), Glauber conditions with fluctuations (Figure 1, panels $c$ and $f$), and Colour Glass Condensate-inspired [5] initial conditions (not shown). We implement initial conditions with either the number of collisions ($\mathrm{N_{coll}}$) or number of participants ($\mathrm{N_{part}}$) scaling. In the future, we plan to introduce a dial-up mechanism for this scaling, so that the initial conditions would scale with a predetermined mix of $\mathrm{N_{coll}}$ and $\mathrm{N_{part}}$, as is closer to experimental observations. The Equation of State (EoS) used as a CHIMERA input is obtained from Lattice QCD calculations by the HotQCD collaboration [6], and agrees well with s95p-vi EoS by Huovinen and Petreczky [7]. This EoS is much softer than the resonance gas EoS predicted by the MIT Bag Model [8], but is stiffer than the lattice-inspired EoS used in VH2 Hydro [9]. All four equations of state are shown in Figure 2.

As the last step before starting viscous hydrodynamics, we employ pre-equilibrium flow, i.e., flow that originates from the non-equilibrated matter at early times (approximately before $\tau = 1$ fm/c), to start the evolution of the heavy ion collision. This is motivated by fits to variables describing the collective behaviour of the fireball, such as Hanbury Brown-Twiss (HBT) radii and collective flow parameters. Pre-equilibrium flow is universal when describing a boost-invariant system with a traceless stress-energy tensor [10]. Since our model utilizes a 2+1 dimensional hydrodynamic evolution (see below), we are well-justified in employing pre-equilibrium flow in our description of heavy ion collisions.

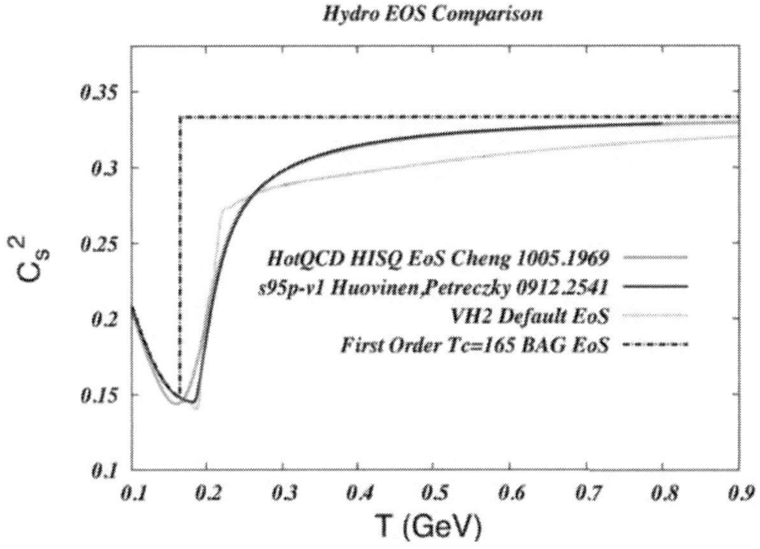


**Figure 2.** Equation of state functions from HotQCD (solid green curve), s95p-v1 (solid black curve), VH2 lattice-inspired EoS (pink curve), and the MIT Bag model EoS with a first order transition.

Using the set up described above, we run VH2 2D+1 viscous hydrodynamics developed by Luzum and Romatschke [9] on the energy-density distribution that resulted from the above assumptions. At the end of the hydrodynamic evolution, we employ Cooper-Frye freezeout [11] with viscous corrections, and then start the UrQMD hadronic cascade [12] to obtain final particle distributions.

After ensuring that the resultant yields and mean $p_T$ ($\langle p_T \rangle$) of the produced particles match those of experimental data, we employ our analytical framework to draw numerical comparisons between our model results and those from experiment. This is done by describing our modelling data using Chebyshev polynomials, and then calculating the $\chi^2$/ndf between these polynomials and experimental data. The $\chi^2$ is calculated taking into account point-to-point correlated, uncorrelated, and scaling uncertainties (A, B, and C-type errors, as described in [13]). Performing this fit and calculation for a grid of initial state parameters, such as viscosity over entropy-density ($\eta/s$) and the maximum temperature of the evolving medium ($T_{central}$), we are then able to do a global comparison of $\chi^2$ minima for a variety of soft observables: particle spectra, elliptic flow coefficients, and femtoscopic measurements.

## 3. Data Analysis

As mentioned in the previous section, randomized Glauber initial conditions for Au+Au collisions at $\sqrt{s_{NN}}$ = 200 GeV with an impact parameter ($b$) of 4.4 fm, shown in panels $b$ and $e$ of Figure 2, are used as an input for this study. The main goal of the analysis is two-fold: first, to study the differences due to inclusion or omission of pre-equilibrium flow. Second, to study the impact of using $N_{coll}$ vs. $N_{part}$ scaling on the accuracy of describing experimental data. Four sets of modelling data were generated: two sets with pre-equilibrium flow turned on (one scaled with $N_{coll}$, the other with $N_{part}$), and two sets with pre-equilibrium flow turned off (also one with $N_{coll}$ scaling and the other with $N_{part}$ scaling). We varied $T_{central}$ (highest temperature of the medium) from 320 to 360 MeV, and $\eta$/s from 0.001 to 0.36. The VH2 EoS was replaced

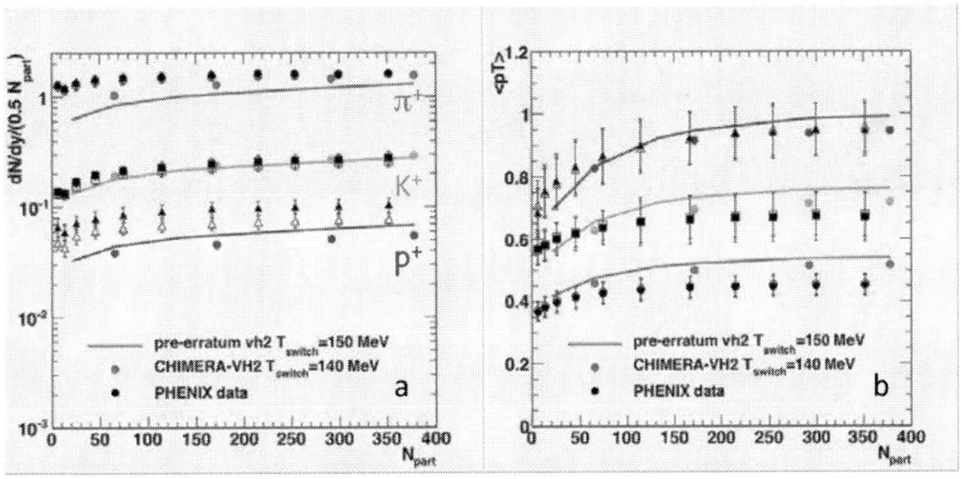


**Figure 3.** $\pi^+$, $K^+$, and $p^+$ dN/dy (panel *a*) and $\langle p_T \rangle$ (panel *b*) distributions from $\sqrt{s_{NN}} = 200$ GeV 10-20% central Au+Au collisions, modelled using CHIMERA (collored filled symbols), VH2 pre-errata data with $T_{switch}$=150 (color curves), and experimental data from PHENIX (black and open symbols).

by HotQCD HISQFIX EoS [6]. Cooper-Fry freeze-out was started when the temperature of the medium reached 165 MeV ($T_{freeze-out}$), roughly corresponding to the temperature of chemical freeze-out derived from the thermal model [14]. UrQMD was started at $T_{switch}$ = 140 MeV.

To check for consistency, we ran a sample of Au+Au 200 GeV data without any centrality selections. We calculate yields per unit of rapidity (dN/dy) and $\langle p_T \rangle$ as a function of $N_{part}$ for the three most common particle species: protons, kaons and pions, and compare these with experimental data, as shown in Figure 3. Since the modelled and experimental data are in a reasonable general agreement, we can now start mapping out the precise correspondence between the experiment and modelling data with variable initial conditions.

For each particle species, and for each of the three observables (HBT radii, spectra, and the elliptic flow coefficient), we obtain a set of data produced with parameters described above. An example of particle spectra, plotted together with experimental data, is shown in Figure 4. The figure illustrates a typical model-data comparison done to show the validity of hydrodynamical codes, and demonstrates a fraction of the available parameters generated by CHIMERA. Note that each set of modelling data is fit using a Chebyshev polynomial (depending on the observable, the polynomial can have four or five parameters).

For each experimental-modelling data pair, we calculate $\chi^2$, to quantify how well a given modelling curve describes a given set of experimental data points, as described in the previous section. Keeping all but two of the initial condition parameters constant, we are able to map out the $\chi^2$/ndf space for particle spectra, flow, and HBT radii, and then use the most probable overlap region to determine the most likely initial states parameter values. In the study performed for this article, the $\chi^2$/ndf was mapped out for $\pi^+$ mesons, with experimental data obtained from both PHENIX [15][16] and STAR [17][18] collaborations. An example, shown in Figure 5, shows a calculation of one map point for each $\chi^2$/ndf map (spectra, flow, and each of the HBT radii). Figure 5 calculations were performed with $N_{part}$ scaled modelling data, with pre-equilibrium flow turned on, at $\eta$/s=0.08, and $T_{central}$=320 MeV. By eye, these parameters represent the data fairly well, but exactly how well can be judged by the $\chi^2$/ndf shown in each panel of the figure.

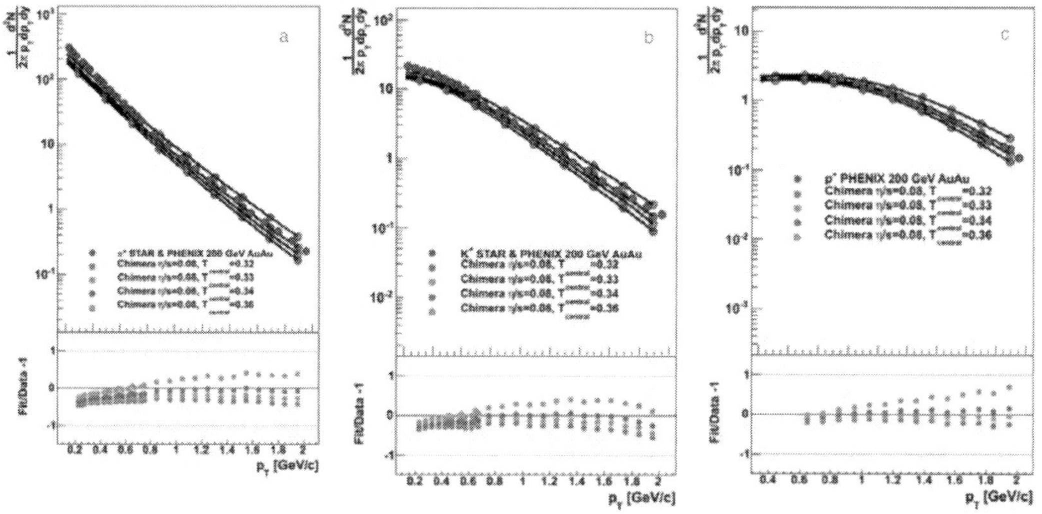

**Figure 4.** $\pi^+$ (panel *a*), $K^+$ (panel *b*), and *p* (panel *c*) spectra distributions from CHIMERA and $\sqrt{s_{NN}} = 200$ GeV 20% central Au+Au RHIC data. $\eta/s$ is kept constant, while $T_{central}$ is varied from 320 to 360 MeV. Bottom sections of each panel show fractional agreement between data and a given set of parameters. Model data are produced with $N_{coll}$ scaling and pre-equilibrium flow off.

**Table 1.** $\chi^2$/ndf minima for each of the three experimental observables: $\pi^+$ meson spectra, flow ($v_2$ coefficient), and the combined HBT radii. The minima have been computed with pre-equilibrium flow turned on and off, and with initial state scaling as the number of binary collisions ($N_{coll}$) or number of participants ($N_{part}$).

| | **Spectra** | | $v_2$ | | **HBT** | |
|---|---|---|---|---|---|---|
| pre-flow | on | off | on | off | on | off |
| $\mathbf{N}_{coll}$ | 18.73 | 11.86 | 8.84 | 9.17 | 4.25 | 12.06 |
| $\mathbf{N}_{part}$ | 5.02 | 5.98 | 9.17 | 26.43 | 8.46 | 27.88 |

## 4. Results

The results of the study are summarized in Table 1 and Figure 6. Table 1 shows the minima from $\chi^2$/ndf distributions for all measurements currently available within CHIMERA made with $\pi^+$ mesons. The $\chi^2$/ndf minima range from 4.25 for $\chi^2$/ndf calculations of the combined HBT radii with $N_{coll}$ scaling and pre-equilibrium flow included, to 26.43 for the $\chi^2$/ndf distributions derived from calculations for $v_2$ without pre-flow and with $N_{part}$ scaling. Trends, evident from the table, are as follows: $\chi^2$/ndf is significantly improved for $v_2$ and HBT measurements by turning on pre-equilibrium flow and by using $N_{coll}$ scaling, while the reverse is true for pion spectra: the best $\chi^2$/ndf value is for $N_{part}$ scaling in simulations with pre-equilibrium flow turned off.

Figure 6 shows $\chi^2$/ndf space distributions for spectra, $v_2$ and HBT radii of $\pi^+$ produced with $N_{coll}$ scaling (panels *a*, *b*, and *c*) and $N_{part}$ scaling (panels *d*, *e*, *f*). The distributions are drawn to display the most likely $\eta/s$ and $T_{central}$ values that would produce the experimental measurements. A global $\eta/s$ and $T_{central}$ value would be at the intersection of the $\chi^2$/ndf minima for all three experimental quantities modelled under the same set of conditions. However, this

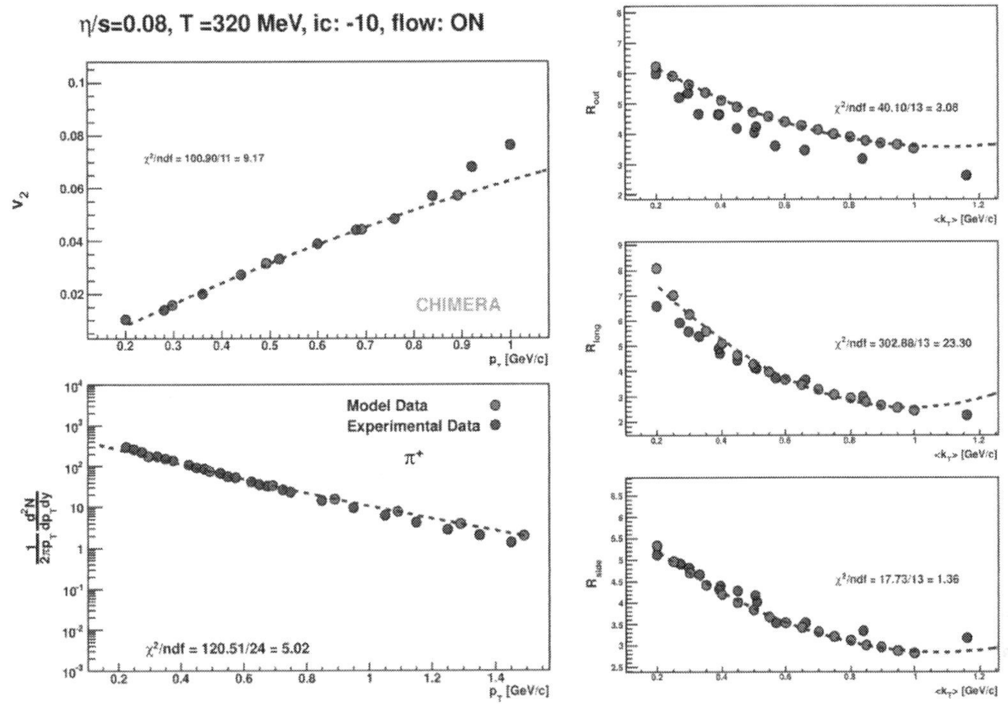


**Figure 5.** An example of $\chi^2/\mathrm{ndf}$ calculations for $\pi^+$ mesons. Each panel shows a comparison between model and experiment for each of the available quantities: $v_2$, spectra, and the three HBT radii (right side of the figure). Pre-equilibrium flow is turned off, and the initial distribution scales as $N_{part}$. Experimental data is shown in purple, and modelling data points – in salmon color filled circles.

is not yet the case. In $N_{coll}$-scaled data, $\eta/s$ and $T_{central}$ tend to higher values for spectra and to some extent for the HBT combined measurement as well, whereas for $v_2$ the most likely $\eta/s$ is much lower ($\sim 0.08$). We also do not see a common minimum in the $N_{part}$-scaled data: there the best $\eta/s$ value for $\pi^+$ spectrum is 0.32, 0.08 for $v_2$, and 0.0001 for HBT. The best $T_{central}$ is the same for spectrum and HBT (320 MeV), but higher for $v_2$. This indicates that the modelling part of CHIMERA is not yet completely physical.

## 5. Summary and outlook

In summary, we presented an outline and the first tests of an evaluative framework for comparing hydrodynamical modelling attempts of the Quark Gluon Plasma to the experimental data available from the existing heavy ion experiments. VH2 viscous hydrodynamical code was used as a base to generate modelling data used in this evaluation. The data comparison was done for 10-20% central $\sqrt{s_{NN}} = 200$ GeV Au+Au collisions at RHIC, and aimed to study the effect of utilizing the initial scaling ($N_{coll}$ vs $N_{part}$). In addition, we analysed possible effects of pre-equilibrium flow on the goodness of model's description of the data. The result of the comparison finds that $N_{coll}$ scaling might not be applicable when attempting to describe the spectra of the data derived from collision's bulk, but does a slightly better job than $N_{part}$ scaling at describing flow and the HBT variables. We also found that pre-equilibrium flow significantly improves the goodness of fit for the HBT radii and $v_2$, as expected, but also undermined the model's description of spectra.

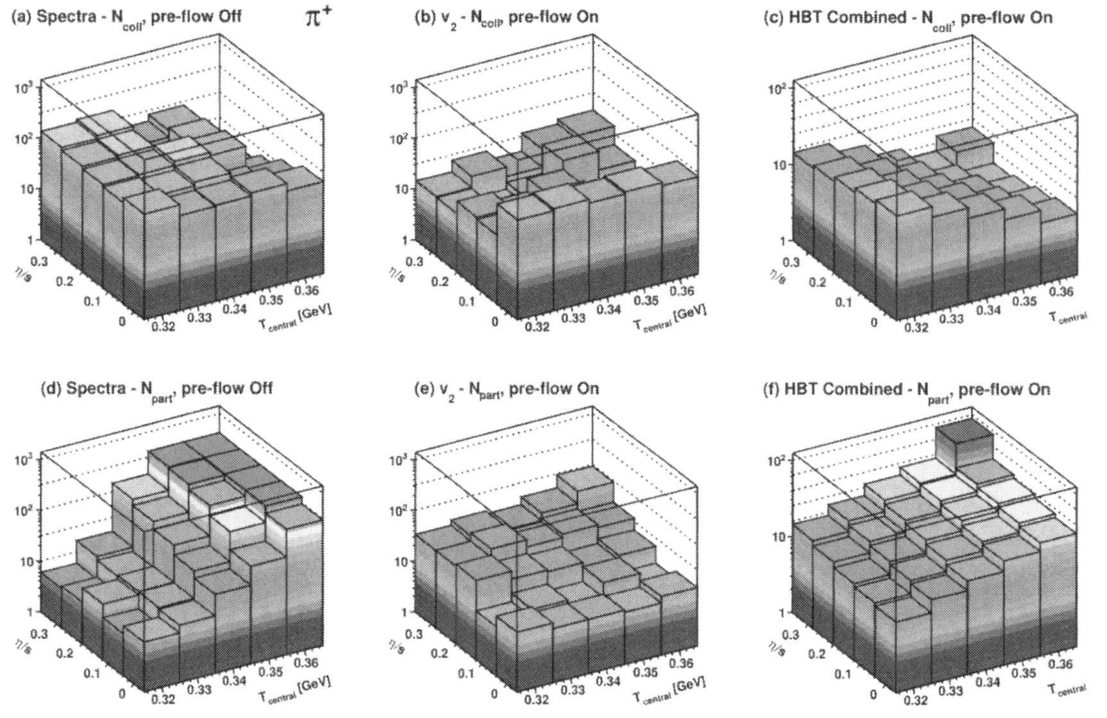


**Figure 6.** Comparison of $\chi^2$/ndf distributions for $N_{coll}$ vs. $N_{part}$ scaling. The top row has been produced with $N_{coll}$ scaling, the bottom row scales with $N_{part}$. Panels *a* and *d* show $\chi^2$/ndf for $\pi^+$ spectra, *b* and *e* – $\pi^+$ flow ($v_2$), and panels *c* and *f* show the distributions for the combined HBT radii. Pre-equilibrium flow was turned on for the four panels on the right, and turned off for the left-most panels.

However, these are initial results, generated primarily as proof of principle. Currently, we plan on implementing a more realistic initial-state energy-density distribution, vary the EoS used for the analysis, and move on to three-dimensional hydro-dynamical modelling. We are also performing $\tau_{start}$ studies and studies to map out $\eta/s$ dependence on temperature. In addition, we are expanding our framework to analyse other collision centralities in Au+Au, other collision species and collision energies. The onslaught of new data from RHIC and the LHC provides us with a unique and exciting opportunity for a comprehensive and detailed study of the matter produced in these collisions. CHIMERA is an important tool for the Heavy Ion community, and we hope it would be used accordingly.

## Acknowledgments

The author thanks Scott Pratt for his invaluable contributions and discussions, Paul Romatschke and Matt Luzum for making their hydrodynamic code available, and Stephen Bass for the availability of UrQMD. Most importantly, without Ron Soltz, Irakli Garishvili, Jason Newby, Andrew Glenn, Michael Cheng, and Alex Linden-Levy there would not be CHIMERA. This work is supported by Lawrence Livermore National Laboratory grant for laboratory-directed research and development, and has been assigned LLNL-PROC-488978.

## References

[1] Adams J *et al* (STAR Collaboration) 2005 *Nucl. Phys. A* **757** 102-83
[2] Adcox K *et al* (PHENIX ollaboration) 2005 *Nucl. Phys. A* **757** 184-283
[3] Alver B, Baker M, Loizides C and Steinberg P 2008 (*Preprint nucl-ex*/0805.4411)
[4] Dumitru A 2011 *Nucl. Phy. A* **84** 71-5
[5] Bazanov A *et al* (HotQCD Collaboration) 2009 *Phys. Rev. D* **80** 014504
[6] Huovinen P and Petreczky P 2010 *Nucl. Phys. A* **837** 26-53
[7] Chodos A, Jaffe R L, Johnson K, Thorn C B and Weisskopf V F 1974 *Phys. Rev. D* **9** 3471
[8] Luzum M and Romatschke P 2008 *Phys. Rev. C* **78** 034915
[9] Vredevoogd J and Pratt S 2009 *Phys. Rev. C* **79** 044915
[10] Cooper F and Frye G 1974 *Phys. Rev. D* **10** 186-9
[11] Bass S *et al* 1998 *Prog. Part. Nucl. Phys.* **41** 255-369
[12] Adare A *et al* (PHENIX Collaboration) 2008 *Phys. Rev. C* **77** 064907
[13] Becattini F 2002 *Nucl. Phys. A* **702** 336-40
[14] Adler S S *et al* (PHENIX Collaboration) 2004 *Phys. Rev. C* **69** 034909-41
[15] Adler S S *et al* (PHENIX Collaboration) 2003 *Phys. Rev. Lett.* **91** 182301
[16] Adams J *et al* (STAR Collaboration) 2005 *Phys. Rev. C* **72** 014904-27
[17] Adams J *et al* (STAR Collaboration) 2005 *Phys. Rev. C* **71** 044906-27

27th Winter Workshop on Nuclear Dynamics (WWND 2011) IOP Publishing
Journal of Physics: Conference Series **316** (2011) 012031 doi:10.1088/1742-6596/316/1/012031

# A derivation of the source term induced by a fast parton from the quark energy-momentum tensor

**R. B. Neufeld**

Los Alamos National Laboratory, Theoretical Division, MS B238, Los Alamos, NM 87545, U.S.A.

E-mail: neufeld@lanl.gov

**Abstract.** The distribution of energy and momentum deposited by a fast parton in a medium of thermalized quarks, or the source term, is evaluated in perturbative thermal field theory. The calculation is performed by directly evaluating the thermal expectation value of the quark energy-momentum tensor. The fast parton is coupled to the medium by adding an interaction term to the Lagrangian. I show that this approach is very general and can be modified to consider more realistic modeling of fast parton propagation, such as a fast parton created in an initial hard interaction or the evolution of a parton shower due to medium induced radiation. For the scenario considered here, it is found that local excitations fall sharply as a function of the energy of the fast parton. These local excitations couple directly to the sound mode in hydrodynamics and are important for generating an observable shockwave structure. This may have implications for the trigger $p_T$ dependence of measurements of azimuthal dihadron particle correlations in heavy-ion collisions. In particular, one would be less likely to observe a conical emission pattern for increasing trigger $p_T$.

## 1. Introduction

The observation of the quark gluon plasma (QGP) created in heavy-ion collisions [1] is one of the most exciting discoveries of the past decade. Among the many interesting experimental results from the heavy-ion programs at the Relativistic Heavy Ion Collider (RHIC) and the Large Hadron Collider (LHC), the suppression in the production rates of energetic, or hard, leading particles and jets relative to a naive superposition of nucleon-nucleon collisions is one of the most striking [2, 3, 4]. These phenomena are often collectively referred to as "jet quenching" and are reflective of the interaction of hard partons with the QGP medium which they traverse.

Another striking result from heavy-ion experiments is that the QGP formed at currently achievable center-of-mass energies may behave as a nearly ideal fluid [5]. In particular, the low energy, or soft particles created in heavy-ion collisions show a strong flow profile consistent with nearly ideal fluidity. The two results cited above lead to the natural question of how the QGP responds to a propagating fast (hard) parton. More generally stated, one would like to understand how the hard and soft physics of the QGP join together.

In this proceedings I will attempt to make some progress in that direction. I will present a new approach for coupling the hard physics of fast parton propagation to the underlying thermal medium. The idea is to couple the fast parton to the QGP via an interaction term in the Lagrangian. The dynamics of the medium in the presence of the fast parton can then be directly obtained by taking the thermal expectation value of the energy-momentum tensor

Journal of Physics: Conference Series **316** (2011) 012031 doi:10.1088/1742-6596/316/1/012031

(EMT), denoted $T^{\mu\nu}$. Although I will here consider an asymptotically propagating fast parton, it should be stressed that the approach used in this proceedings can be easily extended to consider more realistic scenarios. For instance, one can modify the interaction term to consider back to back fast partons created in an initial hard interaction, which is more relevant to phenomenology. Or, equally as relevant, one can incorporate [6] the medium induced evolution of a primary fast parton into a full parton shower through radiative processes [7].

Ideally one would like to evaluate components of the EMT directly, but for the reasons discussed in [8] the focus here will instead be on the divergence of the EMT, $\partial_\mu T^{\mu\nu}$. I will refer to this quantity as the "source term", in keeping with common terminology. The source term carries information about the rate of energy transfer to the medium and acts as a source (hence the name "source term") for the evolution of the underlying medium in the presence of a fast parton. By using an effective theory such as hydrodynamics, one can evaluate the medium response to a fast parton from the source term. A rigorously derived source term thus provides guidance for phenomenological studies of shockwave formation in the QGP [10].

This dual utility of the source term makes it a powerful quantity, particularly when applied to the realistic scenarios discussed above. For instance, the amount and distribution of energy a parton shower loses to the underlying medium can be obtained from the source term. This is useful in the description of jet observables [9], which are a more powerful probe of the QGP than leading particle observables. Additionally, such information may be useful in the experimental/jet medium background separation in heavy-ion experiments [3, 11].

## 2. Formalism and Results

The approach here is to begin with the EMT of the underlying medium. For simplicity I will consider a medium of massless quarks/antiquarks, leaving the inclusion of medium gluons for a forthcoming study. Thus, apart from the coupling strengths, the results will be the same as for a QED plasma. The EMT is given by [12]

$$T^{\mu\nu} = \frac{i}{4}\bar{\psi}\left(\gamma^\mu \overleftrightarrow{D^\nu} + \gamma^\nu \overleftrightarrow{D^\mu}\right)\psi - g^{\mu\nu}\mathscr{L}, \tag{1}$$

where

$$\mathscr{L} = \frac{i}{2}\bar{\psi}\,\overleftrightarrow{\not{D}}\,\psi,\; D^\mu = \partial^\mu - ig\,A^\mu_a t^a \tag{2}$$

and

$$\bar{\psi}\gamma^\mu \overleftrightarrow{D^\nu}\psi = \bar{\psi}\gamma^\mu \overrightarrow{D^\nu}\psi - \bar{\psi}\gamma^\mu \overleftarrow{D^{*\nu}}\psi. \tag{3}$$

In the above equations, $g$ is the strong coupling, $t^a$ are the $SU(3)$ generators in the fundamental representation, and conventional slashed notation is used, $\not{A} = \gamma_\mu A^\mu$, etc. A summation over color, spin, and the active number of quark flavors is implied in the EMT.

As mentioned in the Introduction, one must specify how the fast parton couples to the medium in order to investigate the medium response. The choice that is adopted here is to model the fast parton as an external color current which couples to the Lagrangian:

$$\mathscr{L} \to \mathscr{L} - A^a_\mu\, j^\mu_a, \tag{4}$$

where for the moment I do not specify the explicit form of $j$ ($j^\nu$ should not be confused with the source term, $J^\nu$). Non-Abelian gauge symmetry is preserved by the replacement made in (4) if $D^{ab}_\mu j^\mu_b = 0$.

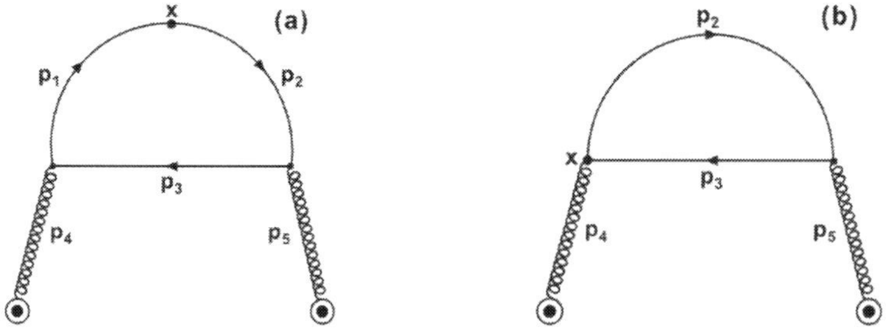


**Figure 1.** The Feynman diagrams contributing to $\langle \partial_\mu T^{\mu\nu} \rangle$ in the presence of the interaction term, $A_\mu^a\, j_a^\mu$. The contribution from the energy-momentum tensor and external current insertions are discussed in the text (see also Figure 2). The diagram in Figure 1(a) can be traced back to terms in the energy-momentum tensor (see equation (1)) which go as $\bar\psi\gamma\partial\psi$, whereas the diagram in Figure 1(b) originates from terms that go as $g\,\bar\psi\gamma A\psi$.

In the presence of the interaction term, $A_\mu^a\, j_a^\mu$, the lowest order Feynman diagrams needed to calculate the thermal expectation $\langle \partial_\mu T^{\mu\nu} \rangle$ are shown in Figure 1. The diagram in Figure 1(a) arises from terms in (1) which go as $\bar\psi\gamma\partial\psi$, whereas Figure 1(b) arises from terms that go as $g\,\bar\psi\gamma A\psi$. Note that a two gluon exchange is necessary to couple to the EMT, which is a color singlet quantity (a two photon exchange is also necessary in QED, from Furry's theorem). The convention used here is that the standard Feynman quark-gluon vertex contributes $ig\gamma^\mu t^a$ and the Feynman gauge is used for gluon propagators.

One must also determine what the correct Feynman rules for the EMT and external current are. To determine these non-standard contributions, which are isolated in Figure 2, one first assigns the appropriate momentum to each derivative. Recalling that one is here interested in $\langle \partial_\mu T^{\mu\nu} \rangle$ and that the final result is Fourier transformed into position space, it is found that the value of Figure 2(a) is [8]

$$\frac{ie^{-ix\cdot(p_1-p_2)}}{4}\Big((p_2^2-p_1^2)\gamma^\nu + p_1^\nu(3\not p_2+\not p_1) - p_2^\nu(3\not p_1+\not p_2)\Big). \tag{5}$$

One applies the same procedure to Figure 2(b) where I choose the convention that the gluon momentum flows away from the external current (or into $x$ and any vertex) in all cases. The result is [8]

$$-ig\,e^{-ix\cdot(p_4+p_3-p_2)}\,(p_4+p_3-p_2)_\mu \frac{(\gamma^\nu j_a^\mu + \gamma^\mu j_a^\nu - 2g^{\mu\nu}\not j_a)\,t^a}{2} \tag{6}$$

which is only valid when the gluon field in Figure 2(b) connects with the source.

Finally, for the external color current in Figure 2(c), one has very generally

$$-i\int d^4z\, j_a^\alpha(z)\, e^{iz\cdot p_4} \tag{7}$$

for the case of an external current which contains one power of $g$. As mentioned in the Introduction, the focus here will be on an asymptotically propagating fast parton. In that case $j_a^\mu = gQ^a(t)U^\mu\,\delta^3(\mathbf{z}-\mathbf{u}t)$ where $\mathbf{u}$ is the fast parton's velocity and $U^\mu = (1,\mathbf{u})$. $gQ^a(t)$ is the charge of a classical particle in QCD (defined by $Q_i^a\,Q_j^a = \delta_{ij}C_{2i}$, with $C_{2i}$ the quadratic Casimir in representation $i$ (3 for a gluon, 4/3 for a quark)) and $Q^a(t)$ evolves in time according to Wong's equations [13]. The time dependence of $Q^a(t)$ is of higher order in $g$ and will not be considered here. For this choice of $j_a^\mu$ Figure 2(b) simplifies to

$$-2\pi i\,g\,Q^a U^\alpha\,\delta(p_4\cdot U). \tag{8}$$

Journal of Physics: Conference Series **316** (2011) 012031 doi:10.1088/1742-6596/316/1/012031

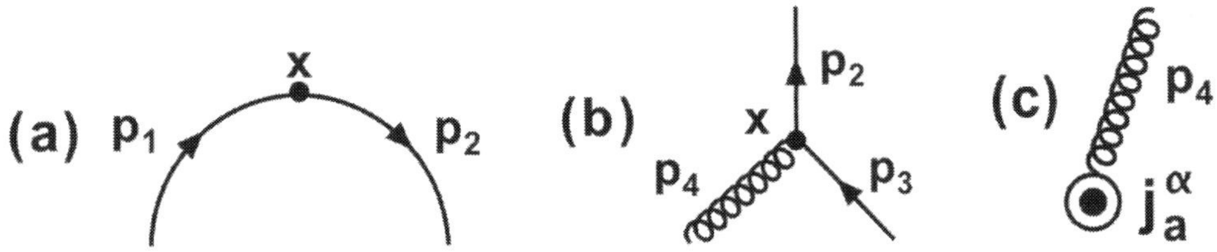


**Figure 2.** Contributions to the Feynman diagrams of Figure 1 which come from inserting the energy-momentum tensor and external current interaction term. Their values and how they are obtained are discussed in the text, see specifically equations (5) - (7).

As discussed in the Introduction, the approach used in this paper can easily be extended to consider more realistic scenarios than the single asymptotic parton considered here. One can modify the interaction term to consider back to back fast partons created in an initial hard interaction or incorporate the medium induced evolution of a primary fast parton into a full parton shower through radiative processes.

To obtain the Green's function structure of Figure 1 I use the conventions of real time thermal field theory as outlined by Das [14]. The details of the calculation will not be presented in this proceedings, but will be presented in a follow-up publication in which medium gluons will also be considered. Using standard Feynman rules for real time thermal field theory in addition to the special rules presented in equations (5) - (7) the result for the source term is obtained as [8]

$$\langle \partial_\mu T^{\mu\nu}(x)\rangle = -4i\, N_F\, C_2\, g^4 \int \frac{d^4p_3\, d^4p_4\, d^4p_5}{(2\pi)^9} e^{-ix\cdot(p_4+p_5)} \delta(p_3^2) n_F(p_3) G_R(p_4) G_R(p_3+p_4) G_R(p_5) \\ \times \delta(p_4\cdot U)\delta(p_5\cdot U) \left[2\,(p_3\cdot U)^2\, p_5^\nu - U^2\, p_3\cdot p_4\, p_5^\nu - U^\nu\,(p_3\cdot U)(2p_3\cdot p_5 + p_4\cdot p_5)\right] \tag{9}$$

where $G_R(p) = (p^2 + i\epsilon p^0)^{-1}$ is the retarded Green's function, $n_F(p) = (e^{|p^0|/T} + 1)^{-1}$ is the Fermi distribution function, $T$ is the temperature, and $N_F$ is the number of active flavors.

Equation (9) is the central result of this proceedings. One can obtain an intuitive picture of the underlying physics it contains by considering the Green's function structure. Reading from left to right in the first line it is seen that a particle from the heat bath (represented by the thermal distribution $\delta(p_3^2)n_F(p_3)$) absorbs a gluon from the external current (represented by $G_R(p_4)$) and then continues to propagate (shown by $G_R(p_3+p_4)$) until it absorbs another gluon from the external current (represented by $G_R(p_5)$). A final propagator representing the reabsorption of the particle by the heat bath (which would be given by $G_R(p_3+p_4+p_5)$) has been cancelled by the momentum structure of $\langle \partial_\mu T^{\mu\nu}\rangle$.

It is useful to consider certain approximations and limits of equation (9). A kinetic theory calculation of the source term generated by an asymptotically propagating fast parton was performed in [15]. Within the hard thermal loop (HTL) approximation equation (9) should reduce to the kinetic theory result (the HTL approximation is formally equivalent to the Vlasov equation [16]). The interested reader can verify (as this author has done) that (9) indeed reproduces the result of [15]. This verification gives confidence to the approach introduced in this proceedings.

I now consider an application of (9) by considering the source term ansatz discussed in [17]. The authors of [17] proposed the following simple form:

$$\langle \partial_\mu T^{\mu\nu}(x)\rangle \approx \frac{dE}{dt}\,(U^\nu - \lambda\partial^\nu)\,\delta(\mathbf{x} - \mathbf{u}t). \tag{10}$$

The utility of (10) is that one can encode basic features of the source term in a compact way. The dimensionfull coefficient $\lambda$ parameterizes local contributions from the source, that is, terms which globally integrate to zero (and for instance do not contribute to $dE/dt$) but may still be important for exciting the medium. It was found in [17] that within linearized hydrodynamics a double peaked structure in the azimuthal emission spectrum associated with the source given by (10) only appeared for rather large values of $\lambda$ (on the order of 0.5 fm or higher for 20 GeV total energy deposited into the medium). In what follows, I will extract $\lambda$ from equation (9).

To extract $\lambda$, one needs the relation [8]

$$\lambda = \frac{-1}{\frac{dE}{dt}} \int d^3x \, x \, \langle \partial_\mu T^{\mu x}(x) \rangle \tag{11}$$

where I have chosen to use the $x$ component of the source to obtain $\lambda$ (according to the ansatz of (10) any component would work). $dE/dt$ here is the collisional energy loss rate, which can be obtained from the source term as

$$\frac{dE}{dt} = \int d^3x \, \langle \partial_\mu T^{\mu 0}(x) \rangle. \tag{12}$$

For this proceedings I will use the HTL limit for the collisional energy loss:

$$\frac{dE}{dt}_{\text{HTL}} = \frac{m_D^2 \, C_2 \, \alpha_s}{2} \left( 1 - \frac{\tanh^{-1}[u]}{\gamma^2 \, u} \right) \ln \frac{p_{max}}{p_{min}} \tag{13}$$

where $\gamma = (1 - u^2)^{-1/2}$ and based on the HTL approximation and physical reasoning $p_{max} \sim T$ and $p_{min} \sim m_D$. For simplicity, I will use the relativistic limit ($\gamma \gg 1$) of (13).

With $dE/dt$ in hand, $\lambda$ can be obtained directly from (9) by using (11) and an integration by parts of the general form:

$$\int d^3x \, d^3p \, x \, e^{i\mathbf{p}\cdot\mathbf{x}} \, f(\mathbf{p}) = i \int d^3x \, d^3p \, e^{i\mathbf{p}\cdot\mathbf{x}} \, \partial_{p_x} f(\mathbf{p}). \tag{14}$$

Care must be taken in using (14) because $\lambda$ has an infrared divergence. The simplest way around this, and what I will adopt here, is to introduce a mass term in the Green's functions $G_R(p_4)G_R(p_5)$ in (9) given by the thermal gluon mass $m^2 = m_D^2/2$.

It is also necessary to consider carefully the momentum scales involved in the integration. For consistency, one should only apply the full result of equation (9) down to some momentum scale $|p_4| \sim q^*$ where $gT \ll q^* \ll T$. When $|p_4| \sim gT$ the full result contains contributions from higher order terms and only the HTL approximation is consistent. From a practical point of view in the calculation of $\lambda$ this means dividing terms in the integration such that

$$\begin{aligned} \lambda &= \lambda_{\text{F}} \text{ when } |p_4| \geq q^*, \\ \lambda &= \lambda_{\text{HTL}} \text{ when } |p_4| \leq q^*, \end{aligned} \tag{15}$$

where $\lambda_{\text{F}}$ is obtained from (9) and $\lambda_{\text{HTL}}$ is from taking the HTL approximation of (9). The final result should be independent of $q^*$ (this type of analysis for the separation of scales in calculations of finite temperature field theory was introduced in [18]).

With these technicalities in place, I will present the result for $\lambda$ and further show that it is independent of $q^*$. The result is presented in Figure 3 as a function of $\gamma = (1 - u^2)^{-1/2}$ for $g = 1$ and $T = 400$ MeV. It is interesting that $\lambda$ falls rather sharply as a function of $\gamma$. As was mentioned above, $\lambda$ parameterizes local contributions from the source and is important for exciting the medium. In particular, $\lambda$ couples directly to sound modes (and not to diffusive

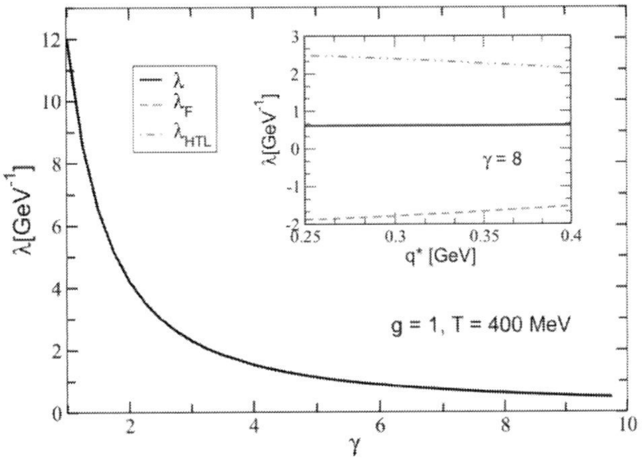


**Figure 3.** In the source term of equation (10) $\lambda$ couples directly to sound modes and is important for generating observable Mach cone signals. The contribution to $\lambda$ from the source term (9) is plotted above for $g = 1$ and $T = 400$ MeV [8]. It is clear that $\lambda$ drops as a function of $\gamma = E/m$ which could have implications for the trigger $p_T$ dependence of azimuthal dihadron correlation measurements (see discussion in text). The inset shows that the result for $\lambda$ is largely independent of the separation parameter, $q^*$ (see text for details).

modes) when (10) is used as a source term for hydrodynamics. It was found that large values of $\lambda$ were crucial to the appearance a conical Mach-like emission spectrum in [17]. The experimental implication of the dependence of $\lambda$ on $\gamma$ as shown in Figure 3 could be found in the trigger $p_T$ dependence of measurements of azimuthal dihadron particle correlations. In particular, a conical emission pattern would be less likely to be observed for increasing trigger $p_T$, which indeed seems to be the case [19]. The inset of Figure 3 shows that the result is largely independent of the separation parameter, $q^*$, introduced above.

## 3. Conclusion

In this proceedings I have presented a new approach for coupling the hard physics of fast parton propagation to the underlying thermal medium. The principle idea is to couple the fast parton to the QGP via an interaction term in the Lagrangian. The dynamics of the medium in the presence of the fast parton are then directly obtained by taking the thermal expectation value of the energy-momentum tensor.

Using this approach the source term for a medium of thermalized quarks in the presence of an asymptotically propagating fast parton was obtained and presented in equation (9). It was then found that local excitations, which may be important for generating observable Mach cone signatures, fall sharply as a function of $\gamma$, as shown in Figure 3. This may have implications for the trigger $p_T$ dependence of measurements of azimuthal dihadron particle correlations in heavy-ion collisions. In particular, a conical emission pattern would be less likely to be observed for increasing trigger $p_T$, which may indeed be the case [19].

A future publication will include medium gluons and present the details of the calculation [20]. The approach introduced here is general and can be modified to consider the medium response to tagged jets [21] or medium induced full parton showering constrained by realistic radiative energy loss calculations [22].

## 4. Acknowledgments
I wish to thank the organizers and participants of the 2011 Winter Workshop on Nuclear Dynamics for a great conference.

## 5. References
[1] I. Arsene *et al.*, Nucl. Phys. A **757**, 1 (2005); B. B. Back *et al.*, Nucl. Phys. A **757**, 28 (2005); J. Adams *et al.*, Nucl. Phys. A **757**, 102 (2005); K. Adcox *et al.*, Nucl. Phys. A **757**, 184 (2005).
[2] K. Adcox *et al.*, Phys. Rev. Lett. **88**, 022301 (2002); C. Adler *et al.*, Phys. Rev. Lett. **89**, 202301 (2002).
[3] S. Chatrchyan *et al.* [ CMS Collaboration ], [arXiv:1102.1957 [nucl-ex]];
[4] K. Aamodt *et al.* [ ALICE Collaboration ], Phys. Lett. **B696**, 30-39 (2011).
[5] H. Song and U. W. Heinz, Phys. Lett. B **658**, 279 (2008).
[6] R. B. Neufeld and I. Vitev, arXiv:1105.2067 [hep-ph].
[7] M. Gyulassy and X. N. Wang, Nucl. Phys. B **420**, (1994) 583; R. Baier, Y. L. Dokshitzer, A. H. Mueller, S. Peigne and D. Schiff, Nucl. Phys. B **484**, (1997) 265; B. G. Zakharov, JETP Lett. **65**, (1997) 615; M. Gyulassy, P. Levai and I. Vitev, Nucl. Phys. B **594**, (2001) 371.
[8] R. B. Neufeld, Phys. Rev. D **83**, 065012 (2011).
[9] I.Vitev, B. W. Zhang, Phys. Rev. Lett. **104**, 132001 (2010); T. Renk, Phys. Rev. C **80**, 044904 (2009); G. Y. Qin and B. Muller, Phys. Rev. Lett. **106**, 162302 (2011).
[10] G. Y. Qin, A. Majumder, H. Song and U. Heinz, Phys. Rev. Lett. **103**, 152303 (2009); R. B. Neufeld and B. Muller, Phys. Rev. Lett. **103**, 042301 (2009). B. Betz, M. Gyulassy, J. Noronha and G. Torrieri, Phys. Lett. B **675**, 340 (2009); R. B. Neufeld, B. Muller and J. Ruppert, Phys. Rev. C **78**, 041901 (2008); R. B. Neufeld, Phys. Rev. C **79**, 054909 (2009);
[11] G. Aad *et al.* [ Atlas Collaboration ], Phys. Rev. Lett. **105**, 252303 (2010).
[12] P. Mathews, V. Ravindran, K. Sridhar, JHEP **0408**, 048 (2004).
[13] S. K. Wong, Nuovo Cim. **A65**, 689-694 (1970).
[14] A. K. Das, *Singapore, Singapore: World Scientific (1997) 404 p*
[15] R. B. Neufeld, Phys. Rev. D **78**, 085015 (2008).
[16] P. F. Kelly, Q. Liu, C. Lucchesi *et al.*, Phys. Rev. **D50**, 4209-4218 (1994).
[17] R. B. Neufeld and T. Renk, Phys. Rev. C **82**, 044903 (2010);
[18] E. Braaten, T. C. Yuan, Phys. Rev. Lett. **66**, 2183-2186 (1991).
[19] A. Adare *et al.* [The PHENIX Collaboration], Phys. Rev. Lett. **104**, 252301 (2010).
[20] R. B. Neufeld and I. Vitev, in preparation.
[21] R. B. Neufeld, I. Vitev, B.-W. Zhang, Phys. Rev. **C83**, 034902 (2011).
[22] G. Ovanesyan and I. Vitev, arXiv:1103.1074 [hep-ph].

# Forward Silicon Vertex Tracker for the PHENIX experiment at RHIC

**Douglas E. Fields for the PHENIX Collaboration**

Department of Physics and Astronomy, 1 University of New Mexico, MSC07-4220 Albuquerque, New Mexico, 87131

fields@unm.edu

**Abstract.** The PHENIX experiment at RHIC is well in the process of constructing a silicon vertex tracker consisting of a central barrel component (already undergoing commissioning), as well as two forward rapidity components. This report gives an overview of the physics goals of the forward vertex tracker (FVTX), and gives the current status of its construction. It is scheduled to be completed this year and commissioned during the next RHIC run.

## 1. Introduction

The main goal of the RHIC heavy ion program is the discovery of the novel ultra-hot high-density state of matter predicted by the fundamental theory of strong interactions and created in collisions of heavy nuclei, the Quark-Gluon Plasma (QGP). From measurements of the large elliptic flow of light mesons and baryons and their large suppression at high transverse momentum $p_T$ that have been made at RHIC, there is evidence that new degrees of freedom, characteristic of a strongly-coupled deconfined QCD medium, drive the dynamics of nucleus-nucleus collisions. It has been recognized, however, that the potential of light quarks and gluons to characterize the properties of the QGP medium is limited and the next phase of the RHIC program calls for the precise determination of its density, temperature, opacity and viscosity using qualitatively new probes, such as heavy quarks.

We are in the construction phase of two Forward Silicon Vertex Trackers (FVTX) for the PHENIX experiment that will directly identify and distinguish charm and beauty decays within the acceptance of the muon spectrometers. The FVTX will provide this essential coverage over a range of forward and backward rapidities ($1.2 < |y| < 2.4$) – a rapidity range coverage which not only brings significantly larger acceptance to PHENIX but which is critical for separating cold nuclear matter effects from QGP effects and is critical for measuring the proton spin contributions over a significant fraction of the kinematic range of interest. In addition, the FVTX will provide greatly reduced background and improved mass resolution for di-muon events, culminating in the first measurements of the $\psi'$ and Drell-Yan at RHIC. These same heavy flavor and di-muon measurements in p+p collisions will allow us to place significant constraints on the gluon and sea quark contributions to the proton's spin and to make fundamentally new tests of the Sivers function universality.

**Figure 1 Half-layout of the PHENIX FVTX showing the four vertical planes of each end cap in the red circles.**

The FVTX detectors will extend the vertex capabilities of the PHENIX Silicon Vertex Tracker (VTX) to cover a much larger rapidity interval, including forward and backward rapidities, and allow displaced track and secondary vertex measurements in conjunction with the PHENIX muon arms. The heavy flavor measurements provided by these detectors will move us from making qualitative statements that heavy flavor is suppressed in heavy ion collisions to being able to distinguish among models which include radiative energy loss, collisional energy loss, and/or collisional dissociation for light and heavy quarks—a distinction which is not possible without the addition of a vertex detector, and without which there remains large uncertainty in what the various suppression mechanisms are for light and heavy quarks and what the extracted properties of the medium should be. By adding vertexing capability covering the muon arms we not only increase the acceptance for physics observables by a factor of 3-10 but also allow coverage in the forward rapidity regions – coverage which is critical to separate cold nuclear matter effects from QGP effects, and critical for understanding the geometric extent of the medium.

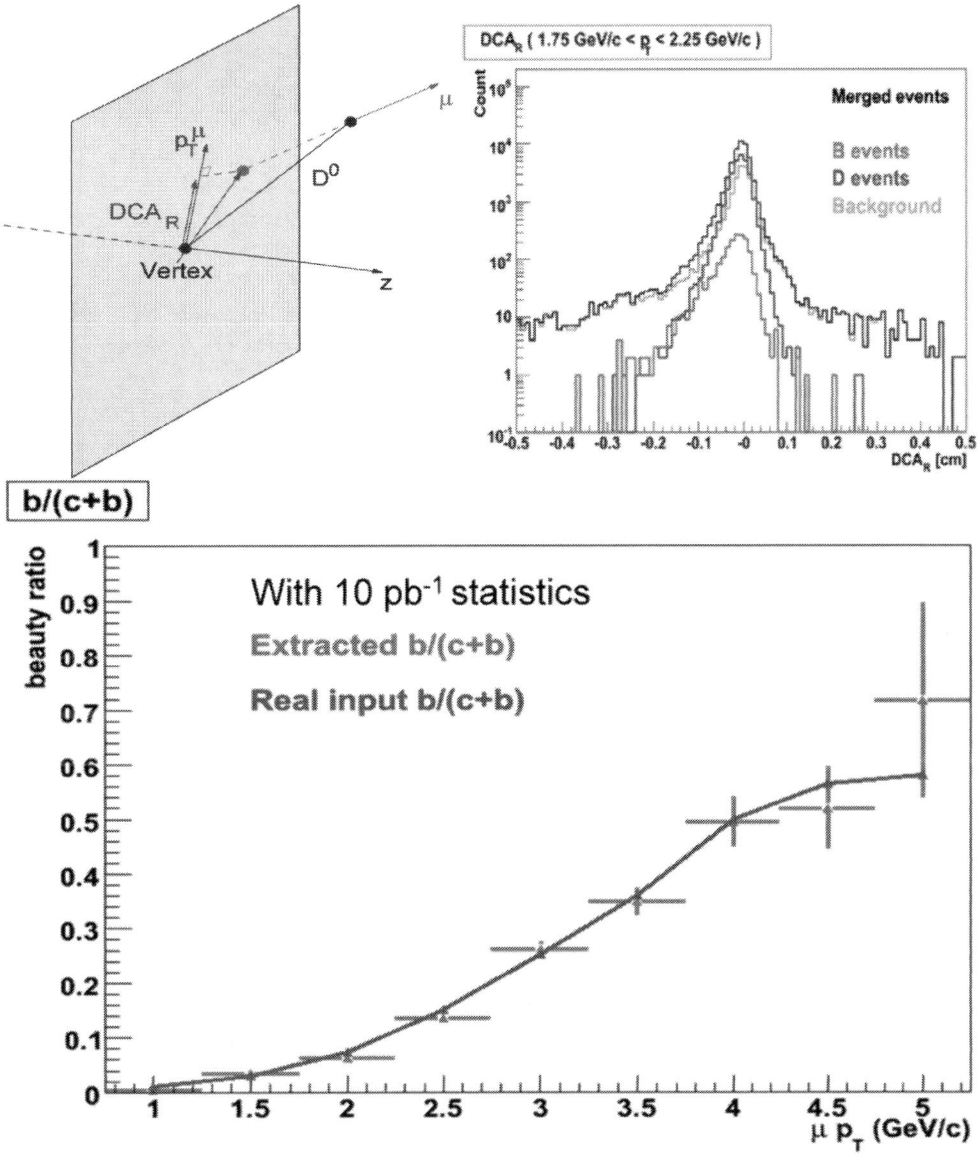


**Figure 2. Upper left panel: Definition of the Radial Distance of Closest Aproach ($DCA_R$). Upper right panel: $DCA_R$ for simulated charm, beauty and background (light meson decay) events. Bottom panel: Extracted beauty/total heavy meson ratio from a fit to the $DCA_R$ distributions, compared to the simulated input ratio.**

J/ψ vector mesons are also considered to be a very sensitive probe of the dense matter created in heavy ion collisions and have been shown to be highly suppressed at RHIC. At SPS the suppression of J/ψ production was one of the measurements given to support the statement that a new state of matter (the QGP) was formed, so the RHIC J/ψ results have been highly anticipated and are of great interest to the heavy-ion community. The baseline PHENIX detector was designed to provide good measurements of inclusive J/ψ production; however, to quantitatively understand the modification of J/ψ production in the medium, these measurements must be coupled with precise open heavy flavor measurements that can be used with recombination models to determine the contribution to J/ψ production that comes from random c and cbar pairs combining to form J/ψs as opposed to prompt J/ψ production. Since J/ψs produced by recombination have different kinematic distributions from prompt J/ψ production, and will *enhance* J/ψ production rather than suppress the production, the recombination contributions must be understood to extract the amount of medium-induced suppression. The FVTX will provide the needed precision open heavy flavor measurements to allow for a much better understanding of J/ψ production via recombination. Together with this, we need to measure production of other vector mesons such as ψ' and $\chi_c$ to understand Debye screening contributions. The different vector mesons have different screening radii, so measuring the difference in their suppression patterns allows one to infer the radius at which Debye screening becomes significant. Additionally, the ψ' and $\chi_c$ contribute to J/ψ production via feed-down, so measuring their suppression pattern allows one to understand how much of the J/ψ suppression is inherited from ψ', $\chi_c$ suppression and how much is suppression of prompt J/ψ production. With the addition of the FVTX detector, we will be able to separate the ψ' and J/ψ production through improved mass resolution and enhanced background rejection. These new measurements, coupled with the J/ψ measurements should significantly advance our understanding of J/ψ suppression in heavy ion collisions.

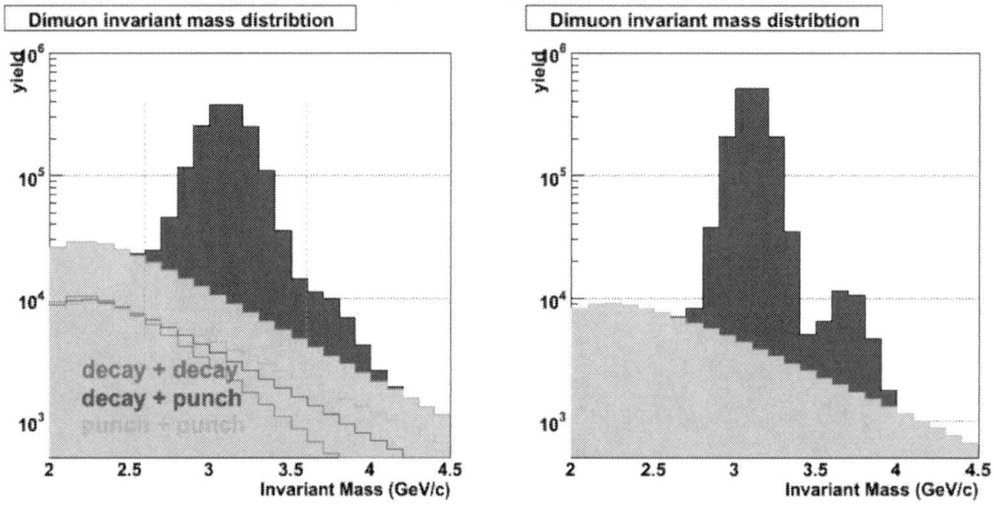


**Figure 3. Di-muon invariant mass spectrum in simulated 200GeV p+p collisions without (left panel) and with (right panel) the use of the FVTX detector's tracking before the magnet steel, demonstrating the improvement in the muon momentum resolution and the reduction in backgrounds.**

The RHIC spin program was developed to determine the gluon and sea quark contributions to the proton's spin and (possibly) put constraints on the angular momentum contributions to the spin. With the addition of the FVTX detectors, we will significantly extend the x coverage for gluon spin measurements in PHENIX by adding precision heavy flavor asymmetry measurements at forward and backward rapidities which are sensitive to the gluon spin distribution. These measurements cannot be made with the baseline detector alone. Measuring the gluon spin distributions over a large x coverage is critical for extracting $\Delta G = \int \Delta G(x)dx$. The FVTX detector will provide precise Drell-Yan measurements (in conjunction with the muon arms), which will allow measurements of sea quark contributions to the proton's spin because the Drell-Yan production asymmetry is sensitive to the ubar polarization. This provides us a unique opportunity to produce a sea quark measurement which is complimentary to the W measurements that are proposed. The FVTX will also provide unique track reconstruction constraints which will help reject hadronic backgrounds which contribute to the single muon spectra. The single muon spectra will be used to measure sea quark contributions to the proton

Journal of Physics: Conference Series **316** (2011) 012032 doi:10.1088/1742-6596/316/1/012032

spin by extracting the W→μX single spin asymmetry at high $p_T$ so background rejection at high $p_T$ is essential to making a precision measurement.

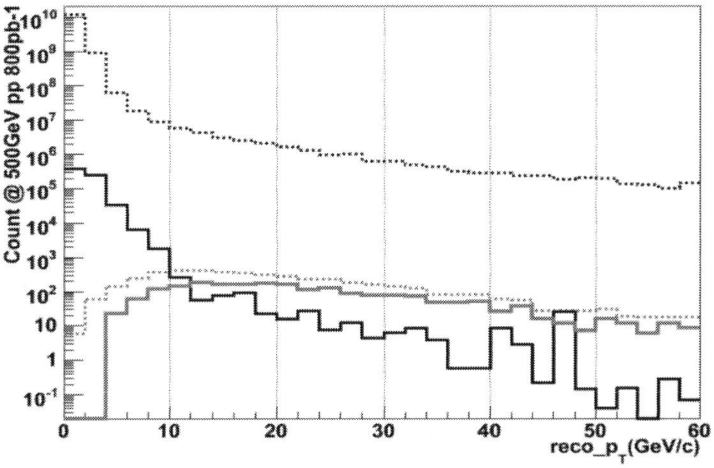


**Figure 4. Background (black lines) and W signal (red lines) without FVTX (dotted lines) and with FVTX (solid lines).**

The Drell-Yan measurements with polarized proton beams will also allow for a new fundamental test of QCD theory. Sivers-type single-spin asymmetry has been observed in semi-inclusive DIS at HERMES and COMPASS very recently. A fundamental prediction of QCD is that such effects will give an opposite sign in the transverse single spin asymmetry in DY production in p+p collisions. Its verification (or not) will be an important milestone in our study of the strong interaction, as it tests all concepts for analyzing hard-scattering reactions that we know of today.

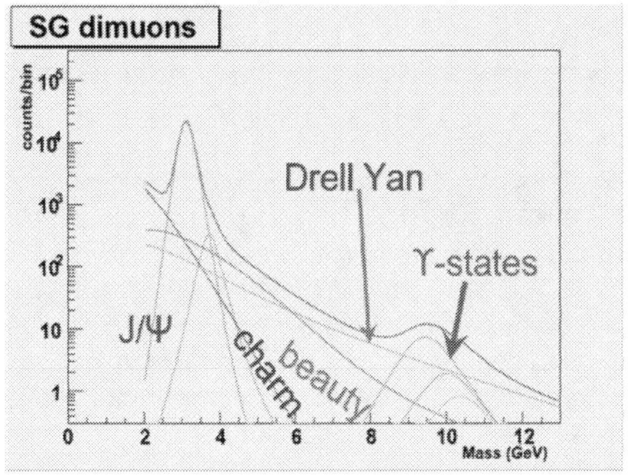


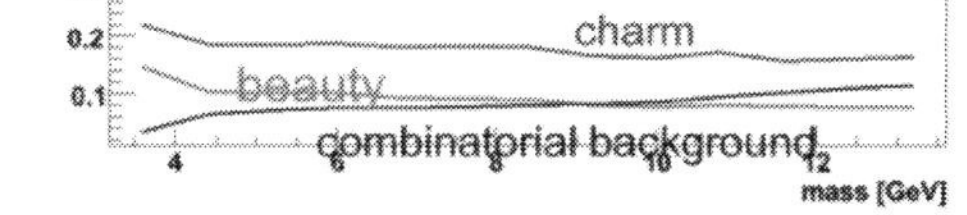


**Figure 5. Left panel: Di-muon invariant mass "cocktail plot" before imposing FVTX cuts. Right panel: Enhancement of the Drell Yan signal over background in the 4 -12 GeV mass range using FVTX cuts.**

The FVTX detectors, covering forward and backward rapidities ($1.2 < |y| < 2.4$) are ideally suited to study the fundamental QCD interactions in d+A collisions at RHIC. Light hadron results indicate that the suppression pattern observed at forward rapidity is suggestive of nuclear shadowing or saturation physics at small values of Bjorken-x. It has been found, however, that such attenuation is not consistent with the measured shadowing in the nuclear structure functions in DIS. Preliminary measurements have instigated recent theoretical activity which suggests that the physics of d+A collisions is much more complex. Nuclear effects that potentially play a role are the Cronin effect, cold nuclear matter energy loss (stopping power of large nuclei) and dynamical shadowing from coherent multiple scattering. The FVTX upgrades will help open new physics channels for precision measurements, such as the Drell-Yan process, where initial- and final state-effects can be separately investigated.

In addition, the FVTX detectors will provide moderate rapidity gap correlation measurements (when combined with the midrapidity PHENIX detectors) and large rapidity gap (when the forward and backward FVTX are used) for studying di-jet correlations. It has been argued in the framework of CGC phenomenology that the dominant particle production mechanism in d+A reactions is 2-to-1 processes, such as gluon fusion and quark scattering on a background gluon field. In contrast, the lowest order process in collinear factorized pQCD is 2-to-2 scattering which would produce a clear di-jet signal. Therefore, to validate or disprove theoretical models of jet production in d+A collisions jet-jet correlation measurements, including heavy quark triggered jets, are of critical importance.

With the present PHENIX detector, heavy-quark production in the forward and backward directions has been measured indirectly via the observation of single muons. The current measurements are inherently limited in accuracy by systematic uncertainties resulting from the large contributions to the single muon spectra from prompt pion and kaon semi-leptonic decays and from pion and kaons which punch through the entire muon system and are mistakenly tagged as muons. In addition, the analysis does not allow for a model-independent separation of the charm and bottom contributions. The FVTX detector will provide vertex tracking with a distance of closest approach (DCA) resolution in r-z that is better than 100 μm over a large coverage in rapidity ($1.2 < |\eta| < 2.2$) and with full azimuthal coverage. This will allow for vertex cuts which separate prompt particles from decay particles and short-lived heavy quark mesons from long-lived light mesons (pions and kaons). In addition, bottom measurements can be made directly via $B \rightarrow J/\psi+X$ by looking for a displaced $J/\psi$ vertex, and this will allow charm and bottom contributions to be separated in semi-inclusive single lepton measurements. Therefore, with this device fundamentally new measurements can be made and current muon measurements will be significantly enhanced.

The precision of the $J/\psi$ and other dimuon measurements in AuAu collisions is currently limited by the large amount of combinatorial background that must be subtracted from the opposite sign dimuon signal obtained with the muon tracker and by the inherent mass resolution. With added rejection power for pion and kaon decays, the significance of all dimuon measurements will be greatly improved. Further improvement in these measurements results from the improved mass resolution, which will be attained because of the more accurate determination of the opening angles of the dimuons. All together, this will result in improved dimuon data as well as provide access to several new measurements: separation of $\psi'$ from $J/\psi$, extraction of a Drell-Yan signal from the dimuon continuum, extraction of $B \rightarrow J/\psi$ and measurement of upsilons at central rapidity.

## 2. Technical description and status of the detector

The FVTX will be composed of two endcaps, with four silicon mini-strip planes each, covering angles ($1.2 < |\eta| < 2.2$) that match the two muon arms. Each silicon plane consists of wedges of mini-strips with 75 μm pitch in the radial direction and 3.75° wide strips in phi, which translates to lengths in the phi direction varying from 2.8 mm at small angles to 12.1 mm at 35 degrees. An r-z DCA resolution of 100 μm can be achieved with a maximum occupancy per strip in central Au+Au collisions of less than 2.8%. In Figure 1, the four stations of the North and South FVTX arms are circled in red, the central support structure for the VTX system can be seen between the two, and the large gray planes surrounding the FVTX sensors are the planes that will hold the readout electronics for the VTX and FVTX systems. A picture of one half-cage of the detector is shown in Figure 6.

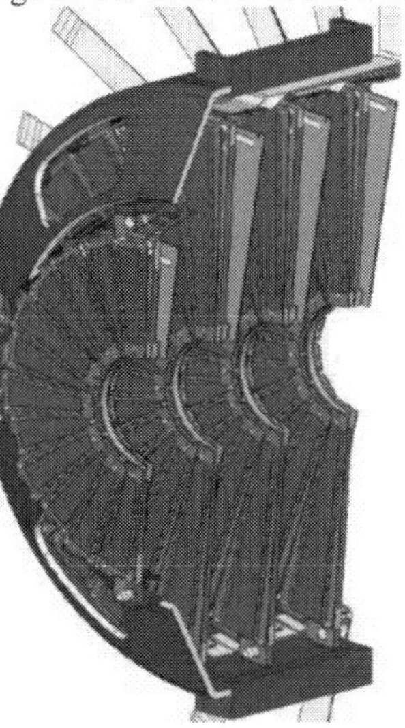

**Figure 6. One half-cage containing the four half-disks of sensor wedges. Each wedge is overlapped sufficiently to give full azimuth coverage.**

The FVTX will consist of approximately 1.1 million mini-strips that will be read out with an IC chip (FPHX) designed by Fermilab, which is wire-bonded directly to the mini-strips. This chip provides analog and digital processing with zero-suppression and produces a digital output which is "data-pushed" at 200 Mbps to intelligent readout boards containing FPGAs. The data are then transformed into the standard PHENIX format and transmitted to the PHENIX DAQ system via fiber optics.

We are using standard p-on-n silicon strip technology, which has been the baseline detector technology for dozens of silicon trackers in nuclear and high-energy physics experiments. In a p-on-n detector, the output signal is generated by the collection of positive charge carriers. The FPHX chip is compatible with positive charge collection. The FPHX is also designed to have leakage current compensation at the front-end, up to 100 nA/ strip. This compensation circuitry allows the possibility of reading out sensors with dc output connections to the FPHX chip, however our sensors are ac-coupled.

The strips are ac-coupled directly on the sensor. The capacitor is made by depositing an oxide layer of approximately 200 nm on top of, and over the entire length of the p-implant. An aluminum metallization is placed on top of the oxide to complete the capacitor and to form the readout conductor. The bias connection to the p-implants is made by a polysilicon resistor of approximately 1.5 MΩ, which is electrically connected to a common bias ring on one end, and to the p-implant on the other end. There is an individual polysilicon resistor for each strip, and three sets of pads on each strip: a spy pad which penetrates the oxide layer to allow probing of the dc characteristics of the strip; probe pads, which are dedicated for probing the ac characteristics of the strips; and bond pads which are only used to wire bond from the detector to the electronics (see Figure 7). All non-metal surfaces of the top side of the sensor are covered with a passivation of silicon-oxide or silicon-nitride. A guard ring is implanted around the perimeter of the wedge between the bias ring and the cut-edge of the sensor to prevent breakdown at the cut-edge of the sensor under a normal range of bias voltage. The sensor breakdown voltage is specified to be ≥200V or ≥50V above operating voltage, whichever is greater. There is no need to specify higher breakdown voltage, or to incorporate multiple guard ring structures, because an integrated 10 year radiation dose of 200 krads does not require it. Leakage current at 20° C is specified to be $\leq$ 160 nA/cm$^2$ at operating voltage. All of these processes and specifications are standard in the industry.

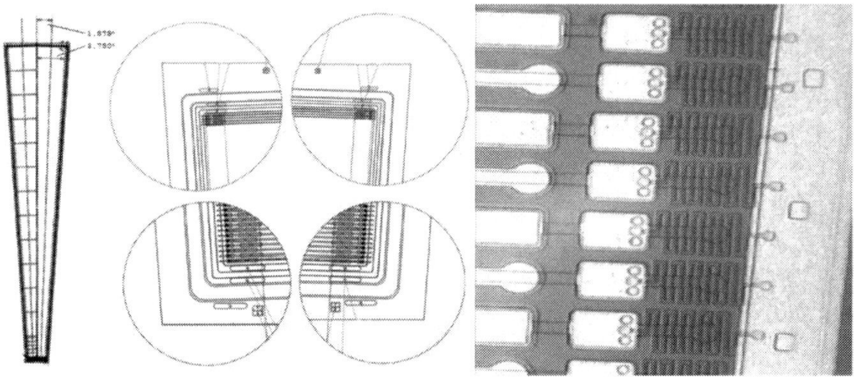

**Figure 7. Left panel: Schematic drawing of the silicon sensors with two rows of strips oriented perpendicular to the radial. Right panel: Close-up view of strips connected to AC-coupled pads, then DC-coupled test pads, the through a poly-silicon resistor to the bias ring (from left to right).**

The sensor readout strips will operate at ground potential, and a positive bias voltage will be applied to the backside of the sensor to fully deplete the sensor volume for efficient charge collection. The large sensor wedge for disks 2, 3, and 4 is approximately 126.5 mm high, 8.7 mm wide at the inner radius, and 25.3 mm wide at the outer radius. The small sensor wedge for disk 1 is approximately 59.5 mm high, 8.7 mm at the inner radius, and 16.5 mm at the outer radius. Our production sensors were produced by Hamamatsu with very good results. Hamamatsu tested and reported leakage current for each strip, and we verified these values on a fraction of the delivered sensors.

The sensor wedge consists of a stack up of a carbon support backing, kapton High-Density Interface (HDI), and sensor and chips. The carbon backing serves as a carrier on which the sensor, HDI, and chips can be mounted separately from the cooling plate. This modular arrangement allows us to fabricate and test all of the wedges separately.The HDI is a seven layer polyamide flexible printed circuit consisting of a bias layer, two signal layers, one ground layer, two power layers, and a component layer. All control lines (which are not active during data taking) are routed under the sensor and all output lines are routed towards the edge of the wedge thus insuring that the output lines will not couple into the signal lines on the sensor. The number of lines required (8 pairs for the control lines and 2 signal pairs per chip for the output lines) fit into the width of the HDI and the line pitch of the HDI challenging, but technically achievable with conventional polyamide manufacturing techniques. Pictures of the unassembled and assembled HDI and wedges are shown in Figure 8.

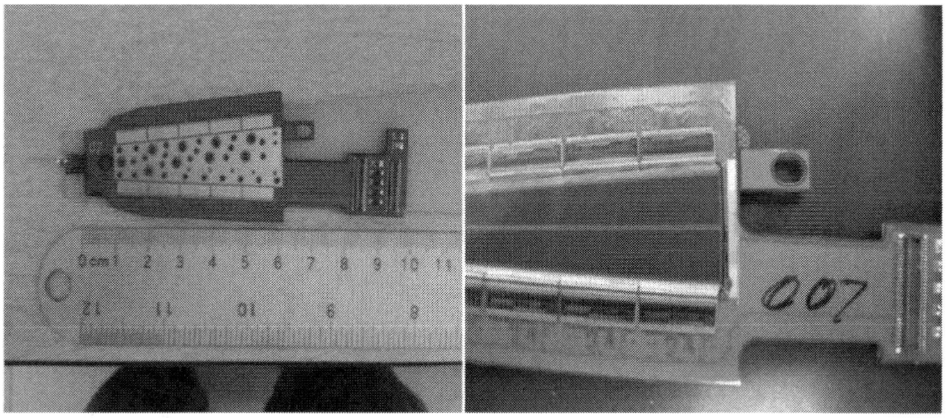


**Figure 8. Left panel: A unassembled production station 1 High-Density Interconnect card (HDI). The central region has the sensor bias plane, with ground planes surrounding it for the placement of the FPHX read-out chips. Right panel: An assembled wedge with the sensor, FPHX chips and wire bonds all in place.**

The FPHX is a custom IC designed by engineers in the Fermi National Accelerator Laboratory Particle Physics Division. The chip design borrows heavily from previously successful IC designs, FPIX2, FSSR, SVX4, etc. The experiment will use a 106 ns beam clock. Each FPHX chip integrates and shapes (CR-RC) signals from 128 channels of mini-strips, digitizes and sparsifies the hit channels each beam crossing, and serially reads out the digitized data. The chip is designed only for positive charge (hole) collection from p-on-n detectors. The strips used by the experiment will have different lengths, resulting in a varying distribution of input capacitance across different chips. For the most part, activity is rare, but when there is an event, it is expected that there will be an average of 2.8% hit pixels per chip, with the possibility of much higher occupancy. Long latency cannot be tolerated, and it is therefore required that four hits will be read out within four beam cross-over periods of an event. A number of FPHX chips will be controlled on a single High Density Interconnect (HDI). Space on the HDI is limited, which is why the data from events must be serialized onto one or two pairs of LVDS digital lines per chip. In addition, the FPHX must be as "self-sufficient" as possible and provide its own internal bias voltages and currents with minimal external support circuitry. The user must be able to control internal parameters and biases, therefore a digital slow control interface is provided on each chip to enable programming.

The FPHX is a mixed-mode chip with two major and distinct sections, the front-end and the back-end (see Figure 9). The mostly analog front-end contains the 128 channels of integrators, shapers, and comparators, in addition to several programmable bias circuits and DACs that are used for setting internal parameters. The output of each front-end channel is simply an 8-bit digital word from 8 comparators (forming one hit discriminator output and seven thermometer-coded ADC outputs, which results in 3-bit magnitude information). Each comparator's threshold is independently programmable, effectively allowing a custom non-linear ADC. The hit and ADC information serves as the input to the Core Logic section of the back-end, which processes the data for readout. Also part of the back-end is the Slow Controller, which accepts a serial data stream to program the chip after power-up.

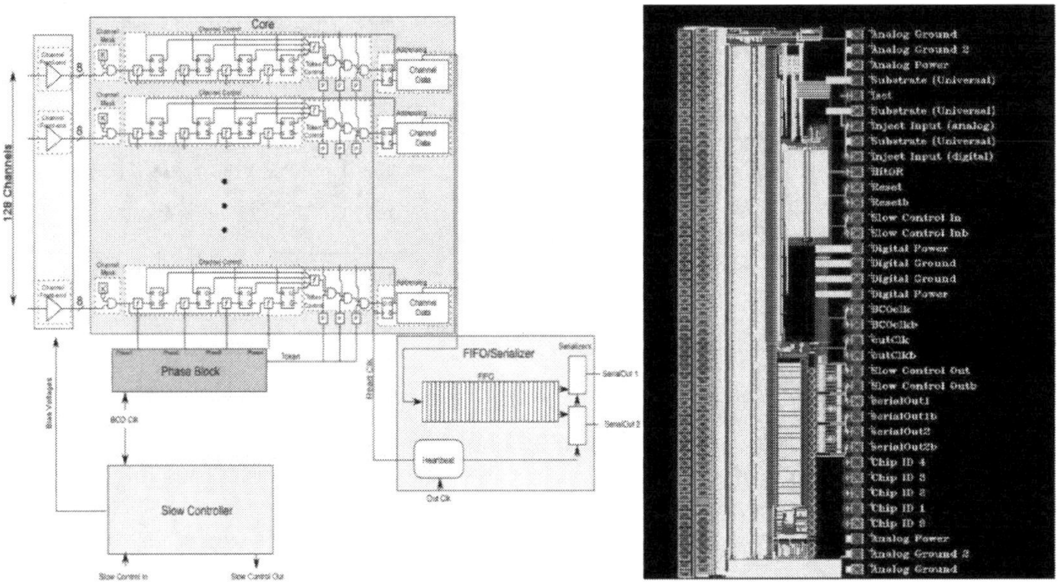


**Figure 9. Left panel: Schematic design of the FPHX read-out chip. Right panel: Layout of FPHX chips with 128 strip input pads (left) and 32 input/output pads (right) for clock, slow control, calibration pulser, power and ground and signal output.**

The overall architecture of the read-out chain is described in Figure 10. Each half-cage has 6 ROCs each connected to 16 sensor wedges. Each ROC is connected via optical fiber to a FEM, and 12 FEMs are grouped in a VME crate with a FEM Controller card. The ROC (Figure 11) will take the continuously streaming data (data-push) from 52 FPHX chips via flexible cables into an FPGA, strip the sync words from the data, combine the data of several chips, serialize it and send it out via fiber to the FEM and the Lvl-1 boards. The time to receive all of the data to pass to the Level 1 trigger is expected to be less than or equal to four beam clocks or 424 nsec. The location of the ROCs will be at the end of the silicon tracker enclosure in the "big wheel" area, as indicated in Figure 1. Twelve ROC boards will be required to service one endcap and these boards will hold a total of 48 FPGAs.

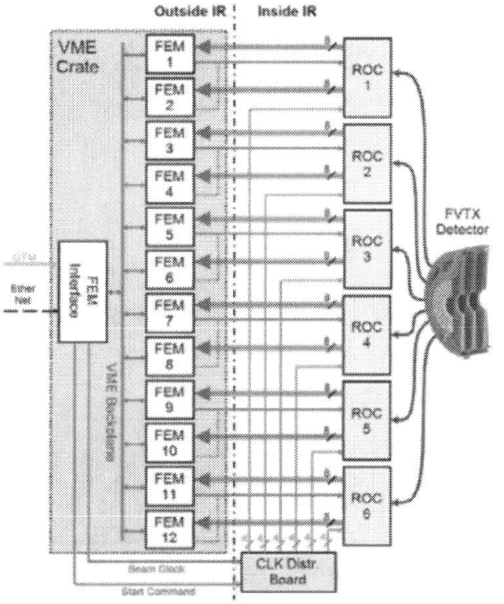


**Figure 10. Schematic of the data-acquisition architecture for the FVTX detector. Each half-cage has 6 ROCs connected to 16 sensor wedges each. Each ROC is connected via optical fiber to a FEM, and 12 FEMs are grouped in a VME crate with a FEM Controller card.**

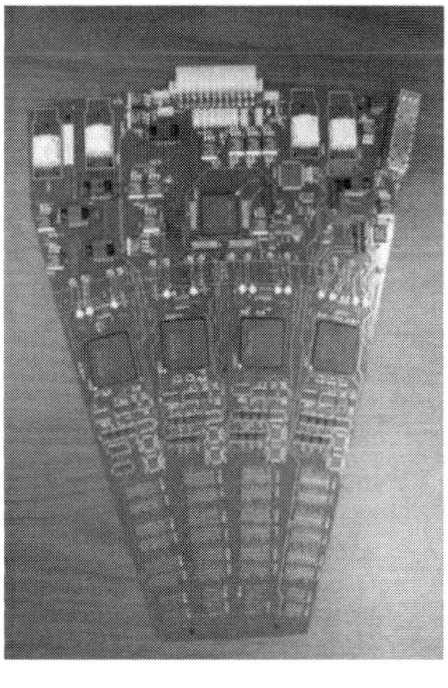

**Figure 11. A production ROC board. Sets of two Hirose connectors at the bottom end are for connections to the sensor wedges. The ROC board also contains bias and power distribution, calibration distribution, clock and slow control distribution, and fiber optic outputs.**

The FEM (Figure 12) will buffer the data for 64 beam clocks (emulating the 64 beam clock analog buffer of current PHENIX detectors), grab the data from the appropriate beam clock upon a Level-1 trigger and reformat the data before it is sent to the PHENIX Data Collection Modules (DCMs). There is a 64 FIFO array, with each FIFO storing the data from a particular beam counter. The FPHX data is routed to a FIFO, selected by the beam clock counter that is embedded in the data stream hits.

The hits from a particular beam crossing stay in the FIFO for not more than 64 clocks, otherwise they are marked as expired. This strategy solves the problem of relatively short beam clock counter wrap around. The FPGA then allows the data from the appropriate beam clock to be sent to the DCM when a LVL-1 trigger accept is received. Upon readout of the FIFO the hits from the expired beam crossing are not being outputted. The existing PHENIX DCMs can be used without modification.

**Figure 12. A production FEM board. It receives 32b parallel data from a single ROC channel over optical fiber at 125 MHz.**

The current FEM channel design has been successfully implemented on an XC4VSX35 Vertex-4 XILINX FPGA and tested at design speed. The buffering block is capable of receiving 32 bit parallel data at up to 300 MHz and reliably sends the data to the output buffer at 300 MHz. Copying of the data from a particular FIFO to the output buffer should be done as fast as possible so that data from one beam crossing are fully emptied before data from another beam crossing arrives. The full design for FVTX readout is implemented on the largest VSX series FPGA XC4VSX55, which has a sufficient number of FIFO blocks to accommodate 4 FEM Channels on a single chip. The data from all 4 channels will be combined on the output in order to reduce the number of DCM channels in readout.

The entire production read-out electronics chain has been tested with a small number of wedges connected, and will soon be tested with a complete half-cage of wedges. We are currently awaiting the delivery of the production extension cables which will allow us to do this final pre-installation testing.

**Figure 13. A production FEM interface board, which controls a VME Crate with 12 FEM boards.**

# Stellar Helium Burning Studied with an Optical Readout TPC (O-TPC) at HI$\gamma$S

**Moshe Gai**
**for the UConn-Yale-Duke-Weizmann-PTB-UCL Collaboration**
Laboratory for Nuclear Science at Avery Point
University of Connecticut, 1084 Shennecossett Rd, Groton, CT 06340-6097, USA
E-mail: moshe.gai@yale.edu

**Abstract.** An Optical Readout Time Projection Chamber (O-TPC) operating with the gas mixture of $CO_2$(80%) + $N_2$(20%) at 100 torr with gamma beams from the HI$\gamma$S facility of TUNL at Duke University were used to study the formation of carbon and oxygen during helium burning. Measurements were carried out with circularly polarized gamma-ray beams at energies: $E_\gamma$ = 9.08, 9.38, 9.58, 9.78, 10.08, 10.38 and 10.68 MeV. We have begun the process of extracting complete angular distributions for the $^{16}O(\gamma,\alpha)^{12}C$ reaction and the $^{12}C(\gamma,3\alpha)$ reaction (eventually with energy bins of approximately 100 keV) in order to determine the values of $S_{E1}$, $S_{E2}$ and the mixing phase $\phi_{12}$ of the $^{16}O(\gamma,\alpha)^{12}C$ reaction. The rate of carbon formation at high temperatures (T > 3 GK) was suggested to increase due to contributions from a higher lying $2^+$ state. We have measured an angular distribution of (essentially) pure E2 transition at $E_\gamma$ = 9.78 MeV of the $^{12}C(\gamma,3\alpha)$ reaction, providing conclusive evidence for the elusive $2_2^+$ state in $^{12}C$.

## 1. Introduction

The outcome of helium burning is the formation of the two elements: carbon and oxygen [1]. The ratio of carbon to oxygen at the end of helium burning has been identified three decades ago as one of the key open questions in Nuclear Astrophysics [1] and it is still today an open question. To solve this problem one must extract the p-wave [$S_{E1}(300)$] and d-wave [$S_{E2}(300)$] cross section factors of the $^{12}C(\alpha,\gamma)^{16}O$ reaction at the Gamow peak (300 keV) with an accuracy of approximately 10% or better. Our current knowledge of these astrophysical cross section factors is not near the required accuracy. In particular the data on the beta-delayed alpha-particle emission of $^{16}N$ does not determine whether the interference between the bound $1^-$ state at 7.117 and the broad ($\Gamma$ = 420 keV) $1^-$ state at 9.585 MeV in $^{16}O$ is constructive or destructive leading to large $S_{E1}$ = 80 keV-b or small $S_{E1}$ = 10 keV-b, respectively, since the measured minima of $\chi^2$ for the $^{16}N$ data are identical for both cases [2] as shown in Fig. 1.

Several new measurements of the $^{12}C(\alpha,\gamma)^{16}O$ reaction using gamma-ray detectors have been reported [3, 4] with energies in the vicinity of 1.0 MeV. However, the S-factors are measured with very low accuracy (±40-80%) and most importantly one cannot rule out a low value (close to 10 keV-b if not zero) of the E1 S-factor [5, 6]. These new experiments use some of the highest intensity alpha-particle beams (10 - 500 $\mu$Amp) with an impressive luminosity of $10^{33}$ [3] and $10^{31}$ $cm^{-2}sec^{-1}$ [4], and a $4\pi$ array of HPGe and BaF gamma-detectors, respectively, that provides large counting statistics. Yet the obtained accuracy of the S-factors is limited due to

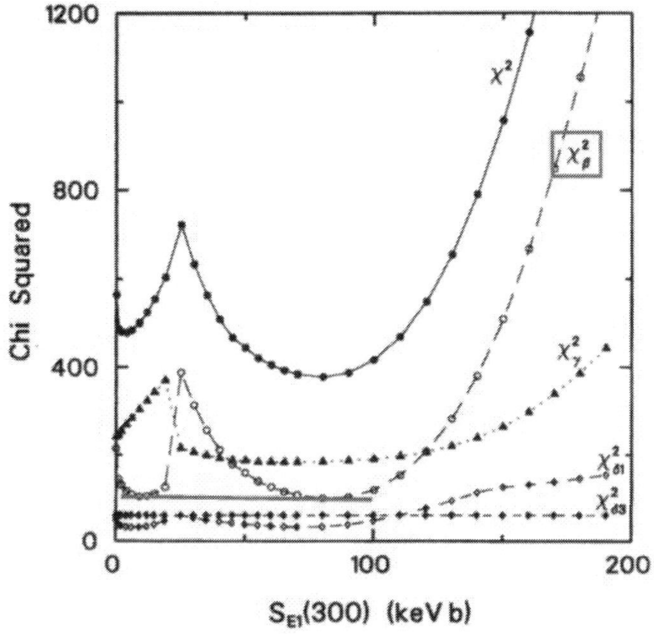


**Figure 1.** The $\chi^2$ obtained from world data as shown in Ref. [2]. The $\chi^2$ for the $^{16}N$ data show two minima at the same low values of $\chi^2$ which does not allow us to distinguish between the destructive and constructive interference solutions as discussed in the text.

the limited accuracy of the measured angular distribution. A major disadvantage of measuring gamma rays is the large background. This large background is not expected in our proposed experiment using a Time Projection Chamber (TPC) and in addition we will measure complete and detailed angular distributions at all angles (including the essential backward angles) and thus we anticipate a large sensitivity to the E2/E1 ratio and the mixing angle $\phi_{12}$.

Carbon is formed during stellar helium burning in the triple-alpha process, the $^8Be(\alpha,\gamma)^{12}C$ reaction, that is mostly governed by the contribution of the $0^+$ Hoyle state at 7.654 MeV. At high temperatures (T > 3 GK) higher lying states in $^{12}C$ may contribute. Indeed a broad ($\Gamma =$ 560 keV, $\Gamma_\gamma = 0.2$ eV) $2^+$ state at 9.11 MeV in $^{12}C$ was included in the NACRE compilation [7] following theoretical prediction [8] for the $2^+$ member of the rotational band built on top of the $0^+$ Hoyle state at 7.654 MeV. It increases the production of carbon at temperatures in excess of 1 GK by up to a factor of 15. A larger production of $^{12}C$ at high temperatures increases the neutron density as required for an r-process, due to the competition between the $^8Be(\alpha,\gamma)$ reaction and the $^8Be(n,\gamma)$ reaction [9, 10]. A $2^+$ member of the rotational band build on top of the Hoyle state is not predicted in the conjectured alpha condensate [11] which predicts a spherical Hoyle state. In this BEC model the $2_2^+$ is predicted to be an alpha-vibration state. An evidence for the broad second $2^+$ state at 9.6 MeV was found in a $^{12}C(\alpha,\alpha')$ [12] and a $^{12}C(p,p')$ measurement [13] but such a state was not observed in the beta-decay of $^{12}B$ and $^{12}N$ [14].

## 2. Measurements of the photo-dissociation of $^{16}O$ and $^{12}C$

We used our Optical-Readout Time Projection Chamber (O-TPC) [15] operating with $CO_2$ gas with circularly polarized gamma-ray beams extracted from the HI$\gamma$S facility of TUNL at Duke University [16] to study the $^{16}O(\gamma,\alpha)$ and the $^{12}C(\gamma,3\alpha)$ reactions at E = 9.08, 9.38, 9.58, 9.78,

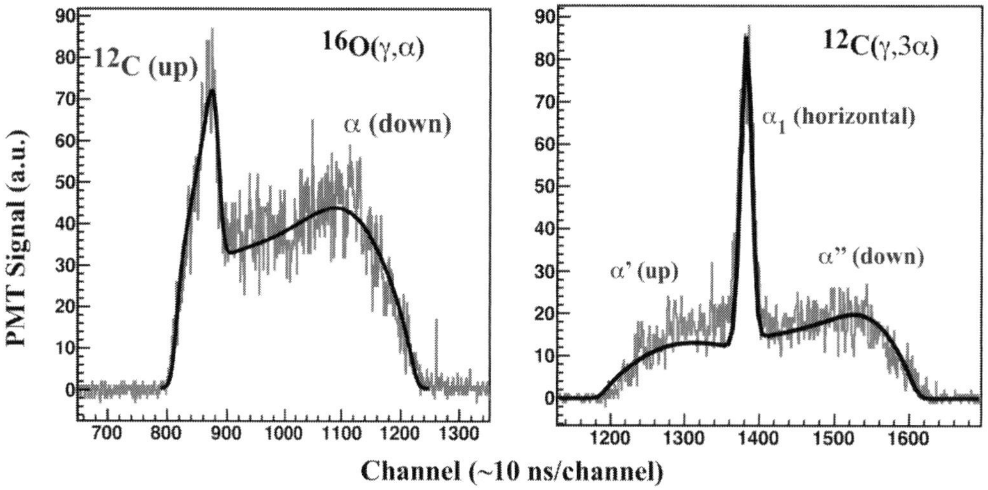


**Figure 2.** The measured PMT signal compared to the predicted line shape for $^{16}O$ (left) and $^{12}C$ (right) dissociation events.

10.08, 10.38 and 10.68 MeV. The outgoing particle resulting from the photo-disscoiation of target nuclei are fully determined by the tracks recorded in the O-TPC in three dimensions. Thus all relevant kinematical variables are measured by the O-TPC [15]. The total energy deposited (grid charge-signal) in the O-TPC detector is determined by the Q-value for the dissociation event; $E_{total} = E_\gamma - Q$; Q = 6.227, 7.162 and 7.275 for the dissociation of $^{18}O$, $^{16}O$ and $^{12}C$, respectively. The Q-values for $^{16}O$ and $^{18}O$ dissociations are sufficiently different (935 keV) and differ considerably more than the beam width of FWHM $\approx$ 300 keV, hence these events are well separated in the total energy spectrum measured by the charge signal. However one major draw back for using $CO_2$ gas is that the difference of Q-values (112 keV) for the dissociation of $^{16}O$ and $^{12}C$ is considerably smaller than the beam width and comparable to the detector resolution of approximately 90 keV [15]. In addition the larger quenching factor for the low energy $^{12}C$ projectiles leads to a smaller grid charge-signal from the dissociation of $^{16}O$ with very similar energy as for the dissociation of $^{12}C$. Hence the total energy deposit cannot be used to separate (and thus identify) $^{12}C$ and $^{16}O$ dissociation events.

To identify and distinguish $^{12}C$ dissociation events we relied on the line shape of the PMT signal which is very well determined by the calculated dE/dX along the track. In Fig. 2 we compare the observed PMT signal to the calculated line shape for a co-linear $\alpha$ + $^{12}C$ event. In this calculation we used the drift velocity of 1.14 cm/$\mu$s (at 100 torr) measured with a well collimated $^{148}Gd$ source [15]. The measured line shape of the PMT signal is compared to 181 calculated functions of the out-of-plane angle (in $\beta$ = 1° increments) to determine the best fit. The calculated line shape shown in Fig. 2 has essentially only one free parameter, the out of plane angle ($\beta$). A good $\chi^2$ is found for the $^{16}O$ dissociation events shown in Fig. 2. The line shape of $^{12}C$ dissociation events arise from a considerably more complicated dE/dx of the three body (non-colinear) decay pattern. The line shape of $^{12}C$ events requires 181x9 = 1,629 functions that are calculated with sufficient angular bin size ($\beta$ = 1°) as well as sufficient bin size (30°) for the $\theta$ and $\phi$ angles of the two alphas ($\alpha'$ and $\alpha''$) from the decay of $^8Be$. For ($\alpha_1$) decay into the excited states, as shown in Fig. 2, we also considered the energy of the excited $^8Be$. A good $\chi^2$ is found for the $^{12}C$ dissociation events shown in Fig. 2.

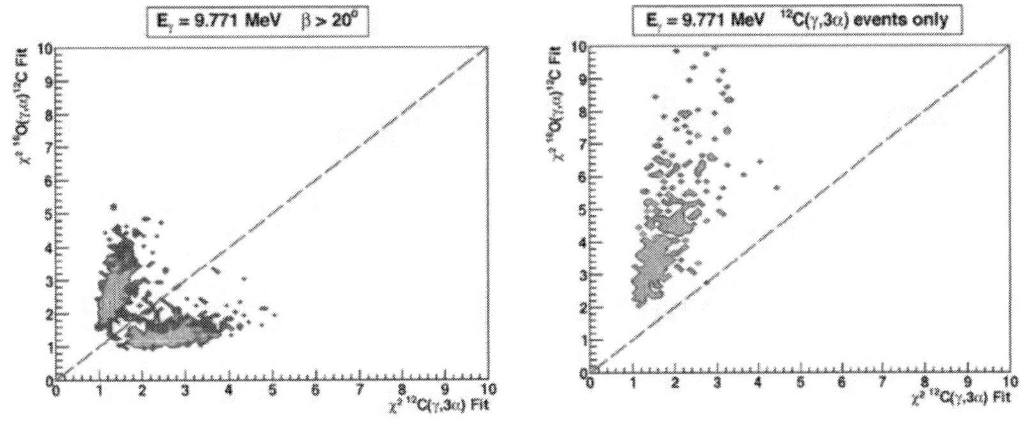


**Figure 3.** The $\chi^2$ obtained by fitting the PMT signal with the line shape predicted for $^{16}O$ and $^{12}C$ events and for selected well identified $^{12}C$ events (right).

Unfortunately the noise level in the CCD camera was too high and it did not permit line shape analysis of the pixel-content. Hence only out of plane events with out of plane angle $\beta$ larger than 20° (approximately 40% of the data) could be analyzed in the current setup. A new cleaner camera has been installed that permits including all data. In addition the resolution of the optical system did not permit resolving the two outgoing alphas emitted from the decay of the ground state of $^8Be$. Due to the poor resolution such decays most of the time appear co-linear in the image recorded by the CCD camera but are clearly distinguished from $^{16}O$ events in the PMT signal. In contrast the two outgoing alphas emitted in the decay of the first excited state of $^8Be^*(3.0)$ are well resolved as shown in Fig. 2. The very low energy of the $^8Be^*(3.0)$ yield a decay pattern which are almost as for a decay in rest.

## 3. A measurement of the $^{12}C(\gamma,\alpha)$ reaction.

The in plane angle ($\alpha$) measured by the track registered in the CCD image and the out-of-plane angle ($\beta$) measured by the Time projection signal of the PMT allow us to deduce for each event the scattering angle ($\theta$) and the azimuthal angle ($\phi$) of the polar coordinate system used in scattering theory: $\cos\theta = \cos\beta$ x $\cos\alpha$ and $\tan\phi = \tan\beta/\sin\alpha$. The so obtained angular distribution is shown in Fig. 4 together with that predicted for a pure $0^+ \rightarrow 2^+$ E2 transition. For these data we used only in plane ($\beta < 20°$) data for which the scattering angle ($\theta$) is determined with high accuracy.

We also measured an excitation curve for the $^{12}C(\gamma,\alpha)$ reaction. The accuracy of these data is limited by the measured beam intensity. A considerably more accurate beam monitoring system has been developed using the known cross section of the photo-dissociation of deuterium. The result of the limited accuracy study is shown in Fig. 5.

## 4. Conclusions

Dissociation events from the $^{12}C(\gamma,3\alpha)$ reaction were identified in our measurement using an O-TPC detector and clear evidence is observed for a pure E2 angular distribution most likely arising from a $2^+$ state just below 10.0 MeV. These data are being remeasured with an improved setup including a CCD camera with lower background. This study is in progress.

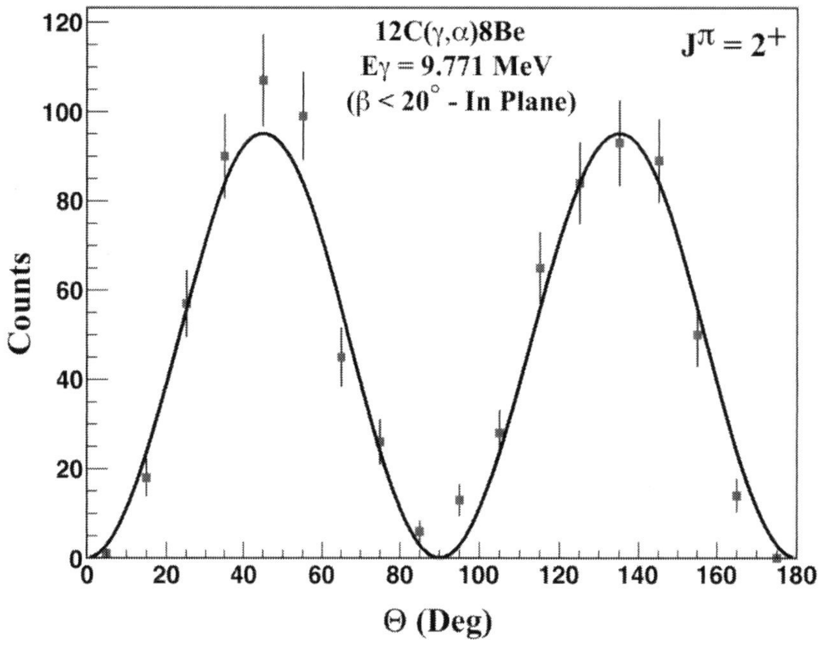


**Figure 4.** Measured angular distribution for in plane ($\beta < 20°$) $^{12}C(\gamma, 3\alpha)$ events compared to the prediction for a pure $0^+ \rightarrow 2^+$ E2 transition.

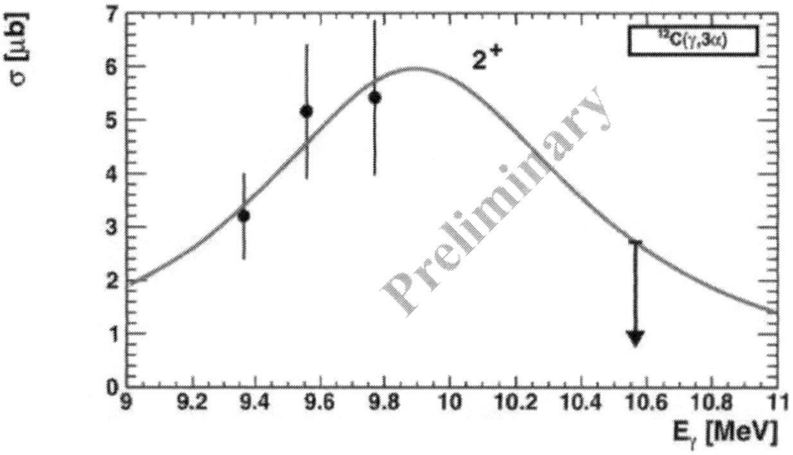


**Figure 5.** Preliminary measured excitation curve for the $^{12}C(\gamma, 3\alpha)$ reaction.

## 5. References

[1] W.A. Fowler, Rev. Mod. Phys. 56(1984)149.
[2] R.E. Azuma *et al.*; Phys. Rev. C**50**(1994)1194.
[3] M. Assuncao *et al.*; Phys. Rev. C**73**(2006)055801.
[4] R. Plag *et al.*; Nucl. Phys. **A758**(2005)415c.
[5] G.M. Hale; Nucl. Phys. **A621**(1997)177c.
[6] L. Gialanella *et al.*; Eur. Phys. J. **A11**(2001)357.
[7] C. Angulo *et al.*; Nucl. Phys. **A656**(1999)3.
[8] Descouvemont P and Baye D 1987 *Phys. Rev. C* **36** 54.
[9] Delano M D and Cameron A G W 1971 *Astr Space Sci.* **10** 203.
[10] Pruet J *et al.* 2005 *Astr. Jour.* **623** 325.
[11] Funaki Y *et al.* 2009 *Phys. Rev. C* **80** 64326 and references therein.
[12] Itoh M *et al.* 2004 *Nucl. Phys.* **A738** 268.
[13] Freer M *et al.* 2009 *Phys. Rev. C* **80** 041303(R).
[14] Hyldegaard S *et al.* 2009 *Phys. Lett.* **B678** 459.
[15] M. Gai, M.W. Ahmed, S.C. Stave, W.R. Zimmerman, A. Breskin, B. Bromberger, R. Chechik, V. Dangendorf, Th. Delbar, R.H. France III, S.S. Henshaw, T.J. Kading, P.P. Martel, J.E.R. McDonald, P.-N. Seo, K. Tittelmeier, H.R. Weller, and A.H. Young; JINST **5**(2010)12004.
[16] H.R. Weller, M.W. Ahmed, H. Gao, W. Tornow, Y. Wu, M. Gai, R. Miskimen; Prog. Part. Nucl. Phys. **62**(2009)257.

# The strangelet saga

**Edwin Norbeck and Yasar Onel**
Department of Physics and Astronomy, University of Iowa, Iowa City, IA 52242 USA

E-mail: edwin-norbeck@uiows.edu

**Abstract.** A strangelet is a small piece of strange quark matter, which consists of roughly equal numbers of u, d, and s quarks. Some have said that strange quark matter could be more stable than ordinary nuclear matter for which the quarks are collected into groups of three as neutrons and protons. At a time when our understanding was largely qualitative it was suggested that a single strangelet could grow without limit by absorbing ordinary nuclei. This raised fears that making a single strangelet in a high energy accelerator could result in the destruction of the earth. Improved theory supplemented with additional experiments, have removed all justification for such catastrophic scenarios. Calculations of color-flavor locking allow for a number of variations of strange quark matter, some of which have a lower energy than any other arrangement of quarks. Such material might be found inside of neutron stars for which the necessary pressure and low temperature are available. Those calculations that predict a density larger than that of compressed neutrons have been ruled out by the recent observation of a neutron star with a mass 2.0 times the mass of the sun.

## 1. Introduction

The term "strange quark matter" refers to a theorized state of matter containing a large number of delocalized quarks (u, d, s) in multiquark droplets. Small pieces of strange quark matter are called strangelets. The smallest strangelet would be the six quark H particle with the quarks of two lambdas in one bag. It has been suggested that the center of neutron stars consists of strange quark matter.

The Alpha Magnetic Spectrometer (AMS-02) was launched and attached to the International Space Station in May, 2011. It will use the unique environment of space to study the universe and its origin by searching for antimatter, dark matter, and strangelets while performing precision measurements of cosmic ray composition and flux [1]. We predict that they will not find strangelets, but nature provides no end of surprises. The papers associated with this project may provide much information about the history and theoretical predictions for strangelets.

Strangelets got into the popular press with the conjecture that a single strangelet could grow and consume the entire earth. There were some who tried to prevent the start of RHIC and LHC because of concerns about this possibility. A search with www.google.com will bring up a number of notes about the alleged danger. In the early days of quark matter studies, there was room for imaginative theorists to formulate disastrous strangelet events, but they are certainly ruled out now by both theory and experiment.

Hypernuclei, consisting mostly of protons, neutrons, and lambdas, can have the same quark content as a strangelet. They can be stable to strong interactions, but the strange baryons in them will decay by weak interactions with lifetimes of the order of 100 ps. It could be difficult to distinguish between strangelets and hypernuclei in measurements made at high altitudes.

The transition between quark matter and hadronic matter takes place by strong interactions and is rapid. It requires only a small rearrangement of the quarks. From this it would follow that all nuclear wave functions have a quark matter component, and that a strangelet and a hypernucleus with the same quark content are different names for the same object.

## 2. The beginning

As soon as quarks became considered as real particles, people started asking why baryons are limited to three quarks. In a system with many quarks the lowering of the Fermi energy by changing some of the u and d quarks to s quarks could, under some circumstances, more than make up for the larger mass of the s quark. Such considerations led to many papers about the possibility of strange quark matter. In 1971 Bodmer [2] suggested that a "collapsed state" of matter would have a lower energy than ordinary nuclei, but that the time required for the transition to the lower energy state could be longer than the age of the universe. As an analogy he noted that the D-D molecule requires a very long time to go into the lower energy state of a He atom.

There were people who wanted to prevent heavy ion studies at RHIC and at the LHC. They claimed that the production of a single piece of such quark matter, if it were neutral or slightly negative, could swallow the earth. Early on, knowledgeable people showed this need not be a concern. E. Whitten, in 1984 [3], pointed out that such an object would have to be positive. "To ensure beta equilibrium, quark matter will have a degenerate electron gas with $\mu_e = \mu_d - \mu_u$. Since this is much more than the electron mass $m_e$, the electron gas is bound only if there is a large electrostatic potential, equal to $\mu_d - \mu_u - m_e = 50$ MeV."

As time went on the calculations became more quantitative. In 1991 Greiner and Stöcker, using the MIT bag model, found that stable quark matter required a small bag constant, $B_{1/4} < 150$ MeV, which would give a very small transition temperature to a quark gluon plasma (QGP), $T_c \sim 100$ MeV [4]. Experiments indicate $T_c > 160$ MeV.

## 3. Possible danger

The logic behind alleged danger was that if a strangelet were the lowest energy state of nuclear matter and had a negative or neutral electric charge it would absorb nearby nuclei and convert them into additional strange quark matter. Once inside the strangelet its quarks would adjust their flavor by weak decays. Since each flavor change would be energetically allowed they could proceed sequentially. The absorption of ordinary nuclei would continue as long as additional nuclei are available—until the entire earth is consumed.

Ordinary nuclear matter does not decay sequentially into strangelets because the initial conversions of u or d quarks to s quarks are not energetically allowed. If the overall energies were such that a bismuth nucleus, for example, would release energy by converting to a strangelet, the time required for the conversion would be extremely long because the decay would require over 60 simultaneous weak decays.

The overlap region in collisions of energetic heavy ion can be considered as an extremely hot strangelet. The quarks are not confined to hadrons and the fraction of strange quarks is large. Many of the anti strange quarks are removed by the process of strangeness distillation [4]. If, but only if, some of this material could cool without forming into hadrons, would it become a strangelet that could be a matter of concern.

## 4. Ice cube from a hot furnace

Ellis has shown that no object even as large as a carbon nucleus could come out of such a hot system [5]. This would include strangelets as well as ordinary nuclei. It has been compared to making ice cubes in a furnace. Andronic et al. [6] make predictions for specific nuclei and hypernuclei and show that their model provides good fits to available data for nucleon-nucleon center of mass energies from 5 to 200 GeV. The maximum production occurs at 5 GeV, and the production of masses larger than 4 falls off rapidly with increasing energy. They find the production ratio for ${}_{\Lambda\Lambda\Xi}{}^{7}\mathrm{He}/\Lambda$ at 5 GeV to be only $5 \times 10^{-12}$. Note that these considerations apply only to production from the hot quark gluon plasma (QGP) and not to the much cooler spectators.

To date there have been no confirmed observations of strangelets. The smallest strangelet would be the H particle (H for hexaquark). This particle was first proposed by Jaffe [7] who noted that the special symmetry of two u, two d, and two s quarks should lead to a stable particle (stable with respect to the strong interactions) with the quark content of two lambdas. He estimated that the mass should be about 80 MeV below the sum of the masses of two lambda particles. Subsequent calculations using a wide variety of models have given mass estimates ranging from values less than the deuteron to unbound states. The H has not been seen in any experiments. The driving interest in the particle lies in its being a new state of matter (six quarks in one bag). In addition, if the mass is near the deuteron mass, it could have a lifetime sufficiently long to have cosmological significance.

The Facility for Antiproton and Ion Research (FAIR) [8] experiment at GSI Darmstadt will study heavy ion reactions with high luminosity at the optimum energy for producing hypernuclei [9]. These studies will provide a more quantitative understanding of multi-hypernuclear systems and possibly strange quark matter. FAIR is projected to deliver its first beams in 2017/2018.

The QGP present in the early universe, shortly after the big bang, is thought to have been a hot soup of delocalized quarks (with roughly equal numbers of u, d, and s quarks) and gluons that is similar to the QGP formed in the interaction region of heavy ion collisions. So for the reasons given above we should not expect to find strangelets, or even medium weight nuclei, left over from the big bang. These considerations do not rule out the possibility of the appearance of strangelets in cosmic rays. We certainly do not understand all possible productions mechanisms by stellar processes. If stable strangelets are possible, nature will find some way to make them. If strange quark matter is stable at zero pressure then compact stars are strange stars, not neutron stars. In that case collisions between strange stars and their partners in binary systems would cause tidal forces that could lead to the ejection of appreciable amount of strange quark matter into the cosmos [1].

## 5. CFL

Recent calculations involving color-flavor locking (CFL) [10] have shown that strange quark matter may be the lowest energy state of matter and that it also could be neutral or even slightly negative. Because of asymptotic freedom, at sufficiently high densities and low temperatures matter can be a Fermi sea of essentially free quarks. The behavior is dominated by quarks near Fermi surface. With an attractive force between them they can form pairs of a bosonic nature that reduces the free energy. Many such pairs will form a bosonic condensate. This is a standard recipe for the BCS type of superconductivity. With color, flavor and spin, many pairing patterns are possible resulting in many varieties of "Cooper pairs" and "color superconductivity" and many different predictions for the density of the resulting strange quark matter. The necessary extreme pressures and adequate cooling times may be found in neutron stars. Pressures near the center are huge, and millions of years are available for cooling, but the pressures may not be sufficient for color-flavor locking even in neutron stars.

## 6. Neutron Stars

There has been considerable interest in the question of strangeness in compact star interiors. For a review and many references, see [11]. Most of the calculations, including CFL calculations, predict that strange quark matter would have a density larger than that of hadronic matter. Those calculations that predict an average density for a star that is much larger than that of compressed neutrons have been ruled out by the recent observation of a neutron star with a mass 2.0 times the mass of the sun. This star, PSR J1614-2230, is a pulsar emitting a radio signal every 3.15 ms [12]. The best calculation of the equation of state for compressed neutrons predicts a radius for the star that is only slightly larger than the Schwarzschild radius. If the average density were larger, the star would be a black hole that could not emit a radio signal. The stellar measurements are compatible with any model of strange quark matter that predicts an equation of state similar to that of compressed neutrons. It is also possible that cold strange quark matter is not the lowest energy state of matter or that the pressure required for the CFL states is more than is found even in neutron stars.

## 7. Hypernuclei from spectators

The only sizable nuclear objects that can be made by high energy heavy ion collisions are the spectators in collisions in which the overlap between the two nuclei is small. For collisions with a larger region of overlap the spectator part disintegrates into individual neutrons and protons. It is suggested that particles from the hot interaction region could get into a spectator and make it into a hypernucleus.

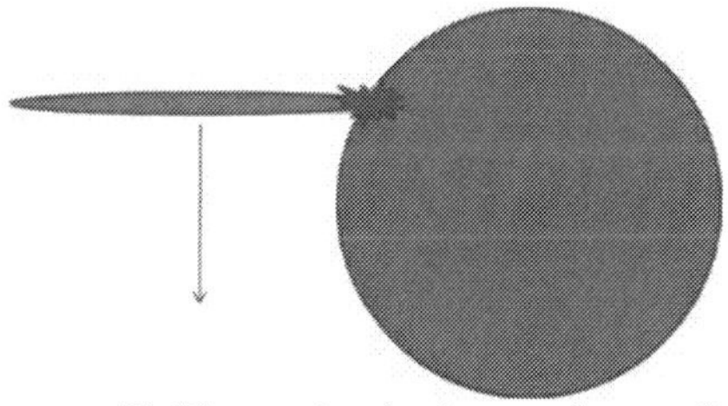

Figure 1. A high energy Pb-Pb reaction in the rest frame of one of the Pb nuclei

Figure 1 shows how strange quarks could develop in a spectator. The figure shows the interacting system in the rest frame of the nucleus that will produce the spectator. The Lorentz contracted incoming nucleus is just making contact. As it passes through the edge of the target nucleus, nucleon-nucleon collisions will produce energetic particles, some of which will pass into the spectator part of the target. PYTHIA version 8.145 was used to calculate the number and type of particles that leave a proton-proton reaction at angles greater than 5°. Reactions involving neutrons should produce particles similar to those from the p-p reactions With Pb-Pb reactions at the LHC in mind, the calculations were done for nucleon-nucleon center of mass energies of 5.5 TeV and 2.75 TeV and then at 0.2 TeV (RHIC energy) to determine the energy dependence. In the center of mass of one nucleus this required the incoming Pb nuclei to have energies of 16000 TeV/nucleon (for 5.5 TeV) and 4000 TeV/nucleon (for 2.75 TeV). For each energy, 30 p-p reactions were calculated and the number of reaction products summed. This is an approximation to the number of nucleon-nucleon reactions that would occur during a semiperipheral Pb-Pb reaction.

There are two types of reaction products of interest. There are particles containing strange quarks and the much larger number of pions. The pions mostly have energies suitable for initiating reactions that result in strange quarks. The neutral pions are included along with the charged pions. The lifetime of a $\pi^0$ is much too long for it to decay into gammas during its passage through the spectator nucleus. Also included in the pion counts are a small number of $\eta$, $\rho$, and $\omega$ mesons. There were also a few neutrons and protons with angles greater than 5°; these were not included in the count. In order of decreasing energy the pion counts (for 30 pp reactions) are 148, 141, and 115, which shows an increase with energy. The numbers for particles carrying strangeness are smaller, 1 $K^-$, 2 $\Lambda^o$, 3 anti$K^o$, and 2 $K^o$ for the highest energy and 1 $K^-$, 1 $\Lambda^o$, 2 anti$K^o$, and 2 $K^o$ for 2.74 TeV. For 0.2 TeV the numbers are larger, 3 $K^-$, 4 $\Lambda^o$, 4 anti$K^o$, and 5 $K^o$. The larger numbers at the lowest energy reflects the peak in the production cross section not far above threshold.

With these numbers one could expect that LHC experiments would produce some spectators with one $\Lambda^0$ but few with more than one. It would have been useful to have extended these calculations to the much higher energies available in cosmic rays, but there was doubt that Pythia would give reliable results at such energies.

We speculate that at sufficiently high energies a large number of strange quarks would reach the spectators and produce multi-hypernuclei. A large nucleus, Fe or Pb, colliding with an atmospheric nitrogen nucleus would always leave considerable spectator matter.

It would be good to make a thorough study of spectators from Pb-Pb reactions at LHC energies to see how well the "universal" distribution function [13] works at the much larger energy. A careful study would also determine the fraction of the spectators that are hypernuclei. The experiments would be difficult, and so far no one has accepted the challenge.

## 8. All nuclei have a quark matter component

There is generally assumed dichotomy between matter that is composed of nucleons and quark matter for which the quarks are not localized in nucleons. However, in an ordinary nucleus the force binding the nucleons is the result of quarks and gluons being somewhat delocalized. A good place to look for a quark matter component in the wave function is in the $^4$He nucleus. With two flavors, two spin states, and three colors, up to 12 quarks, 6 u and 6 d, can all be in the ground state. A detailed calculation shows that 12 quarks do indeed provide an energy minimum as a function of quark number [14]. Mosallem and Uzhinskii [15] found that it was impossible to describe simultaneously the p-$^4$He and d-$^4$He elastic scattering cross sections using the same set of N-N amplitude parameters. The data was fitted well using a $^4$He wave function of which 10.5% was a 12 quark bag. For $_{\Lambda\Lambda}{}^6$He it could be expected that the delocalized quark component would be an even larger part of the wave function. This would be an extremely tightly bound nucleus, but with a mass still larger than normal $^6$He so that it would decay by weak interactions.

From this point of view the nature of the ground state of a nuclear object depends only on the number of u, d, and s quarks. The terms hyper nucleus and strangelet refer to the endpoints of a continuum. Starting from the other end of this continuum, consider a large box containing quarks. The forces between the quarks will ensure no concentrations of color charge. The quarks will be associated in color-neutral triplets. Only a small amount of motion of the quarks, or a reduction in temperature, is required for the triplets to become normal baryons.

Returning to the H particle, Jaffe [7] contrasted a particle with 6 quarks in one bag with two bags with 3 quarks each. Nature does not make such a distinction. The H particle could be expected to be mostly a deuteron like object consisting of two $\Lambda^o$s. While the two s quarks in the six quark component could, in principle, allow two $\Lambda^o$s to be bound when two neutrons are not, this appears to not happen.

It might be expected that in neutron stars there is a delocalized quark component and that some of the quarks are strange. The magnitude of these components would depend on the pressure (depth). The question is not whether there is strange quark matter or not in a neutron star, but rather the relative numbers of u, d, and s quarks and the extent of delocalization. In neutron stars there is the additional feature that a CFL phase would require some of all of the quarks to be combined as Cooper pairs [10].

## 9. CASTOR and ZDC

Many unusual cosmic-ray events have been recorded at high altitude observatories [16]. Some of these may have been caused by strangelets [17]. The AMS-02 experiment will see such objects if they are part of the primary cosmic rays. If they are produced by reactions of heavy ions with the upper atmosphere, it is likely that they would also be formed by Pb-Pb reactions at the LHC and may be seen by the CASTOR (Centauro And STrange Object Research) detector [18]. CASTOR is a small angle detector covering the angles from 0.7° down to 0.08° ($5.2<\eta<6.6$) from the beam direction that has sufficient granularity to identify and characterize unusual nuclear objects. It is divided into two parts, a short electromagnetic part (EM) and a long part (HAD) for studying objects with the penetrating potential of hadrons. Its total length provides 10.3 nuclear interaction lengths. The showers are sampled by observing Čerenkov light from thin quartz plates between tungsten plates. It is divided into 14 longitudinal sections, each with 16 azimuthal divisions, for a total of 224 analog signals. Figure 2 shows the structure. There have been no reports yet from CASTOR about strange objects. The detailed calibration has been difficult because CASTOR, with 224 photomultiplier tubes (PMTs), is located in a part of CMS [19] that has a large and irregular magnetic field.

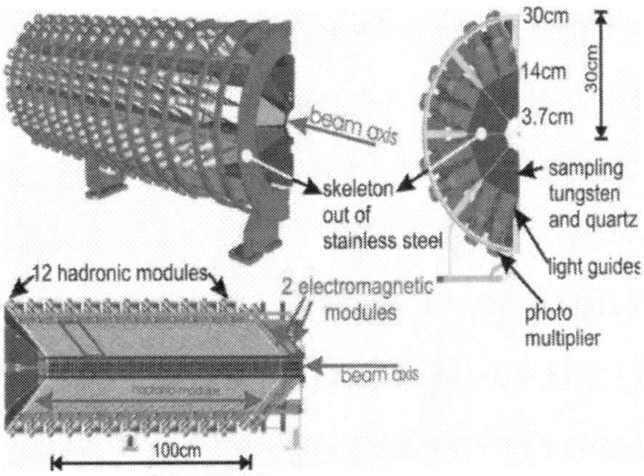


**Figure 2.** Details of the structure of CASTOR

CASTOR cannot see spectator matter because it follows the outgoing beam through the middle of CASTOR. It is not expected that strangelets will come from the hot quark gluon plasma at LHC energies. However it is likely that detailed small angle studies with CASTOR will lead to a better understand of the unusual cosmic ray events.

For detailed studies of reaction products at even smaller angles there are only the Zero Degree Calorimeters (ZDCs). These are located at the closest point for which the incoming and outgoing beams are in separate beam pipes. This is 140 m from the interaction point for CMS. A ZDC must be small because the distance between the beam pipes is only 10 cm. Figure 3 shows the design of a ZDC for CMS [20]. With heavy ion reactions a ZDC sees mostly spectator neutrons. A spectator hypernucleus would get out of the beam pipe and into the ZDC only if it had a Z/A ratio less than 0.2. The Pythia calculations, present earlier, appear to rule out this possibility.

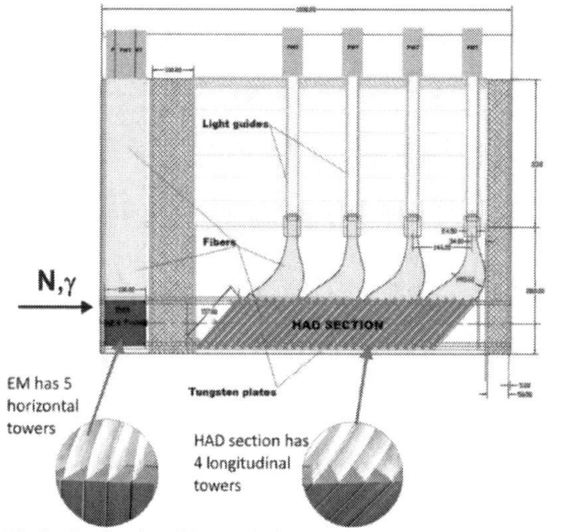


**Figure 3.** Design of ZDC. Both the EM and the HAD section consist of tungsten plates as the absorbing material. The Čerenkov light from quartz fibers between the metal plates is collected by PMTs at the top.

From the experimental point of view, a ZDC should see any nearly neutral heavy object that leaves the interaction point in the same direction as the original beam and has a cross section for producing charged particles in tungsten metal that is within an order of magnitude of that of hadrons. Figure 4 shows the energy spectrum from one of the CMS ZDCs gated on a signal from elsewhere in CMS for Pb-Pb at 2.75 TeV/nucleon. The sharp peak on the left is from single neutrons released from a Pb nucleus by the intense Coulomb field of the other Pb nucleus. The broad peak is from spectator neutrons from actual Pb-Pb collisions. The number of spectator neutrons is a function of the impact parameter (the distance of closest approach of the centers of the two nuclei), which can be determined from the total number of particles in CASTOR or other parts of CMS. Heavy, nearly neutral objects would appear as deviations from the predicted number.

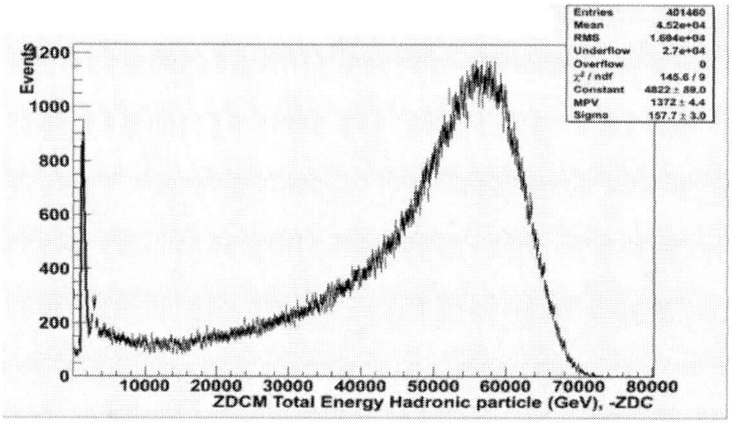


**Figure 4.** ZDC energy spectrum gated on other CMS signals showing the single (and small double) neutron peaks from ultra-peripheral collisions and a broad peak of spectator neutrons from actual Pb-Pb collisions at a center of mass energy of 2.75 TeV/nucleon.

The interacting Coulomb fields can produce particles with masses up to about 200 GeV. Perhaps some of the unusual objects seen in high altitude cosmic rays studies will be produced by this process. They should be seen in CASTOR or other parts of CMS in coincidence with the one neutron peak. All of the single neutrons in figure 4 are in coincidence with other objects seen by CMS. If the ZDC had supplied the gate, the area under the single neutron peak would have been an order of magnitude larger than the area under the broad multi-neutron peak.

## 10. Conclusions

- There is no longer any reason to be concerned that strangelet production at an accelerator could destroy the earth.
- To date no strangelets have been observed.
- The distinction between quark matter and hadronic matter may be spurious. All nuclear matter may be a mixture of both.
- High energy, heavy ion cosmic rays may produce multi-hyper nuclear objects by the spectator part of the heavy ion being infiltrated with strange quarks from the hot quark gluon plasma.

## References

[1] Sandweiss J 2004 *J. Phys.* G: *Nucl. Part. Phys.* **30** S51-9
[2] Bodmer A R 1971 *Phys. Rev.* D **4** 1601-6
[3] Whitten E 1984 Cosmic separation of phases *Phys. Rev.* D **30** 272-85
[4] Greiner C and Stöcker H 1991 *Phys. Rev.* D **44** 3517
[5] Ellis J *et al* 2008 *J. Phys.* G **35** 115004
[6] Andronic A, Braun-Munzinger P, Stachel J and Stöcker H 2011 Production of light nuclei, hypernuclei and their antiparticles in relativistic nuclear collisions *Phys. Lett.* B **697** 203-7
[7] Jaffe R L 1977 *Phys. Rev. Lett.* **38** 195
[8] Henning W F 2008 FAIR: recent developments and status *Nucl Phys.* A **805** 502
[9] Steinheimer J, Sturm C, Schramm S and Stöcker H 2010 Strange perspectives at FAIR *J. Phys. G Nucl. Part. Phys.* **37** 094026
[10] Alford M G, Schmitt A, Rajagopal K and Schäfer T 2008 *Rev. Mod. Phys.* **80** 1455-1515 (arXiv:0709.4635v2 [hep-ph])
[11] Blaschke D, Klähn T, Lastowiecki R and Sandin F 2010 *J. Phys. G Nucl. Part. Phys.* **37** 094063
[12] Demorest P B, Pennucci T, Ransom S M, Roberts M S E and Hessels J W T 2010 A two-solar-mass neutron star measured using Shapiro delay *Nature.* **467** 1081-3
[13] Schüttauf A *et al* 1996 *Nucl. Phys.* A **607** 457-86
[14] Neudatchin V G, Obukhovsky I T, Smirnov Yu F and Tkalya E V 1983 *Z. Phys.* A **313** 357-66
[15] Mosallem A M, Uzhinskii V V, 2002 arXiv:nucl-th/0208043v1
[16] Gladysz-Dziadus E 2003 *Phys. Part. Nucl.* **34** 285-347
[17] Gladysz-Dziadus E 2006 *Acta Physica Polonica* B **37** 153-160
[18] http://cmsdoc.cern.ch/cms/castor/html/
[19] CMS Collaboration 2008 *JINST* **3** S08004
[20] Grachov O A *et al* (CMS Collaboration) 2006 *AIP Conf. Proc.* **867** 258-65 (nucl-ex/0608052)

# AUTHOR INDEX